Springer Proceedings in Physics

Volume 171

More information about this series at http://www.springer.com/series/361

Leonid Bulavin · Nikolai Lebovka
Editors

Physics of Liquid Matter: Modern Problems

Proceedings, Kyiv, Ukraine, 23–27 May 2014

Springer

Editors
Leonid Bulavin
Faculty of Physics
Taras Shevchenko National University
Kyiv
Ukraine

Nikolai Lebovka
Head of Department of Physical Chemistry of Disperse Minerals
Biocolloidal Chemistry Institute
Kyiv
Ukraine

ISSN 0930-8989 ISSN 1867-4941 (electronic)
Springer Proceedings in Physics
ISBN 978-3-319-20874-9 ISBN 978-3-319-20875-6 (eBook)
DOI 10.1007/978-3-319-20875-6

Library of Congress Control Number: 2015943449

Springer Cham Heidelberg New York Dordrecht London

Printed on acid-free paper

Springer International Publishing AG Switzerland is part of Springer Science+Business Media (www.springer.com)

Preface

These proceedings comprise invited and contributed papers presented at the 6th International Conference “Physics of Liquid Matter: Modern Problems” (PLMMP-2014) which was held during 23–27 May 2014 in Kyiv, Ukraine. Understanding of Liquid Matter properties is the foundation stone for many scientific and engineering disciplines including physics, chemistry, chemical engineering, and materials science. It provides the basis for research in different fields from biological objects to supernovae behavior. The combination of experiments, theory, and simulation allows scientists to study the complex systems on molecular and macroscopic levels. The PLMMP-2014 Conference was organized by the Taras Shevchenko National University of Kyiv, Institute for Safety Problems of Nuclear Power Plants and Bogolyubov Institute for Theoretical Physics of the National Academy of Sciences of Ukraine. The scientific program covered the most recent developments in the broader field of liquid state, including interdisciplinary problems. The format of the conference was based on invited lectures, oral presentations, and posters. The presented papers are divided into four parts: (i) *Structure of Liquids in Confined Systems*, (ii) *Phase Transitions, Supercritical Liquids & Glasses*, (iii) *Colloids*, and (iv) *Medical and Biological Aspects*. The presented objects include liquids, solutions, soft polymers, colloids, amphiphiles, micelles, emulsions, and liquid crystals. On behalf of the organizing committee we would like to thank all the plenary and invited speakers for their valuable contributions. We would like also to thank the International Advisory Board and Local Organizing Committee for helping to organize the scientific program and operation of the PLMMP-2014 Conference.

Kyiv, Ukraine

Leonid Bulavin
Nikolai Lebovka

Contents

Part III Colloids

Part IV Medical and Biological Aspects

Contributors

Evgeny Apfelbaum Joint Institute for High Temperatures of Russian Academy of Sciences, Moscow, Russia

Sergey Artemenko Institute of Refrigeration, Cryotechnologies, and Eco-Power Engineering (Former Academy of Refrigeration), Odessa National Academy of Food Technologies, Odessa, Ukraine

Shyam Badu M2NeT Laboratory, MS2 Discovery Interdisciplinary Research Institute, Wilfrid Laurier University, Waterloo, ON, Canada

Olexandr Bakai The National Scientific Center Kharkiv Institute of Physics and Technology (Formerly the Ukrainian Institute of Physics and Technology), Kharkiv, Ukraine

M. Bombrowski Department of Physical Chemistry, TU Bergakademie Freiberg, Freiberg, Germany

Taras Bryk Institute for Condensed Matter Physics of the National Academy of Sciences of Ukraine, Lviv, Ukraine

Leonid Bulavin Department of Molecular Physics, Faculty of Physics, Taras Shevchenko National University, Kyiv, Ukraine

Alexander V. Chalyi Department of Medical and Biological Physics, Bogomolets National Medical University, Kyiv, Ukraine

Vadim Cheplak Department of Theoretical Physics, Odessa National University, Odessa, Ukraine

S.T. Goga Department of Physical Chemistry, Kharkov V. N. Karazin National University, Kharkiv, Ukraine

Federico Gorelli Istituto Nazionale di Ottica, CNR-INO, and LENS, Sesto Fiorentino, Italy

William T. Heller Neutron Scattering Directorate, Oak Ridge National Laboratory, Biology and Soft Matter Science Division, Oak Ridge, TN, USA

Myroslav Holovko Institute for Condensed Matter Physics of the National Academy of Sciences of Ukraine, Lviv, Ukraine

N.N. Kamneva Department of Physical Chemistry, Kharkov V. N. Karazin National University, Kharkiv, Ukraine

Valeriy Klepko Institute of Macromolecular Chemistry NAS of Ukraine 48, Kyiv, Ukraine

Vitaly K. Koltover Institute of Problems of Chemical Physics, Russian Academy of Sciences, Chernogolovka, Moscow Region, Russian Federation

A.P. Kryshtal Department of Physics and Technology, Kharkov V. N. Karazin National University, Kharkiv, Ukraine

Vladimir L. Kulinskii Department of Theoretical Physics, Odessa National University, Odessa, Ukraine

Nikolai Lebovka Department of Physical Chemistry of Disperse Minerals, Biocolloidal Chemistry Institute named after F. D. Ovcharenko, Kyiv, Ukraine

Longin Lisetski Institute for Scintillation Materials of STC "Institute for Single Crystals", NAS of Ukraine, Kharkiv, Ukraine

Eduard Lysenkov V.O. Sukhomlyns'kyi Mykolayiv National University, Mykolayiv, Ukraine

Sidney T. Malak School of Materials Science and Engineering, Georgia Institute of Technology, Atlanta, GA, USA

A.I. Marynin National University of Food Technologies, Kyiv, Ukraine

Victor Mazur Institute of Refrigeration, Cryotechnologies, and Eco-Power Engineering (Former Academy of Refrigeration), Odessa National Academy of Food Technologies, Odessa, Ukraine

N.O. Mchedlov-Petrossyan Department of Physical Chemistry, Kharkov V. N. Karazin National University, Kharkiv, Ukraine

Yuri B. Melnichenko Neutron Scattering Directorate, Oak Ridge National Laboratory, Biology and Soft Matter Science Division, Oak Ridge, TN, USA

Roderick Melnik M2NeT Laboratory, MS2 Discovery Interdisciplinary Research Institute, Wilfrid Laurier University, Waterloo, ON, Canada

Iryna Melnyk Faculty of Physics, Taras Shevchenko National University, Kyiv, Ukraine

H.-J. Mögel Department of Physical Chemistry, TU Bergakademie Freiberg, Freiberg, Germany

Axel H.E. Müller Makromolekulare Chemie II and Bayreuther Zentrum für Kolloide und Grenzflächen, Universität Bayreuth, Bayreuth, Germany; Institut für Organische Chemie, Johannes Gutenberg-Universität Mainz, Mainz, Germany

E. Ōsawa Faculty of Textile Science and Technology, AREC (Asama Research Extension Center), Nano Carbon Research Institute Ltd, Shinshu University, Ueda, Nagano, Japan

Taras Patsahan Institute for Condensed Matter Physics of the National Academy of Sciences of Ukraine, Lviv, Ukraine

Felix A. Plamper Institute of Physical Chemistry, RWTH Aachen University, Aachen, Germany

Sanjay Prabhakar M2NeT Laboratory, MS2 Discovery Interdisciplinary Research Institute, Wilfrid Laurier University, Waterloo, ON, Canada

Giancarlo Ruocco Dipartimento di Fisica, Universita di Roma La Sapienza, Rome, Italy; Center for Life Nano Science @Sapienza, Istituto Italiano di Tecnologia, Rome, Italy

Mario Santoro Istituto Nazionale di Ottica, CNR-INO, and LENS, Sesto Fiorentino, Italy

P. Schiller Department of Physical Chemistry, TU Bergakademie Freiberg, Freiberg, Germany

Tullio Scopigno Dipartimento di Fisica, Universita di Roma La Sapienza, Rome, Italy

Volodymyr Shmotolokha Institute for Condensed Matter Physics of the National Academy of Sciences of Ukraine, Lviv, Ukraine

Marat Soskin Institute of Physics, NAS of Ukraine, Kyiv, Ukraine

Christopher V. Synatschke Makromolekulare Chemie II and Bayreuther Zentrum für Kolloide und Grenzflächen, Universität Bayreuth, Bayreuth, Germany

V.V. Tkachenko Department of Physical Chemistry, Kharkov V. N. Karazin National University, Kharkiv, Ukraine

Vladimir V. Tsukruk School of Materials Science and Engineering, Georgia Institute of Technology, Atlanta, GA, USA

Olena Vasilieva Kyiv National University of Trade and Economics, Kyiv, Ukraine

Vladimir Vorob'ev Joint Institute for High Temperatures of Russian Academy of Sciences, Moscow, Russia

M. Wahab Department of Physical Chemistry, TU Bergakademie Freiberg, Freiberg, Germany

Weinan Xu School of Materials Science and Engineering, Georgia Institute of Technology, Atlanta, GA, USA

Part I
Structure of Liquids in Confined Systems

Chapter 1
Thermodynamics of Molecular Liquids in Random Porous Media: Scaled Particle Theory and the Generalized Van der Waals Equation

Myroslav Holovko, Volodymyr Shmotolokha and Taras Patsahan

Abstract A new approach to the theoretical description of molecular liquids confined in random porous media is proposed in order to study their thermodynamic properties. The models applied in our study are characterized by the intermolecular interactions consisting of repulsive and attractive parts, both of which are of the anisotropic nature. To take into account an anisotropy of the repulsion the scaled particle theory (SPT) is extended for the system of a hard convex body (HCB) fluid in a quenched matrix of hard particles forming a random porous medium. A contribution of the anisotropic attractive interaction is considered on the level of the mean-field or Van der Waals approximation. Therefore, combining the obtained analytical results within the framework of the perturbation theory the equation of state for confined liquids is derived. On the basis of the developed approach we can consider a fluid in a random matrix using various models. A reliability of the SPT theory is proved on the examples of hard sphere and hard spherocylinder fluids in different matrices. For a spherocylinder fluid with attractive intermolecular interaction the phase transition diagrams are constructed to study a vapour-liquid-nematic equilibrium and the effect of confinement on it. It is shown that a matrix porosity decrease leads to decreasing of the critical temperature and the critical density of vapour-liquid phase transition. In the case of long spherocylinders ($L_1/D_1 = 10$) the vapour-liquid transition of a fluid in a matrix can disappear completely being suppressed by the isotropic-nematic phase transition. On the other hand the coexistence between vapour and nematic phases is observed for a spherocylinder fluid at the conditions comparable to the Onsager limit

M. Holovko (✉) · V. Shmotolokha · T. Patsahan
Institute for Condensed Matter Physics of the National Academy of Sciences of Ukraine,
1 Svientsitskii Str, Lviv 79011, Ukraine
e-mail: holovko@icmp.lviv.ua

V. Shmotolokha
e-mail: shmotolokha@icmp.lviv.ua

T. Patsahan
e-mail: tarpa@icmp.lviv.ua

L. Bulavin and N. Lebovka (eds.), *Physics of Liquid Matter: Modern Problems*,
Springer Proceedings in Physics 171, DOI 10.1007/978-3-319-20875-6_1

($L_1/D_1 = 80$). The anisotropy of attractive potential causes the broadening of the liquid-nematic coexistence region and in the case of essentially high rates of anisotropy the vapour-liquid transition vanishes. It is noticed that the presence of porous medium enhances this effect. The presented review is aimed to illustrate an application of the SPT approach which developed recently for fluids of non-spherical molecules confined in random porous media.

1.1 Introduction

Many different materials such as silicas and zeolites, activated carbons and clays, cements and ceramics, metal foams and others can be considered as porous media. Molecular fluids confined in such porous materials with pore sizes ranging from a few nanometers to hundreds nanometers can undergo drastic modifications in their physical and physicochemical properties. For example, it is well established that confinement can induce drastic shifts of phase equilibria, e.g. narrowing of the vapour-liquid coexistence curve, lowering of the pore critical temperature, decreasing of the critical density and the appearance of new types of phase transitions, which are not observed in the bulk [1]. Besides its fundamental interest, a thorough understanding of the influence of confinement on the physical and chemical properties of fluids is highly useful in many areas of applied science and engineering, geosciences, biophysics, material science etc. Different porous materials are widely used in the chemical, oil and gas, food and pharmaceutical industries for pollution control, mixture separation, and as catalyst or catalyst support for chemical reactions. In parallel with experimental studies a lot of different molecular models were introduced for the investigation of the properties of fluids in porous media within the framework of computer simulations and theoretical approaches. Atomistic molecular simulations methods have been widely used to characterize adsorption in porous materials [2, 3], including the structure and dynamics of adsorbed phases, thermodynamics of adsorption, the influence of structural heterogeneity of pores and chemical heterogeneity of pore surface [4, 5], the influence of porous media on the isotropic-nematic phase transition like in confined liquid crystals fluids [6, 7]. Also numerous computer simulation studies have been devoted to the same problems, but in isolated pores of slit-like or cylindrical shapes. Theoretical approaches used for the description of fluids in porous media are mostly based on the method of Ornstein-Zernike equations [8–11] and the method of density functional theory [12, 13]. In the present review we consider fluids confined in porous materials with a random or disordered structure, in which pores are formed by randomly distributed solid particles. In these materials the pore shapes and sizes are not well defined, they are not isolated but build a network with very complex topology. The systems of such a kind cannot be described by a single pore model, thus in theoretical approaches as well as in computer simulations one should take into account the whole variety of statistically probable configurations of pores.

Much theoretical efforts have been devoted within the framework of statistical mechanical methods to a study of fluids in random porous media during the last three decades starting from the pioneering work of Madden and Gland [8]. In this work a porous medium is presented as a quenched configuration of randomly distributed spherical particles that form so-called matrix [8]. The specific of description of fluids in such porous media is connected with the double quenched-annealed averages: the annealed average is taken over all fluid configurations and the additional quenched average should be taken over all realizations of the matrix. One standard approach to solve this problem is based on the replica method. It consists in the description of a fluid in a random porous medium as the $(s+1)$ component equilibrium mixture of a matrix and s replicated copies of a fluid, which do not interact with each other, and then the limit $s \to 0$ is to be taken. Using the replica Ornstein-Zernike (ROZ) integral equation theory [9], the statistical mechanics approach of liquid state was extended to a description of different models of a fluid confined in random porous matrices [14, 15] including the chemical reacting fluids adsorbed in porous media [16, 17]. However, unlike bulk fluids, no analytical result have been obtained from the ROZ integral equations approach even for the simplest model such as a hard-sphere fluid in a hard sphere matrix.

The first rather accurate analytical results for a hard sphere fluid in hard-sphere (HS) and overlapping hard-sphere (OHS) matrices were obtained quite recently [18–20] by extending the classic scaled particle theory (SPT) [21–23]. The SPT approach is based on a combination of the exact treatment of a point scaled particle in a HS fluid with the thermodynamic consideration of a finite size scaled particle. The exact result for a point scaled particle in a HS fluid confined in a random matrix was obtained in [18]. However, the approach proposed in [18] and named as SPT1 contains a subtle inconsistency appearing when a size of matrix particles is essentially larger than a size of fluid particles. Later, this inconsistency was eliminated in a new approach named as SPT2 [20].

The expressions obtained in SPT2 include two types of porosities. One of them is defined by a pure geometry of porous medium (geometrical porosity ϕ_0 characterizing the free volume for a fluid) and the second one is defined by the chemical potential of a fluid in the limit of infinite dilution (probe particle porosity ϕ characterizing the adsorption of a fluid in an empty matrix). On the basis of SPT2 approach the approximation SPT2b was proposed, and it was shown that it reproduces the computer simulation data with a very good accuracy at small and intermediate fluid densities. However, the expressions obtained in the SPT2 and SPT2b approximations contain a divergence at the packing fraction of fluid equal to the probe particle porosity ϕ. Consequently, the prediction of thermodynamic properties at high densities of a fluid can be inaccurate, especially when it reaches close packing conditions. An accuracy of SPT2 and SPT2b approximations also becomes worse when fluid and matrix particles are of comparable sizes [20]. Later, in the investigation of one-dimensional hard rode fluid in a random porous medium [24] a series of the new approximations SPT2b1, SPT2b2 and SPT2b3 were proposed, which are free of the mentioned drawbacks. Two last approximations contain the third type of porosity ϕ^* defined by the maximum value of packing fraction of fluid in a porous medium. It was shown that these new approximations essentially improve the SPT predictions at high fluid densities. The application of the SPT theory and generalization of the SPT2b1 approximation to the case of HS fluid confined in random matrices were reviewed recently in [25], where thermodynamic properties of fluid were calculated. A comparison of obtained results with computer simulation data proved applicability and high reliability of the SPT for the wide range of fluid densities and different matrix parameters.

A remarkable feature of the SPT theory is a possibility of its generalization for the description of non-spherical hard convex body fluids in the bulk, which can be done using one [26, 27] and two [28] scaling parameters. The SPT theory can also be applied for the description of nematic ordering in hard convex body (HCB) fluids [28–30]. Recently, the SPT theory was extended for HCB fluids confined in random matrix [31]. The generalization of the SPT theory for a HCB fluid confined in a random porous medium with a use of two scaling parameters was presented, and the effect of porous media on the orientational ordering in a HCB fluid was studied [32].

In this chapter the extension of SPT theory for the description of thermodynamic properties of non-spherical molecular fluids confined in random porous media is reviewed. First we present the generalization of SPT theory for a HCB fluid in random porous media. Then we consider the SPT theory with two scaling parameters for the description of a hard spherocylinder fluid in a random matrix. After that a system of hard spherocylinders in a matrix is used as the reference system [33, 34] in the generalization of Van der Waals equation for anisotropic fluids confined in random porous media. Finally, the derived equations are applied to investigate the effect of porous media on the vapour-liquid-nematic phase equilibria in molecular fluids.

1.2 HCB Fluids in Random Porous Media: SPT with One Scaling Parameter

Hard convex body (HCB) particles are characterized by three geometrical parameters—the volume V, the surface area S and the mean curvature R with a factor $1/4\pi$. For example, for a frequently considered case of a system of spherocylindrical rods with the length L and the diameter D, these parameters are

$$V = \frac{1}{4}\pi D^2 L + \frac{1}{6}\pi D^3, \quad S = \pi DL + \pi D^2, \quad R = \frac{1}{4}L + \frac{1}{2}D. \tag{1.1}$$

The basic idea of the SPT approach is an insertion of an additional scaled particle of a variable size into a fluid. To this aim we introduce the scaling parameter λ_s in such a way that the volume V_s, the surface area S_s and the curvature R_s of scaled particle are modified as

$$V_s = \lambda_s^3 V_1, \quad S_s = \lambda_s^2 S_1, \quad R_s = \lambda_s R_1, \tag{1.2}$$

where V_1, S_1 and R_1 are the volume, the surface area and the mean curvature of a fluid particle respectively. Hereafter, we use the conventional notations [9, 14, 15], where the index "1" is used to denote fluid component and the index "0" denotes matrix particles. For scaled particles the index "s" is used.

Procedure of insertion of the scaled particle into a fluid is equivalent to a creation of cavity, which is free of any other fluid particles. The key point of considered reformulation of the SPT theory consists in a derivation of the excess chemical potential of a scaled particle μ_s^{ex}, which is equal to a work needed to create the corresponding cavity. In the presence of a porous medium the expression of excess chemical potential for a small scaled particle in a HCB fluid can be written in the form

$$\begin{aligned} \beta\mu_s^{ex} = \beta\mu_s - \ln(\rho_1 \Lambda_1^3 \Lambda_{1R}) = \ln p_0(\lambda_s) \\ - \ln\left[1 - \frac{\eta_1}{p_0(\lambda_s)}\left(1 + 3\lambda_s\alpha_1 + 3\lambda_s^2\alpha_1 + \lambda_s^3\right)\right], \end{aligned} \tag{1.3}$$

where $\beta = 1/(k_B T)$, k_B is the Boltzmann constant, T is the temperature, $\eta_1 = \rho_1 V_1$ is the fluid packing fraction, ρ_1 is the fluid density, Λ_1 is the fluid thermal wave length, the quantity Λ_{1R}^{-1} is the rotational partition function of a single molecule [35], and $\alpha_1 = \frac{R_1 S_1}{3V_1}$ is the non-sphericity parameter of a fluid particle. The term $p_0(\lambda_s) = \exp(-\beta\mu_s^0)$ is defined by the excess chemical potential of the scaled particle confined in an empty matrix μ_s^0, and it has a meaning of probability to find a cavity created by the scaled particle in the matrix in the absence of fluid particles.

For the large scaled particle the excess chemical potential is given by the thermodynamical expression for the work needed to create a macroscopic cavity inside a fluid confined in a porous medium. It can be presented as follows

$$\beta\mu_s^{ex} = w(\lambda_s) + \frac{\beta P V_s}{p_0(\lambda_s)}, \tag{1.4}$$

where P is the pressure of fluid, V_s is the volume of scaled particle. The multiplier $1/p_0(\lambda_s)$ appears due to an excluded volume occupied by matrix particles, which can be considered as a probability to find a cavity created by a scaled particle in the matrix in the absence of fluid particles. This probability is related directly to two different types of the porosities [20]. The first one corresponds to the case of $\lambda_s = 0$ and gives the geometrical porosity

$$\phi_0 = p_0(\lambda_s = 0), \tag{1.5}$$

which depends only on a structure of matrix and it is related to the volume of a void between matrix particles. The second type of porosity corresponds to the case of $\lambda_s = 1$ and gives the probe particle porosity [20]

$$\phi = p_0(\lambda_s = 1), \tag{1.6}$$

which is defined by the excess chemical potential of a fluid in the limit of infinite dilution μ_1^0. Thus, it depends also on a nature of fluid under study.

According to the ansatz of the SPT theory [18–25], $w(\lambda_s)$ can be presented in the form of expansion

$$w(\lambda_s) = w_0 + w_1\lambda_s + \frac{1}{2}w_2\lambda_s^2. \tag{1.7}$$

The coefficients of this expansion can be found from the continuity of μ_s^{ex} and its corresponding derivatives $\partial\mu_s^{ex}/\partial\lambda_s$ and $\partial^2\mu_s^{ex}/\partial\lambda_s^2$ at $\lambda_s = 0$. Consequently, one derives the following expressions [25–31]:

$$\begin{aligned} w_0 &= -\ln(1 - \eta_1/\phi_0), \\ w_1 &= \frac{\eta_1/\phi_0}{1 - \eta_1/\phi_0}\left(3\alpha_1 - \frac{p_0'}{\phi_0}\right), \\ w_2 &= \frac{\eta_1/\phi_0}{1 - \eta_1/\phi_0}\left[6\alpha_1 - 6\alpha_1\frac{p_0'}{\phi_0} + 2\left(\frac{p_0'}{\phi_0}\right)^2 - \frac{p_0''}{\phi_0}\right] \\ &\quad + \left(\frac{\eta_1/\phi_0}{1 - \eta_1/\phi_0}\right)^2\left(3\alpha_1 - \frac{p_0'}{\phi_0}\right)^2, \end{aligned} \tag{1.8}$$

where $p_0' = \frac{\partial p_0(\lambda_s)}{\partial\lambda_s}$ and $p_0'' = \frac{\partial^2 p_0(\lambda_s)}{\partial\lambda_s^2}$ at $\lambda_s = 0$.

After setting $\lambda_s = 1$ the expression (1.4) leads to the relation between the pressure P and the excess chemical potential μ_1^{ex} of a fluid in a matrix

$$\begin{aligned} \beta(\mu_1^{ex} - \mu_1^0) &= -\ln(1 - \eta_1/\phi_0) + A\frac{\eta_1/\phi_0}{1 - \eta_1/\phi_0} + B\frac{(\eta_1/\phi_0)^2}{(1 - \eta_1/\phi_0)^2} \\ &\quad + \frac{\beta P}{\rho_1}\frac{\eta_1}{\phi}, \end{aligned} \tag{1.9}$$

where the coefficients A and B define the porous medium structure and the expressions for them are as follow:

$$\begin{aligned} A &= 3\alpha_1 - \frac{p_0'}{\phi_0} + \frac{1}{2}\left[6\alpha_1 - 6\frac{p_0'}{\phi_0}\alpha_1 + 2\left(\frac{p_0'}{\phi_0}\right)^2 - \frac{p_0''}{\phi_0}\right], \\ B &= \frac{1}{2}\left(3\alpha_1 - \frac{p_0'}{\phi_0}\right)^2. \end{aligned} \tag{1.10}$$

Using the Gibbs-Duhem equation

$$\left(\frac{\partial P}{\partial \rho_1}\right)_T = \rho_1 \left(\frac{\partial \mu_1}{\partial \rho_1}\right)_T \tag{1.11}$$

one derives the fluid compressibility

$$\begin{aligned}
\beta\left(\frac{\partial P}{\partial \rho_1}\right)_T = {} & \frac{1}{(1-\eta_1/\phi)} + (1+A)\frac{\eta_1/\phi_0}{(1-\eta_1/\phi)(1-\eta_1/\phi_0)} \\
& + (A+2B)\frac{(\eta_1/\phi_0)^2}{(1-\eta_1/\phi)(1-\eta_1/\phi_0)^2} \\
& + 2B\frac{(\eta_1/\phi_0)^3}{(1-\eta_1/\phi)(1-\eta_1/\phi_0)^3},
\end{aligned} \tag{1.12}$$

which makes it possible to obtain the total chemical potential, $\beta\mu_1 = \ln(\rho_1 \Lambda_1^3 \Lambda_{1R}) + \beta\mu_1^{ex}$, and the pressure of a fluid, and a result of integration of (1.12) over the fluid density leads to [31]:

$$\begin{aligned}
\beta(\mu_1^{ex} - \mu_1^0) = {} & -\ln(1-\eta_1/\phi) + (A+1)\frac{\phi}{\phi-\phi_0}\ln\frac{1-\eta_1/\phi}{1-\eta_1/\phi_0} \\
& + (A+2B)\frac{\phi}{\phi-\phi_0}\left(\frac{\eta_1/\phi_0}{1-\eta_1/\phi_0} - \frac{\phi}{\phi-\phi_0}\ln\frac{1-\eta_1/\phi}{1-\eta_1/\phi_0}\right) \\
& + 2B\frac{\phi}{\phi-\phi_0}\left[\frac{1}{2}\frac{(\eta_1/\phi_0)^2}{(1-\eta_1/\phi_0)^2} - \frac{\phi}{\phi-\phi_0}\frac{\eta_1/\phi_0}{1-\eta_1/\phi_0}\right. \\
& \left. + \frac{\phi^2}{(\phi-\phi_0)^2}\ln\frac{1-\eta_1/\phi}{1-\eta_1/\phi_0}\right],
\end{aligned} \tag{1.13}$$

$$\begin{aligned}
\frac{\beta P}{\rho_1} = {} & -\frac{\phi}{\eta_1}\ln\frac{1-\eta_1/\phi}{1-\eta_1/\phi_0} + (1+A)\frac{\phi}{\eta_1}\frac{\phi}{\phi-\phi_0}\ln\frac{1-\eta_1/\phi}{1-\eta_1/\phi_0} \\
& + (A+2B)\frac{\phi}{\phi-\phi_0}\left[\frac{1}{1-\eta_1/\phi_0} - \frac{\phi}{\eta_1}\frac{\phi}{\phi-\phi_0}\ln\frac{1-\eta_1/\phi}{1-\eta_1/\phi_0}\right] \\
& + 2B\frac{\phi}{\phi-\phi_0}\left[\frac{1}{2}\frac{\eta_1/\phi_0}{(1-\eta_1/\phi_0)^2} - \frac{2\phi-\phi_0}{\phi-\phi_0}\frac{1}{1-\eta_1/\phi_0}\right. \\
& \left. + \frac{\phi}{\eta_1}\frac{\phi^2}{(\phi-\phi_0)^2}\ln\frac{1-\eta_1/\phi}{1-\eta_1/\phi_0}\right].
\end{aligned} \tag{1.14}$$

The expressions (1.13) and (1.14) are the result of SPT2 approach. At high fluid densities the obtained expressions have two divergences, which appear in $\eta_1 = \phi$ and $\eta_1 = \phi_0$ respectively. Since $\phi < \phi_0$ the first divergence in $\eta_1 = \phi$ occurs at lower densities than the second one. From geometrical point of view such a divergence should appear at higher densities close to the maximum value of fluid packing fraction $\eta_1^{\max}$ available for a fluid in a given matrix. The different corrections and improvements of the SPT2 approach were proposed in [20, 24, 25, 31]. First corrections were given in [20], where on the basis of SPT2 four approximations were developed. One of them called SPT2b can be derived if ϕ is replaced by ϕ_0 everywhere in (1.12) except the first term. As a result, (1.13) and (1.14) can be rewritten in the following form

$$\begin{aligned}\beta(\mu_1^{ex} - \mu_1^0)^{SPT2b} &= -\ln(1-\eta_1/\phi) + (1+A)\frac{\eta_1/\phi_0}{1-\eta_1/\phi_0} \\ &\quad + \frac{1}{2}(A+2B)\frac{(\eta_1/\phi_0)^2}{(1-\eta_1/\phi_0)^2} + \frac{2}{3}B\frac{(\eta_1/\phi_0)^3}{(1-\eta_1/\phi_0)^3},\end{aligned} \tag{1.15}$$

$$\begin{aligned}\left(\frac{\beta P}{\rho_1}\right)^{SPT2b} &= -\frac{\phi}{\eta_1}\ln\left(1-\frac{\eta_1}{\phi}\right) + \frac{\phi_0}{\eta_1}\ln\left(1-\frac{\eta_1}{\phi_0}\right) + \frac{1}{1-\eta_1/\phi_0} \\ &\quad + \frac{A}{2}\frac{\eta_1/\phi_0}{(1-\eta_1/\phi_0)^2} + \frac{2B}{3}\frac{(\eta_1/\phi_0)^2}{(1-\eta_1/\phi_0)^3}.\end{aligned} \tag{1.16}$$

The second approximation proposed in [24, 25] is called SPT2b1 and it corrects SPT2b by removing the divergence at $\eta_1 = \phi$ through an expansion of the logarithmic term in (1.15)

$$-\ln(1-\eta_1/\phi) \approx -\ln(1-\eta_1/\phi_0) + \frac{\eta_1(\phi_0-\phi)}{\phi_0\phi(1-\eta_1/\phi_0)}. \tag{1.17}$$

Therefore, one obtains the expressions for the chemical potential and pressure within the SPT2b1 approximation as follows

$$\begin{aligned}\beta(\mu_1^{ex} - \mu_1^0)^{SPT2b1} &= -\ln(1-\eta_1/\phi_0) + (1+A)\frac{\eta_1/\phi_0}{1-\eta_1/\phi_0} + \frac{\eta_1(\phi_0-\phi)}{\phi_0\phi(1-\eta_1/\phi_0)} \\ &\quad + \frac{1}{2}(A+2B)\frac{(\eta_1/\phi_0)^2}{(1-\eta_1/\phi_0)^2} + \frac{2}{3}B\frac{(\eta_1/\phi_0)^3}{(1-\eta_1/\phi_0)^3},\end{aligned} \tag{1.18}$$

$$\begin{aligned}\left(\frac{\beta P}{\rho_1}\right)^{SPT2b1} &= \frac{1}{1-\eta_1/\phi_0}\frac{\phi_0}{\phi} + \left(\frac{\phi_0}{\phi}-1\right)\frac{\phi_0}{\eta_1}\ln\left(1-\frac{\eta_1}{\phi_0}\right) \\ &\quad + \frac{A}{2}\frac{\eta_1/\phi_0}{(1-\eta_1/\phi_0)^2} + \frac{2B}{3}\frac{(\eta_1/\phi_0)^2}{(1-\eta_1/\phi_0)^3}.\end{aligned} \tag{1.19}$$

Two other approximations called SPT2b2 and SPT2b3 contain the third type of porosity ϕ^* defined by the maximum value of packing fraction of a fluid in a porous medium and provide the more correct description of thermodynamic properties of a fluid in the high-density region, which corresponds to the close packing condition.

To introduce ϕ^* in the expression for the chemical potential (1.15) the logarithmic term is modified in the following way

$$-\ln(1-\eta_1/\phi) \approx -\ln(1-\eta_1/\phi^*) + \frac{\eta_1(\phi^*-\phi)}{\phi^*\phi(1-\eta_1/\phi^*)}. \tag{1.20}$$

Consequently, the SPT2b2 approximation is derived as

$$\begin{aligned}\beta\left(\mu_1^{ex}-\mu_1^0\right)^{SPT2b2} &= -\ln(1-\eta_1/\phi^*) + \frac{\eta_1/\phi_0}{1-\eta_1/\phi_0}(1+A) \\ &\quad + \frac{\eta_1(\phi^*-\phi)}{\phi^*\phi(1-\eta_1/\phi^*)} + \frac{1}{2}(A+2B)\frac{(\eta_1/\phi_0)^2}{(1-\eta_1/\phi_0)^2} + \frac{2}{3}B\frac{(\eta_1/\phi_0)^3}{(1-\eta_1/\phi_0)^3},\end{aligned} \tag{1.21}$$

$$\begin{aligned}\left(\frac{\beta P}{\rho_1}\right)^{SPT2b2} &= -\frac{\phi^*}{\eta_1}\ln\left(1-\frac{\eta_1}{\phi^*}\right) + \frac{\phi_0}{\eta_1}\ln\left(1-\frac{\eta_1}{\phi_0}\right) + \frac{1}{1-\eta_1/\phi_0} \\ &\quad + \frac{\phi^*-\phi}{\phi}\left[\ln(1-\eta_1/\phi^*) + \frac{\eta_1/\phi^*}{1-\eta_1/\phi^*}\right] \\ &\quad + \frac{A}{2}\frac{\eta_1/\phi_0}{(1-\eta_1/\phi_0)^2} + \frac{2}{3}B\frac{(\eta_1/\phi_0)^2}{(1-\eta_1/\phi_0)^3}.\end{aligned} \tag{1.22}$$

Finally, the SPT2b3 approximation can be obtained similar to the SPT2b2 approximation through an expansion of the logarithmic term $\ln(1-\eta_1/\phi^*)$ in the expression (1.21) for the chemical potential. As a result we obtain

$$\begin{aligned}\beta\left(\mu_1^{ex}-\mu_1^0\right)^{SPT2b3} &= -\ln(1-\eta_1/\phi_0) + \frac{\eta_1/\phi^*}{1-\eta_1/\phi_0} + \frac{\eta_1(\phi^*-\phi)}{\phi^*\phi(1-\eta_1/\phi^*)} \\ &\quad + A\frac{\eta_1/\phi_0}{1-\eta_1/\phi_0} + \frac{1}{2}(A+2B)\frac{(\eta_1/\phi_0)^2}{(1-\eta_1/\phi_0)^2} + \frac{2}{3}B\frac{(\eta_1/\phi_0)^3}{(1-\eta_1/\phi_0)^3},\end{aligned} \tag{1.23}$$

$$\begin{aligned}\left(\frac{\beta P}{\rho_1}\right)^{SPT2b3} &= \frac{\phi^*-\phi}{\phi}\left[\ln\left(1-\frac{\eta_1}{\phi^*}\right) + \frac{\eta_1/\phi^*}{1-\eta_1/\phi^*}\right] + \frac{1}{1-\eta_1/\phi_0} \\ &\quad + \frac{\phi_0-\phi^*}{\phi^*}\left[\ln(1-\eta_1/\phi_0) + \frac{\eta_1/\phi_0}{1-\eta_1/\phi_0}\right] \\ &\quad + \frac{A}{2}\frac{\eta_1/\phi_0}{(1-\eta_1/\phi_0)^2} + \frac{2}{3}B\frac{(\eta_1/\phi_0)^2}{(1-\eta_1/\phi_0)^3}.\end{aligned} \tag{1.24}$$

In [24] it was shown on the example of one-dimensional system of a fluid in a random matrix that ϕ^* is related to ϕ_0 and ϕ with the following relation

$$1/\phi^{*} = (1/\phi - 1/\phi_0)/\ln(\phi_0/\phi). \tag{1.25}$$

Since (1.25) is presented in the general form and does not depend directly on the dimensionality of the system, we extend its application to the three-dimensional case. In the bulk all the porosities are equal $\phi = \phi_0 = \phi^{*} = 1$. Therefore, the expressions for pressure and chemical potential of bulk hard sphere (HS) and HCB fluids have the same divergence at $\eta_1 = 1$ [33, 34]. In the case of a fluid in a porous medium one gets an inequality $\phi < \phi^{*} < \phi_0$. It should be noted that in the case when matrix particles size is essentially larger than a size of fluid particles, i.e. a size ratio of fluid to matrix particles tends to zero, hence the porosities tend to the same value $\phi \approx \phi_0 \approx \phi^{*}$ and all the considered approximations lead to the same result, which is equivalent to a bulk fluid with the effective density $\hat{\eta}_1 = \eta_1/\phi_0$.

An application of the developed theory is illustrated for two models of porous medium in [31]. The first model is a HCB matrix and the second model is an overlapping hard convex body (OHCB) matrix. The geometrical porosities for these models have the form [31]

$$\phi_0 = 1 - \eta_0 \tag{1.26}$$

for a HCB matrix and

$$\phi_0 = \mathrm{e}^{-\eta_0} \tag{1.27}$$

for an OHCB matrix, where $\eta_0 = \rho_0 V_0$, $\rho_0 = \frac{N_0}{V}$, N_0 is the number of matrix particles, V_0 is the volume of a matrix particle and V is the total volume of system.

Using the SPT theory [27] the following expression for the probe particle porosity ϕ is derived

$$\begin{aligned}\phi = e^{-\beta\mu_1^0} = (1-\eta_0)\exp\Bigg[-\Bigg(\frac{\eta_0}{1-\eta_0}\left(3\alpha_0\left(\frac{R_1}{R_0}+\frac{S_1}{S_0}\right)+\frac{V_1}{V_0}\right)\\ +\frac{3\alpha_0\eta_0^2}{2(1-\eta_0)^2}\left(3\alpha_0\frac{R_1^2}{R_0^2}+2\frac{V_1}{V_0}\right)+\frac{3\alpha_0^2\eta_0^3}{(1-\eta_0)^3}\frac{V_1}{V_0}\Bigg)\Bigg]\end{aligned} \tag{1.28}$$

for the case of a HCB matrix and

$$\phi = e^{-\beta\mu_1^0} = \exp\left[-\eta_0\left(1+3\alpha_0\frac{R_1}{R_0}+3\alpha_0\frac{S_1}{S_0}+\frac{V_1}{V_0}\right)\right] \tag{1.29}$$

for the case of an OHCB matrix, where S_0 and R_0 are the surface area and the mean curvature of matrix particles respectively, $\alpha_0 = R_0 S_0/3V_0$ is the parameter of non-sphericity of matrix particles.

The probability to find a place for the scaled particle in a HCB matrix is equal to

$$p_0(\lambda_s) = 1 - \eta_0\left(1 + 3\alpha_0\frac{R_1\lambda_s}{R_0} + 3\alpha_0\frac{\lambda_s^2 S_1}{S_0} + \lambda_s^3\frac{V_1}{V_0}\right). \tag{1.30}$$

The corresponding derivatives from (1.30) used for the SPT ansatz are

$$p_0' = -3\eta_0\alpha_0\frac{R_1}{R_0}, \quad p_0'' = -6\eta_0\alpha_0\frac{S_1}{S_0}. \tag{1.31}$$

Similarly, in the case of OHCB matrix one can obtain an expression for the probability $p_0(\lambda_s)$:

$$p_0(\lambda_s) = \exp\left[-\eta_0\left(1 + 3\alpha_0\frac{R_1}{R_0}\lambda_s + 3\alpha_0\frac{S_1}{S_0}\lambda_s^2 + \frac{V_1}{V_0}\lambda_s^3\right)\right] \tag{1.32}$$

as well as the expressions for its derivatives

$$p_0' = -3\eta_0\alpha_0\frac{R_1}{R_0}\phi_0, \quad p_0'' = -3\eta_0\alpha_0\left(2\frac{S_1}{S_0} - 3\eta_0\alpha_0\left(\frac{R_1}{R_0}\right)^2\right)\phi_0. \tag{1.33}$$

Therefore, using (1.10) the coefficients A and B can be derived as

$$\begin{aligned} A &= 3\left[2\alpha_1 + \alpha_0\frac{\eta_0}{(1-\eta_0)}\frac{R_1}{R_0}(1 + 3\alpha_1) + \alpha_0\frac{S_1}{S_0}\frac{\eta_0}{(1-\eta_0)} + 3\alpha_0^2\frac{R_1^2}{R_0^2}\frac{\eta_0^2}{(1-\eta_0)^2}\right], \\ B &= \frac{9}{2}\left(\alpha_1 + \alpha_0\frac{R_1}{R_0}\frac{\eta_0}{1-\eta_0}\right)^2 \end{aligned} \tag{1.34}$$

for a HCB fluid in a HCB matrix and as

$$\begin{aligned} A &= 3\left[2\alpha_1 + \alpha_0\eta_0\frac{R_1}{R_0}(1 + 3\alpha_1) + \alpha_0\frac{S_1}{S_0}\eta_0 + 3\alpha_0^2\frac{R_1^2}{R_0^2}\eta_0^2\right], \\ B &= \frac{9}{2}\left(\alpha_1 + \alpha_0\frac{R_1}{R_0}\eta_0\right)^2 \end{aligned} \tag{1.35}$$

for a HCB fluid in an OHCB matrix.

An accuracy of the approximations proposed in [31] were tested by a comparison with computer simulation data obtained from the grand-canonical Monte Carlo (GCMC) simulations [36]. In this study a model of a HS fluid in a OHCB matrix was investigated. Four different OHCB matrices with parameters presented in Table 1.1 were considered. Using the different approximations the chemical potential depending on the fluid density (packing fraction) is calculated (Fig. 1.1). As one can see at low fluid densities all approximations are correct except the SPT2

Table 1.1 Parameters and characteristics of matrices for the systems A, B, C and D

System	L_0	D_0	η_0	ϕ_0	ϕ	α_0
A	2.0	3.0	0.282	0.754	0.556	1.111
B	3.0	2.0	0.271	0.762	0.493	1.346
C	5.0	2.0	0.099	0.9052	0.781	1.658
D	10.0	1.0	0.167	0.846	0.491	4.125

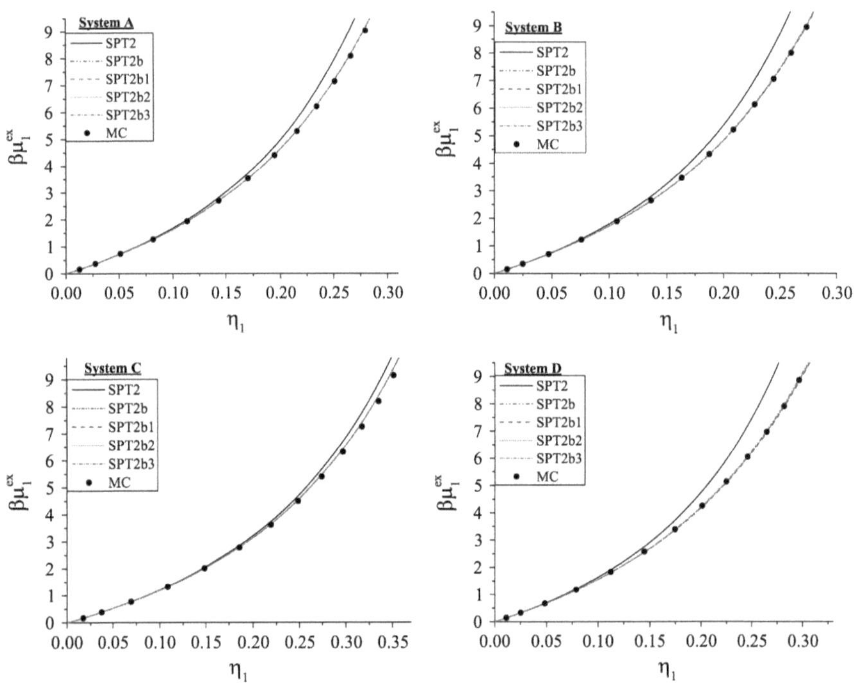

Fig. 1.1 The excess chemical potential $\beta\mu_1^{ex}$ versus the fluid packing fraction η_1 for a HS fluid in random OHCB matrices. The parameters of matrices are presented in Table 1.1. A comparison of the different approximations (*lines*) with the GCMC simulation results (*symbols*)

approach, which overestimates the chemical potential at intermediate densities in a comparison with the GCMC data. On the other hand, the SPT2b approximation improves essentially the results for intermediate fluid densities. In the most considered cases the results of SPT2b coincide with other approximations (SPT2b1, SPT2b2 and SPT2b3) and the observed deviations are comparable with the statistical errors of the simulations (0.5 %). However, as it was shown in [31] in the case of a large difference between the porosities ϕ and ϕ_0, at high fluid densities the approximations SPT2b1, SPT2b2 and SPT2b3 give better results than the SPT2b approach.

1.3 Hard Spherocylinder Fluid in Random Porous Media: SPT with Two Scaling Parameters

Generally, if a shape of hard convex body particles is not strictly specified or it is rather complex, it is acceptable to restrict oneself by one scaling parameter. On the other hand, particles of spherocylindrical shape are characterized only by two measures, i.e. by their length and diameter. Therefore, applying the SPT formalism it is reasonable to change sizes of scaled particle along exactly these two measures. We consider two scaling parameters generalization of the SPT theory for the description of thermodynamic properties of a hard spherocylinder (HSC) fluid in random matrices. We also apply these results to a study of the effect of porous medium on the isotropic-nematic orientational transition appearing in this fluid.

As it was mentioned above a hard spherocylinder system is defined by the volume V, the surface area S and the mean curvature R given by (1.1). According to the general idea of the SPT theory [28, 29, 32] we introduce in a confined HSC fluid an additional hard spherocylinder with the scaling diameter D_s and the scaling length L_s:

$$D_s = \lambda_s D_1, \quad L_s = \alpha_s L_1, \tag{1.36}$$

where D_1 and L_1 are the diameter and the length of fluid spherocylinders respectively. The excess chemical potential for the small scaled particle in a HSC fluid confined in a matrix can be written in the form

$$\begin{aligned} \beta\mu_s^{ex} = & -\ln p_0(\alpha_s, \lambda_s) - \ln\Bigg[1 - \frac{\eta_1}{V_1 p_0(\alpha_s, \lambda_s)}\bigg(\frac{\pi}{6}D_1^3(1+\lambda_s)^3 \\ & + \frac{\pi}{4}D_1^2 L_1(1+\lambda_s)^2(1+\alpha_s) \\ & + \frac{\pi}{4}D_1 L_1^2(1+\lambda_s)\alpha_s \int f(\Omega_1)f(\Omega_2)\sin\vartheta_{12}d\Omega_1 d\Omega_2\bigg)\Bigg], \end{aligned} \tag{1.37}$$

where $\eta_1 = \rho_1 V_1$ is the fluid packing fraction and ρ_1 is the fluid density; $p_0(\alpha_s, \lambda_s)$ is the probability to find a cavity created by a scale particle in the empty matrix and it is defined by the excess chemical potential μ_s^0 of the scale particle in the limit of infinite dilution of a fluid; $\Omega = (\vartheta, \varphi)$ is the orientation of particles defined by the angles ϑ and φ; $d\Omega = \frac{1}{4\pi}\sin\vartheta d\vartheta d\varphi$ is the normalized angle element; ϑ_{12} is the angle between orientational vectors of two molecules; $f(\Omega)$ is the single orientational distribution function normalized in such a way that

$$\int f(\Omega)d\Omega = 1. \tag{1.38}$$

For the large scale particle the excess chemical potential is given by a thermodynamic expression, which can be presented in the form similar to (1.4):

$$\beta\mu_s^{ex} = w(\alpha_s, \lambda_s) + \beta P V_s / p_0(\lambda_s, \alpha_s), \tag{1.39}$$

where $w(\alpha_s, \lambda_s)$ is the following:

$$w(\lambda_s, \alpha_s) = w_{00} + w_{10}\lambda_s + w_{01}\alpha_s + w_{11}\alpha_s\lambda_s + \frac{w_{20}\lambda_s^2}{2}. \tag{1.40}$$

According to the ansatz of SPT theory [28, 29, 32] the coefficients of the expansion (1.40) can be found from the continuity of the excess chemical potential given in (1.37) and (1.39), as well as from the corresponding derivatives $\partial\mu_s^{ex}/\partial\lambda_s$, $\partial\mu_s^{ex}/\partial\alpha_s$, $\partial^2\mu_s^{ex}/\partial\alpha_s\partial\lambda_s$ and $\partial^2\mu_s^{ex}/\partial\lambda_s^2$. As a result one derives the following coefficients

$$w_{00} = -\ln(1 - \eta_1/\phi_0), \tag{1.41}$$

$$w_{10} = \frac{\eta_1/\phi_0}{1 - \eta_1/\phi_0}\left(\frac{6\gamma_1}{3\gamma_1 - 1} - \frac{p'_{0\lambda}}{\phi_0}\right), \tag{1.42}$$

$$w_{01} = \frac{\eta_1/\phi_0}{1 - \eta_1/\phi_0}\left(\frac{3(\gamma_1 - 1)}{3\gamma_1 - 1} + \frac{3(\gamma_1 - 1)^2}{3\gamma_1 - 1}\tau(f) - \frac{p'_{0\alpha}}{\phi_0}\right), \tag{1.43}$$

$$\begin{aligned} w_{11} = \frac{\eta_1/\phi_0}{1 - \eta_1/\phi_0}\Bigg(&\frac{6(\gamma_1 - 1)}{3\gamma_1 - 1} + \frac{3(\gamma_1 - 1)^2\tau(f)}{3\gamma_1 - 1} - \frac{p''_{0\alpha\lambda}}{\phi_0} \\ &+ 2\frac{p'_{0\alpha}p'_{0\lambda}}{\phi_0^2} - \frac{3(\gamma_1 - 1) + 3(\gamma_1 - 1)^2\tau(f)}{3\gamma_1 - 1}\frac{p'_{0\lambda}}{\phi_0} - \frac{6\gamma_1}{3\gamma_1 - 1}\frac{p'_{0\alpha}}{\phi_0}\Bigg) \\ &+ \left(\frac{\eta_1/\phi_0}{1 - \eta_1/\phi_0}\right)^2\left(\frac{6\gamma_1}{3\gamma_1 - 1} - \frac{p'_{0\lambda}}{\phi_0}\right) \\ &\times \left(\frac{3(\gamma_1 - 1)}{3\gamma_1 - 1} + \frac{3(\gamma_1 - 1)^2\tau(f)}{3\gamma_1 - 1} - \frac{p'_{0\alpha}}{\phi_0}\right), \end{aligned} \tag{1.44}$$

$$\begin{aligned} w_{20} = \frac{\eta_1/\phi_0}{1 - \eta_1/\phi_0}&\left(\frac{6(1 + \gamma_1)}{3\gamma_1 - 1} - \frac{12\gamma_1}{3\gamma_1 - 1}\frac{p'_{0\lambda}}{\phi_0} + 2\left(\frac{p'_{0\lambda}}{\phi_0}\right)^2 - \frac{p''_{0\lambda\lambda}}{\phi_0}\right) \\ &+ \left(\frac{\eta_1/\phi_0}{1 - \eta_1/\phi_0}\right)^2\left(\frac{6\gamma_1}{3\gamma_1 - 1} - \frac{p'_{0\lambda}}{\phi_0}\right)^2, \end{aligned} \tag{1.45}$$

where

$$\gamma_1 = 1 + \frac{L_1}{D_1}, \tag{1.46}$$

$$\tau(f) = \frac{4}{\pi}\int f(\Omega_1)f(\Omega_2)\sin\vartheta_{12}d\Omega_1 d\Omega_2. \tag{1.47}$$

If both the scale parameters equal to zero $\alpha_s = \lambda_s = 0$, the probability p_0 is equivalent to the geometrical porosity:

$$\phi_0 = p_0(\alpha_s = \lambda_s = 0), \tag{1.48}$$

and in the case of HSC matrix or overlapping HSC matrix it is defined by the relations (1.26) or (1.27) respectively.

Setting $\alpha_s = \lambda_s = 1$ in the (1.39) leads to the expression similar to (1.9), the chemical potential μ_1^{ex} of a fluid in a matrix. However, now the constants A and B have more general and complicated form

$$\begin{aligned} A(\tau(f)) = {} & 6 + \frac{6(\gamma_1 - 1)^2\tau(f)}{3\gamma_1 - 1} \\ & - \frac{p'_{0\lambda}}{\phi_0}\left(4 + \frac{3(\gamma_1 - 1)^2\tau(f)}{3\gamma_1 - 1}\right) - \frac{p'_{0\alpha}}{\phi_0}\left(1 + \frac{6\gamma_1}{3\gamma_1 - 1}\right) \\ & - \frac{p''_{0\alpha\lambda}}{\phi_0} - \frac{1}{2}\frac{p''_{0\lambda\lambda}}{\phi_0} + 2\frac{p'_{0\alpha}p'_{0\lambda}}{\phi_0^2} + \left(\frac{p'_{0\lambda}}{\phi_0}\right)^2, \end{aligned} \tag{1.49}$$

$$\begin{aligned} B(\tau(f)) = {} & \left(\frac{6\gamma_1}{3\gamma_1 - 1} - \frac{p'_{0\lambda}}{\phi_0}\right) \\ & \times \left(\frac{3(2\gamma_1 - 1)}{3\gamma_1 - 1} + \frac{3(\gamma_1 - 1)^2\tau(f)}{3\gamma_1 - 1} - \frac{p'_{0\alpha}}{\phi_0} - \frac{1}{2}\frac{p'_{0\lambda}}{\phi_0}\right), \end{aligned} \tag{1.50}$$

where $p'_{0\lambda} = \frac{\partial p_0(\alpha_s,\lambda_s)}{\partial\lambda_s}, p'_{0\alpha} = \frac{\partial p_0(\alpha_s,\lambda_s)}{\partial\alpha_s}, p''_{0\alpha\lambda} = \frac{\partial^2 p_0(\alpha_s,\lambda_s)}{\partial\alpha_s\partial\lambda_s}, p''_{0\lambda\lambda} = \frac{\partial^2 p_0(\alpha_s,\lambda_s)}{\partial\lambda_s^2}$ are the corresponding derivatives at $\alpha = \lambda = 0$. Also the probe particle porosity ϕ can be obtained from

$$\phi = p_0(\alpha_s = \lambda_s = 1). \tag{1.51}$$

In order to derive expressions for the chemical potential we repeat the calculations presented in the previous section. Using the Gibbs-Duhem equation we get an expression for the compressibility in the form (1.12). After integration of this expression over the fluid density we obtain the excess chemical potential μ_1^{ex} and the pressure of a fluid in the form similar to the SPT2 approximation (1.13)–(1.14). On the basis of SPT2 result we construct the SPT2b approximation in the form (1.15) to (1.16). Following to the the scheme presented in the previous section we also derive the expressions for the SPT2b1, SPT2b2 and SPT2b3 approximations in the form like (1.18)–(1.19), (1.21)–(1.22) and (1.23)–(1.24) respectively. The only difference in the new expressions of the chemical potential in a comparison with ones obtained in the previous section is an additional entropic term $\sigma(f)$:

$$\sigma(f) = \int f(\Omega)\ln f(\Omega)d\Omega. \tag{1.52}$$

As an example we present here the expression for the chemical potential of a fluid confined in a matrix using the SPT2b approximation:

$$\begin{aligned}\beta(\mu_1^{ex} - \mu_1^0)^{SPT2b} &= \sigma(f) - \ln(1 - \eta_1/\phi) + (1 + A(\tau(f)))\frac{\eta_1/\phi_0}{1 - \eta_1/\phi_0} \\ &+ \frac{1}{2}(A(\tau(f)) + 2B(\tau(f)))\frac{(\eta_1/\phi_0)^2}{(1 - \eta_1/\phi_0)^2} \\ &+ \frac{2}{3}B(\tau(f))\frac{(\eta_1/\phi_0)^3}{(1 - \eta_1/\phi_0)^3}.\end{aligned} \tag{1.53}$$

From the thermodynamic relationship

$$\frac{\beta F}{V} = \beta\mu_1\rho_1 - \beta P \tag{1.54}$$

one can obtain the expression for the free energy. Within the SPT2b approximation the free energy of a confined fluid is the following

$$\begin{aligned}\beta V^{-1}F^{SPT2b} &= \rho_1\sigma(f) + \rho_1(\ln(\Lambda_1^3\rho_1) - 1) + \beta\mu_1^0\rho_1 - \rho_1\ln(1 - \eta_1/\phi) \\ &+ \frac{\rho_1\phi}{\eta_1}\ln(1 - \eta_1/\phi) - \frac{\rho_1\phi_0}{\eta_1}\ln(1 - \eta_1/\phi_0) \\ &+ \rho_1\frac{A(\tau(f))}{2}\frac{\eta_1/\phi_0}{1 - \eta_1/\phi_0} + \rho_1\frac{B(\tau(f))}{3}\left(\frac{\eta_1/\phi_0}{1 - \eta_1/\phi_0}\right)^2.\end{aligned} \tag{1.55}$$

Now we return to the orientational distribution function $f(\Omega)$ introduced in the beginning of this section. Distribution function $f(\Omega)$ can be determined from a minimization of the free energy with respect to variations in this distribution. This procedure leads to the nonlinear integral equation

$$\ln f(\Omega_1) + 1 + C\int f(\Omega_2)\sin\vartheta_{12}d\Omega_2 = 0, \tag{1.56}$$

where

$$C = \frac{\eta_1/\phi_0}{1-\eta_1/\phi_0}\left[\frac{3(\gamma_1-1)^2}{3\gamma_1-1}\left(1-\frac{p'_{0\lambda}}{2\phi_0}\right)\right.$$
$$\left.+\frac{\eta_1/\phi_0}{(1-\eta_1/\phi_0)}\frac{(\gamma_1-1)^2}{3\gamma_1-1}\left(\frac{6\gamma_1}{3\gamma_1-1}-\frac{p'_{0\lambda}}{\phi_0}\right)\right]. \tag{1.57}$$

The (1.56) should be solved together with the normalization condition (1.38). The solution of the (1.56) can be calculated numerically using an iteration procedure according to the algorithm proposed in [37]. We should note that the (1.56) for the singlet distribution function $f(\Omega)$ has the same structure as the corresponding equation obtained by Onsager [38] for the hard spherocylinder fluid in the limit $L\to\infty$, $D_1\to 0$, while the dimensionless density of fluid $c=\frac{1}{4}\pi L_1^2 D_1\rho_1$ is fixed. Therefore, in the Onsager limit one has

$$C\to c=\frac{1}{4}\pi L_1^2 D_1\rho_1. \tag{1.58}$$

This result within the framework of the SPT theory was generalized for a HSC fluid with the finite value of L_1 and D_1 [39, 40], and in this case

$$C=\frac{\eta_1}{1-\eta_1}\left[\frac{3(\gamma_1-1)^2}{3\gamma_1-1}+\frac{\eta_1}{1-\eta_1}\frac{6\gamma_1(\gamma_1-1)^2}{(3\gamma_1-1)^2}\right]. \tag{1.59}$$

It is not difficult to show that the expression (1.59) corresponds to the bulk case $(\phi_0=1, p'_{0\lambda}=0)$ of our result (1.57).

From the bifurcation analysis of (1.56) it is found that this equation has two characteristic points c_i and c_n [41]. For the Onsager model in the bulk [41, 42]

$$c_i=3.290,\quad c_n=4.191, \tag{1.60}$$

where c_i corresponds to high densities of stable isotropic fluid and c_n is related with the minimal density of stable orientational ordering, i.e. a nematic state of fluid.

In the presence of a porous medium for the Onsager model we obtain

$$c_i/\phi_0=3.290,\quad c_n/\phi_0=4.191. \tag{1.61}$$

It means that the isotropic-nematic phase transition in the presence of a matrix shifts to lower densities of a fluid.

For finite values of L_1 and D_1 we can put

$$C_i=3.290,\quad C_n=4.191, \tag{1.62}$$

where C_i and C_n are defined from (1.57). The values in (1.62) define the isotropic-nematic phase diagram for a HSC fluid in a matrix depending on the ratio L_1/D_1 and the parameters of a matrix.

As an example we consider a porous medium formed by a HS matrix. The probability to find scaled spherocylinder in an empty HS matrix is equal to

$$p_0(\alpha_s, \lambda_s) = 1 - \eta_0 \frac{1}{V_0} \frac{\pi}{2} \left[\frac{1}{3} (D_0 + \lambda_s D_1)^3 + \frac{1}{2} \alpha_s L_1 (D_0 + \lambda_s D_1)^2 \right]. \tag{1.63}$$

From (1.63) one can find the derivatives needed for the description of thermodynamic properties of a confined fluid:

$$\begin{gathered} p'_{0\alpha} = -3 \frac{D_1}{D_0} \eta_0, \quad p'_{0\alpha} = -\frac{3}{2} \eta_0 \frac{L_1}{D_0}, \quad p''_{0\alpha\lambda} = -3\eta_0 \frac{L_1}{D_0} \frac{D_1}{D_0}, \\ p''_{0\lambda\lambda} = -6\eta_0 \frac{D_1^2}{D_0^2}, \end{gathered} \tag{1.64}$$

where η_0 is the packing fraction of HS matrix particles. The probe particle porosity in this case is equal to [32]

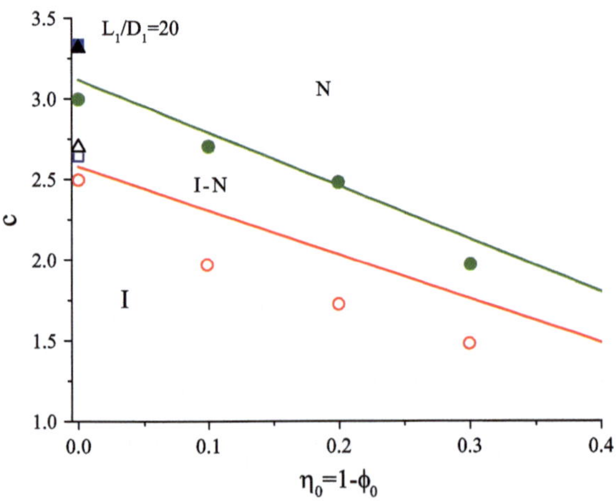

Fig. 1.2 Coexistence lines of isotropic-nematic phases of a hard spherocylinder fluid in a hard sphere matrix for $L_1/D_1 = 20$ and $D_0 = L_1$ presented as a dependence of the spherocylinder fluid density $c = \frac{1}{4}\pi\rho_1 L_1^2 D_1$ on the matrix packing fraction η_0. The GEMC simulation results taken from [7] are shown as *circles*, from [43] are shown as *squares* and *triangles* (GDI). The isotropic phase are denoted by open symbols and the nematic phase—by filled symbols. Solid lines corresponds to the SPT theory. The notations "I" and "N" mean isotropic and nematic phases respectively

$$\phi = (1-\eta_0)\exp\left[-\frac{\eta_0}{1-\eta_0}\frac{D_1}{D_0}\left(\frac{3}{2}(\gamma_1+1)+3\gamma_1\frac{D_1}{D_0}\right)-\frac{\eta_0^2}{(1-\eta_0)^2}\frac{9}{2}\gamma_1\frac{D_1^2}{D_0^2}\right.$$
$$\left.-\frac{\eta_0}{(1-\eta_0)^3}(3\gamma_1-1)\frac{1}{2}\frac{D_1^3}{D_0^3}(1+\eta_0+\eta_0^2)\right]. \tag{1.65}$$

The analysis of the phase diagrams of a fluid in a matrix [32] obtained according to (1.57) and (1.62) shows that the isotropic-nematic phase coexistence shifts to smaller densities with decreasing of the value of L_1/D_1 as well as with decreasing of the matrix porosity ϕ_0. In Fig. 1.2 one can observe how the matrix porosity affects the phase coexistence in the case of the HSC fluid with $L_1/D_1 = 20$ and the HS matrix with $L_1/D_0 = 1$. For comparison in this figure it is also presented the results of computer simulation of Schmidt and Dijkstra [7] obtained by the method of Gibbs ensemble Monte Carlo (GEMC). For the bulk case ($\eta_0 = 0$) the results of Bolhuis and Frenkel [43] are shown in Fig. 1.2 as well. These results were obtained using the common GEMC method and GEMC combined with the modified Gibbs-Duhem integration (GDI) method. As one can see in Fig. 1.2 our theory overestimates the effect of porous medium, especially it is noticeable for the isotropic branch of phase coexistence. On the other hand, the nematic branch looks rather satisfactory.

We should note that for isotropic-nematic coexistence lines can also be found from the condition of thermodynamic equilibrium. According to this the isotropic and nematic phases have the same pressure and the same chemical potential:

$$P_i(c_i) = P_n(c_n), \quad \mu_i(c_i) = \mu_n(c_n). \tag{1.66}$$

In [41] it was shown for the Onsager model in the bulk case that the results obtained from bifurcation analysis and from the thermodynamic consideration coincide exactly. Evidently, we can expect the same for the Onsager model in the case of the porous medium presence. We observed in [32] that for the finite value of L_1/D_1 there is some deviation between the results obtained from the thermodynamic and bifurcation analysis, which increases slightly with increasing of the ratio L_1/D_1.

1.4 Generalization of Van der Waals Equation for Anisotropic Fluid in Random Porous Media

It is well established that the short-range order in simple and molecular liquids is determined by the repulsive part of intermolecular interaction [33, 34]. Such a short-range structure is essentially related to the packing of hard core particles, which can be modeled by hard spheres (HS) in the case of simple fluid or hard convex bodies (HCB) system in the case of molecular fluid. Similar to the bulk case [33, 34, 44] the results obtained from the SPT theory for HS and HCB fluids confined in random matrices can be used as the reference system within the perturbation theory of fluids. In this section as the first step we consider the possibility of an extension of the Van der Waals Equation of state to a simple and anisotropic molecular fluid in a random porous medium.

Considering the case of a simple fluid with a pair potential of interaction, which consists of a HS repulsive and an attractive parts, we start from the well-known Kac potential [45, 46]

$$U^{att}(r) = \gamma^3 U(\gamma r), \tag{1.67}$$

where r is a distance between two particles. In the same way as for a bulk fluid in the limit $\gamma \to 0$ the pressure of a confined fluid can be presented in the form [44, 45]

$$\frac{\beta P}{\rho_1} = \left(\frac{\beta P}{\rho_1}\right)_{HS} - 12a\eta_1\beta, \tag{1.68}$$

where $\left(\frac{\beta P}{\rho_1}\right)_{HS}$ is a contribution of a HS interaction, which can be obtained from (1.16) within the SPT2b approximation, and $\beta = 1/k_B T$. Therefore, a HS fluid confined in a random matrix is taken as a reference system. The second term in (1.68) is a contribution of the attractive interaction defined by the constant a, which can be calculated from the following expression:

$$a = -\frac{1}{\phi_0 D_1^3} \int_0^\infty \gamma^3 U(\gamma r) r^2 dr, \tag{1.69}$$

where the factor $1/\phi_0$ excludes the volume occupied by matrix particles, since this volume does not contribute to the fluid attraction. We also introduce a size of hard core of fluid particles D_1, which means a diameter of HS particles.

As an example, we substitute an attractive pair potential $U^{att}(r)$ with the Lennard-Jones potential [44] in the form

$$U^{att}(r) = \begin{cases} 4\varepsilon_1 \left[\left(\frac{D_1}{r}\right)^{12} - \left(\frac{D_1}{r}\right)^6 \right], & r \geq D_1 \\ 0, & r < D_1 \end{cases}. \tag{1.70}$$

Using the Gibbs-Duhem relationship one can derive the expression of chemical potential from (1.68). Having the analytical expressions for the equation of state and the chemical potential one can build the liquid-vapour phase diagram in coordinates $\eta_1 - T^*$, where T^* is dimensionless temperature $T^* = kT/\varepsilon_1 = 1/\beta\varepsilon_1$.

In Fig. 1.3 the liquid-vapour coexistence curves are presented for a simple fluid in HS matrices of the different porosities ϕ_0. One can see that the coexistence curves shift toward lower temperatures and lower fluid densities if the matrix

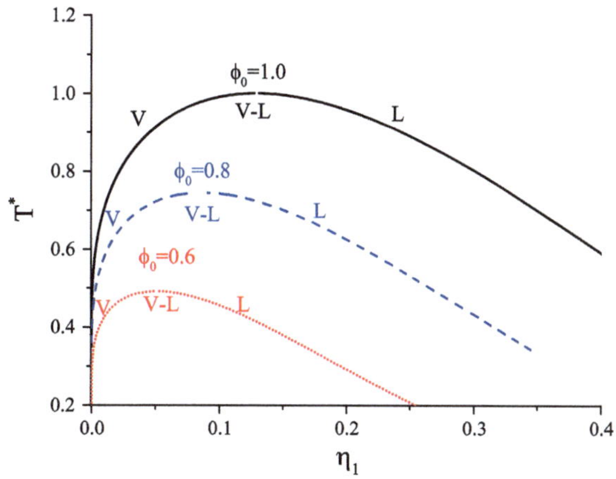

Fig. 1.3 Liquid-vapour coexistence curves ($T^* = kT/\varepsilon_1$ is a reduced temperature, η_1 is a fluid packing fraction) calculated from a generalized Van der Waals (1.68) for a simple fluid in a HS matrix of the different porosity ϕ_0

porosity decreases. Therefore, the critical density η_c and critical temperature T_c^* decrease with matrix porosity decreasing. This behavior is very common for fluids in random confinements [11, 25, 47].

It is worth mentioning that the interpretation of experimental results for the phase behaviour of fluid in random porous media [48] is enough controversial. From one point of view a fluid in a quenched disorder or in a random matrix can be considered as experimental realization of the random-field Ising model [49]. Within this model the random field describes the spatially varying preference of the porous media for different fluid phases. From the other point of view the behaviour of a fluid in a porous medium can be described in terms of the wetting states of the two phases in a single pore of ideal geometry [50]. We do not focus here specially on the influence of porous media on the behaviour of fluid near the critical point. However, since the conventional Van der Waals Equation of state for the bulk fluid gives the mean field description, one can consider that our analog of the Van der Waals equation for a fluid in a random porous matrix leads to the same critical exponents.

A description of molecular fluids requires a corresponding generalization of (1.68), which takes into account an anisotropic nature of the interaction between molecules. An extension of the expression (1.68) to the case of a system non-spherical particles with orientations starts from the following

$$\frac{\beta P}{\rho_1} = \left(\frac{\beta P}{\rho_1}\right)_{HCB} - 12a\eta_1\beta, \tag{1.71}$$

where $\left(\frac{\beta P}{\rho_1}\right)_{HCB}$ is the hard convex body contribution of the reference system. In our study the HCB particles are considered as hard spherocylinders (HSC). Also we restrict ourselves to the approximation SPT2b.

The attraction term depends on the constant a which is expressed in the general form as

$$a = -\frac{1}{\phi_0 V_1}\int f(\Omega_1)f(\Omega_2)U^{att}(r_{12}\Omega_1\Omega_2)d\bar{r}_{12}d\Omega_1 d\Omega_2, \tag{1.72}$$

where V_1 is the volume of a spherocylindrical molecule. One can see that the attractive pair potential $U^{att}(r_1\Omega_1\Omega_2)$ in (1.72) is orientational dependent. Therefore, except the anisotropic repulsive interaction of HSC particles the attractive part of intermolecular interaction is anisotropic as well. We introduce an orientational dependence for the potential U^{att} by modifying the Lennard-Jones potential in the following way

$$U^{att}(r_{12}\Omega_1\Omega_2) = U_{LJ}\left(\frac{\sigma(\Omega_1\Omega_2\Omega_r)}{r_{12}}\right)[1 + \chi P_2(\cos\vartheta_{12})], \tag{1.73}$$

$$U_{LJ}\left(\frac{\sigma(\Omega_1\Omega_2\Omega_r)}{r_{12}}\right) = \begin{cases} 4\varepsilon_1\left[\left(\frac{\sigma(\Omega_1\Omega_2\Omega_r)}{r_{12}}\right)^{12} - \left(\frac{\sigma(\Omega_1\Omega_2\Omega_r)}{r_{12}}\right)^{6}\right], & r_{12} \geq \sigma(\Omega_1\Omega_2\Omega_r) \\ 0, & r_{12} < \sigma(\Omega_1\Omega_2\Omega_r) \end{cases} \tag{1.74}$$

where $P_2(\cos\vartheta_{12}) = \frac{1}{2}(3\cos^2\vartheta_{12} - 1))$ is the second Legendre polynomial, the relative orientation ϑ_{12} corresponds to the angle between the principal axes of the two molecules. $\sigma(\Omega_1\Omega_2\Omega_r)$ is the contact distance between molecules, and it depends on the orientations of two interacting molecules as well as on the orientation of a distance vector $\vec{r}_{12}$ between their centers. It is worth noting that in the case of the repulsive part of the potential is spherically symmetric (σ is fixed) the expression for the potential (1.73) reduces to the Maier-Saupe potential [51].

One can see that (1.73) is a sum of two Lennard-Jones potentials, where the first one is related to the isotropic attraction and another one corresponds to the anisotropic attraction. The ratio of the well depths of these two potentials $\chi = \varepsilon_2/\varepsilon_1$ specifies a rate of anisotropy in the attraction of the resulted potential (1.73).

Following the traditional scheme [52], taking into account that $d\bar{r} = r^2 dr d\Omega_r$ and using a dimensionless intermolecular distance $r^* = r/\sigma(\Omega_1\Omega_2\Omega_r)$ one obtains

$$\begin{aligned} a = -\frac{1}{\phi_0 V_1}\int d\Omega_1 d\Omega_2 f(\Omega_1)f(\Omega_2)[1 + \chi P_2(\cos\vartheta_{12})] \\ \times V_1^{exc}(\Omega_1\Omega_2)3\int_0^\infty r^{*2}dr^*\beta U_{LJ}(r^*), \end{aligned} \tag{1.75}$$

where

$$V_{exc}(\Omega_1\Omega_2) = \frac{1}{3}\int d\Omega_r[\sigma(\Omega_1\Omega_2\Omega_r)]^3 \tag{1.76}$$

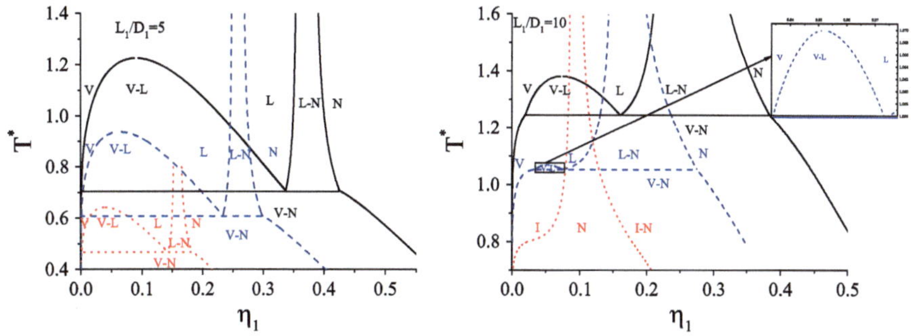

Fig. 1.4 The temperature-density phase diagram ($T^* = kT/\varepsilon_1$ is a reduced temperature, η_1 is a fluid packing fraction) calculated from the generalized Van der Waals (1.71) for the attractive spherocylinder fluid in a bulk and in HS matrices of porosity $\phi_0 = 1.0$ (*solid lines*), $\phi_0 = 0.8$ (*dashed lines*) and $\phi_0 = 0.6$ (*dotted lines*). The *horizontal lines* correspond to the vapour-liquid-nematic three-phase coexistence separating the vapour-liquid (VL), liquid-nematic (L-N) and vapour-nematic (V-N) regions. Two ratios of length to diameter of spherocylinder molecules are considered: $L_1/D_1 = 5$ (*left panel*) and $L_1/D_1 = 10$ (*right panel*)

is excluded volume formed by two hard spherocylinders with the orientations Ω_1 and Ω_2.

In order to study the anisotropy effect only in the repulsive part of intermolecular interaction, we put $\chi = 0$. For this case the phase diagrams for the system of HSC fluid with $L_1/D_1 = 5$ and $L_1/D_1 = 10$ in the bulk ($\phi_0 = 1.0$) and in the porous matrices ($\phi_0 < 1.0$) are presented in Fig. 1.4. Similar as it was shown for the bulk [52], for our model three regions of liquid phase equilibria are apparent. At low and intermediate densities the phase equilibrium between vapour (*V*) and isotropic liquid (*L*) states is observed. The coexistence region between the isotropic liquid (*L*) and the anisotropic nematic (*N*) states appears at high densities. The isotropic-nematic (I-N) transition is related mainly to the non-spherical shape of molecules, hence the position of this transition does not change with the temperature. In the high-temperature limit the liquid-nematic (L-N) transition vanishes and the system tends to that for a HSC fluid in a matrix. In contrast to this the temperature decrease leads to the L-N region becomes broader. At sufficiently low temperatures, the region merges into the continuous vapour-liquid region at the vapour-liquid-nematic (V-L-N) triple point. Below the triple point temperature only the vapour-nematic (V-N) coexistence is seen. In the porous matrix presence all the phase diagrams shift to the region of lower temperatures and lower densities similar as it was observed in the case of simple fluids (Fig. 1.3).

With increasing of the ratio L_1/D_1 to 10 the liquid-nematic region becomes much more extensive (Fig. 1.4). The vapour-liquid region for $L_1/D_1 = 10$ is essentially narrower than in the case of $L_1/D_1 = 5$, while the liquid-nematic region covers a wide range of densities. As one can see in Fig. 1.4 a porous medium for $L_1/D_1 = 10$ can modify the phase behaviour of a fluid qualitatively. For instance, for the porosity $\phi_0 = 0.8$ the vapour-liquid coexistence region in the case of $L_1/D_1 = 10$ is very small. Therefore, for such a fluid in matrices with porosities lower than $\phi_0 = 0.8$

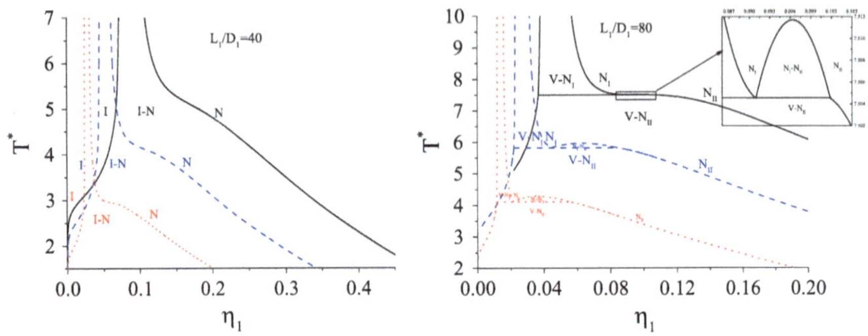

Fig. 1.5 The same as in Fig. 1.4, but for the ratios of length to diameter of spherocylinder molecules $L_1/D_1 = 40$ (*left panel*) and $L_1/D_1 = 80$ (*right panel*). For $L_1/D_1 = 80$ the coexisting region of nematic vapour and nematic liquid $(N_I - N_{II})$ with corresponding critical point is observed. The horizontal lines correspond to the vapour-nematic-nematic three phase coexistence, which separates the vapour-nematic $(V - N_I)$, nematic I-nematic II $(N_I - N_{II})$ and vapour-nematic II $(V - N_{II})$ regions

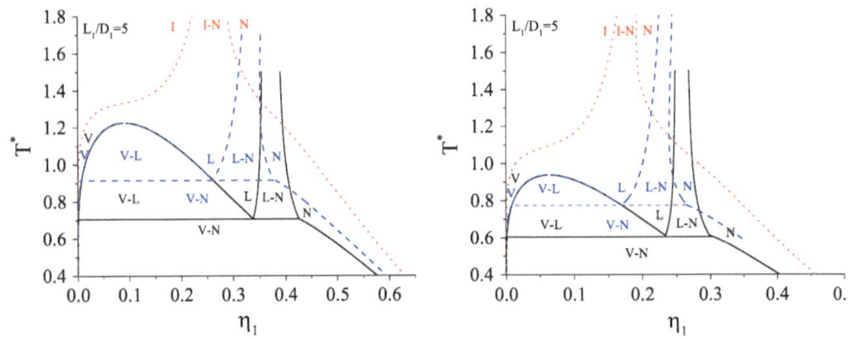

Fig. 1.6 The temperature-density phase diagram ($T^* = kT/\varepsilon_1$ is a reduced temperature, η_1 is a fluid packing fraction) calculated from the generalized Van der Waals (1.69) for the spherocylinder fluid with $L_1/D_1 = 5$ in a bulk (*left panel*) and in a HS matrix of porosity $\phi = 0.8$ (*right panel*). The different anisotropic rates of the attractive potential are considered: $\chi = 0.0$ (*solid lines*), $\chi = 0.1$ (*dashed lines*) and $\chi = 0.4$ (*dotted lines*)

the vapour-liquid coexistence region can disappear completely. Such a situation is observed for $\phi_0 = 0.6$ in Fig. 1.4.

It was also observed that for considerably high values of $L_1/D_1 = 40$ the vapour-liquid coexistence region disappears even in the bulk and the presence of porous medium for these cases does not change this (Fig. 1.5, left panel). Only the isotropic-nematic transition (I-N) is observed for the ratio $L_1/D_1 = 40$. One can see that the porosity decrease leads to the narrowing of the I-N region and it shifts to the lower densities. Since in the Onsager limit $L_1 \to \infty$, $D_1 \to 0$ the isotropic-nematic transition shifts to the lower densities we can expect an appearance of vapour-liquid coexistence in nematic region. Such a situation is observed for $L_1/D_1 = 80$ and it is presented in Fig. 1.5 (right panel). It is seen that below the triple point temperature of vapour-nematic-nematic transition there is a relatively broad region of V-N coexistence. A decrease of the matrix porosity causes the vapour-nematic region to be more pronounced.

Finally, we focus on the effect of anisotropic attractive interaction on the fluid phase behaviour in a random confinement. The temperature-density projection of the phase behaviour in the case of $L_1/D_1 = 5$ spherocylinder fluid with the different rates of anisotropic attractive interactions χ is presented in Fig. 1.6. As it can be expected the introduction of the anisotropic attractive interaction enhances an ability of the fluid to form orientationally ordered states. The liquid-nematic region broadens out significantly if χ increases, while the vapour-liquid coexistence curve remains unchanged. As a consequence, one can see an increase of the vapour-liquid-nematic triple point. For sufficiently large anisotropy ($\chi = 0.4$) the triple and critical points merge, and as a result only the I-N phase behaviour is found. The presence of porous medium as usually shifts the phase diagram to lower densities and temperatures. Although quantitatively the phase behaviour in the bulk (Fig. 1.6, left panel) and in the presence of porous medium with the porosity $\phi_0 = 0.8$ (Fig. 1.6, right panel) are practically the same.

1.5 Conclusions

The development and the application of the scaled particle theory (SPT) for a study of the thermodynamic properties of molecular liquids in random porous media are reviewed in this chapter. Within the proposed approach a series of different approximations are considered and tested by a comparison with computer simulations. It is shown that the SPT2b approximation fits the simulation results with a good accuracy at low and intermediate fluid densities. The SPT2b1 approximation improves the description at high densities and for the case when sizes of fluid and matrix particles are comparable. For a hard spherocylinder fluid confined in a random matrix the SPT2 approach is extended with a use of two scaling parameters. The results obtained for a hard spherocylinder fluid with two scaling parameters make it possible to study the effect of a porous medium on the isotropic-nematic phase transition. The proposed theory predicts that this transition is of the first order and a decrease of porosity shifts the phase diagrams toward lower fluid densities and temperatures. This prediction is supported by computer simulations [7] and also at least for enough large pores by experimental results [53, 54].

We have demonstrated that the results obtained within the framework of the SPT theory for a system of hard convex body fluid and in particular for a hard spherocylinder fluid can be used as a reference system for an extension of the Van der Waals equation of state to the case of molecular anisotropic fluid in a random porous matrix. Starting from this generalization the Van der Waals equation has made it possible to examine the evolution of vapour-liquid-nematic phase equilibrium depending on the anisotropy of fluid molecules and the porosity of a random confinement. As it is known [33, 34] the principal defect of the Van der Waals equation is related with a neglect of a fluid structure. Recently [55], we applied the Barker-Henderson perturbation theory [44] for the description of the liquid-vapour phase transition of a simple fluid in a random matrix. The pair distribution function of a hard sphere fluid in a porous medium needed for this theory was taken from numerical calculations of the replica Ornstein-Zernike equations [8, 9, 14, 15]. However, it is observed that the effect of porous medium on the vapour-liquid phase diagram qualitatively comparable with the prediction obtained from the Van der Waals equation. We plan in future to generalize the Barker-Henderson perturbation theory for anisotropic molecular fluids in random porous media.

Another way of a development of the presented theory is related to taking into account of association effects in molecular fluids confined in random porous media. Recently [47], within the framework of the SPT theory we have obtained the analytical expression for the contact value of a pair distribution function of a hard sphere fluid in random matrices. This allowed us to apply the thermodynamic perturbation theory [56, 57] to the treatment of association effects and to construct the phase diagrams of network-forming fluids confined in random matrices. We hope to generalize this approach for molecular anisotropic fluids in random porous media as well.

The developed theory opens new possibilities for the modeling of porous media. Within the framework of the SPT theory a porous medium can be presented as quenched HCB or OHCB particles, thus one can get a description of fluids in wider range of porous structures than those considered in other studies before. However, we should note that a structure of real porous materials can be much more complicated. As it was noted in [1], sometimes the simulation and theory can be used to study the behaviour of adsorbates confined in hypothetical porous materials that do not necessarily correspond to real materials. In order to establish the relation between such simple models of porous medium and more realistic ones we can use the morphological principle of mapping between the thermodynamic properties of a fluid in various matrices [25]. According to this principle the fluid in two different matrices has the same thermodynamic properties if the both matrices have the same probe particle porosity ϕ, the specific pore area s, the mean curvature and the Gaussian curvature. It was shown in [25] that these four morphological measures of a porous medium is enough to make a prediction of the thermodynamic properties of a confined fluid.

Finally, we should emphasize that in this review we focus on the consideration of the effect on fluids caused by porous media exceptionally with a random structure. However, major conclusions made for the case of random porous materials are valid for the case of regular porous materials as well, at least for materials with a high porosity. An understanding of the effect of a non-regular porous structure on confined fluid in a comparison with ordered porous materials needs deeper analysis and it will be studied elsewhere.

References

1. L.D. Gelb, K.E. Gubbins, R. Radhakrishnan, M. Sliwinska-Bartkowiak, Rep. Prog. Phys. **62**, 1573 (1999)
2. C.M. Lastoskie, K.E. Gubbins, Adv. Chem. Eng **28**, 203 (2001)
3. P.A. Monson, Adsorption **11**, 29 (2005)
4. P.A. Monson, J. Chem. Phys **128**, 084701 (2008)
5. Q.-T. Doan, G. Lefevre, O. Hurisse, F.-X. Coudert, Mol. Simulat. **40**, 16 (2014)
6. J. Ilnytskyi, S. Sokolowski, O. Pizio, Phys. Rev. E **59**, 4161 (1999)
7. M. Schmidt, M. Dijkstra, J. Chem. Phys. **121**, 12067 (2004)
8. W.G. Madden, E.D. Glandt, J. Stat. Phys. **51**, 537 (1988)
9. J.A. Given, G. Stell, J. Chem. Phys. **97**, 4573 (1992)
10. T. Patsahan, A. Trokhymchuk, M. Holovko, J. Mol. Liq. **92**, 117 (2001)
11. T. Patsahan, A. Trokhymchuk, M. Holovko, J. Mol. Liq. **105**, 227 (2003)
12. M. Schmidt, Phys. Rev. E **66**, 041108 (2002)
13. M. Schmidt, J. Phys.: Condens. Matter **17**, S3481 (2005)
14. M.L. Rosinberg, C. Caccamo, J.P. Hansen, G. Stell (eds.), *New Approaches to Problems in Liquid State Theory. NATO Science series C*, vol. 529 (Kluwer Dordrecht (Holland), 1999), pp. 245–278
15. O. Pizio, M. Borowko (eds.), *Computational Methods in Surface and Colloidal Science. Surfactant Science Series*, vol. 89 (Kluwer, Marcell Deker, New York 2000), p. 293
16. A.D. Trokhymchuk, O. Pizio, M.F. Holovko, S. Sokolowski, J. Phys. Chem. **100**, 17004 (1996)
17. A.D. Trokhymchuk, O. Pizio, M.F. Holovko, S. Sokolowski, J. Chem. Phys. **106**, 200 (1997)
18. M. Holovko, W. Dong, J. Phys. Chem B **113**, 6360 (2009)
19. W. Chen, W. Dong, M. Holovko, X.S. Chen, J. Phys. Chem. B **114**, 1225 (2010)
20. T. Patsahan, M. Holovko, W. Dong, J. Chem. Phys. **134**, 074503 (2011)
21. H. Reiss, H.L. Frisch, J.L. Lebowitz, J. Chem. Phys. **31**, 369 (1959)
22. H. Reiss, H.L. Frisch, E. Helfand, J.L. Lebowitz, J. Chem. Phys. **32**, 119 (1960)
23. J.L. Lebowitz, E. Helfand, E. Praestgaard, J. Chem. Phys. **43**, 774 (1965)
24. M. Holovko, T. Patsahan, W. Dong, Condens. Matter Phys. **15**, 23607 (2012)
25. M. Holovko, T. Patsahan, W. Dong, Pure Appl. Chem. **85**, 115 (2013)
26. R.M. Gibbons, Mol. Phys. **17**, 81 (1969)
27. T. Boublik, Mol. Phys. **27**, 1415 (1974)
28. M.A. Cotter, D.E. Martire, J. Chem. Phys. **52**, 1970 (1909)
29. M.A. Cotter, Phys. Rev. A **10**, 625 (1974)
30. G. Lasher, J. Chem. Phys. **53**, 4141 (1970)
31. M. Holovko, V. Shmotolokha, T. Patsahan, J. Mol. Liq. **189**, 30 (2014)
32. M. Holovko, V. Shmotolokha, T. Patsahan, (in preparation)
33. I.R. Yukhnovskyi, M.F. Holovko, *Statistical theory of Classical Equilibrium Systems* (Naukova Dumka, Kyiv, 1980)
34. J.P. Hansen, I.R. McDonald, *Theory of Simple Liquids* (Academic Press, London, 2006)
35. C.G. Gray, K.E. Gubbins, *Theory of Molecular Fluids* (Clarendon Press, Oxford, 1984)
36. D. Frenkel, B. Smith, *Understanding Molecular Simulations* (Academic, San Diego, 1995)
37. J. Herzfeld, A.E. Berger, J.W. Wingate, Macromolecules **17**, 1718 (1984)
38. L. Onsager, Ann. N. Y. Acad. Sci. **51**, 627 (1949)
39. M. Cotter, D.C. Wacker, Phys. Rev. A **18**, 2669 (1978)
40. R. Tuinier, T. Taniguchi, H.H. Wensink, Eur. Phys. J. E **23**, 355 (2007)
41. R.F. Kayser Jr., H.J. Raveche, Phys. Rev. A **17**, 2067 (1978)
42. G.J. Vroege, H.N.W. Lekkerkerker, Rep. Prog. Phys. **55**, 1241 (1992)
43. P. Bolhuis, D. Frenkel, J. Chem. Phys. **107**, 666 (1997)
44. J.A. Barker, D. Henderson, Rev. Mod. Phys. **48**, 587 (1976)
45. M. Kac, G.E. Uhlenbeck, P.C. Hammer, J. Math. Phys. **4**, 216 (1963)

46. J.L. Lebowitz, O. Penrose, J. Math. Phys. **7**, 98 (1966)
47. Y.V. Kalyuzhnyi, M. Holovko, T. Patsahan, P. Cummings, J. Phys. Chem. Lett. **5**, 4260 (2014)
48. A.P.Y. Wong, S.B. Kim, W.J. Goldburg, M.H.W. Chan. Phys. Rev. Lett. **70**, 954 (1993)
49. F. Brochard, P.G. de Gennes, J. Phys. Lett (Paris) **44**, 785 (1983)
50. L. Monette, A. Liu, G.S. Grest, Phys. Rev. A. **46**, 7664 (1992)
51. W. Maier, A. Saupe, Z. Naturforsch **14a**, 882 (1959)
52. M. Franco-Melgar, A.J. Haslam, G. Jackson, Mol. Phys. **107**, 2329 (2009)
53. M.D. Dadmun, M. Muthukumar, J. Chem. Phys. **98**, 4850 (1993)
54. G.S. Iannacchione, S. Qian, D. Finotello, F.M. Aliev, Phys. Rev. E **56**, 554 (1997)
55. M. Holovko, T. Patsahan, V. Shmotolokha, Cond. Matter Phys. **18**, 13607 (2015)
56. M.S. Wertheim, J. Stat. Phys. **35**, 19 (1984), **35**, 34 (1984)
57. M.S. Wertheim, J. Stat. Phys. **42**, 459 (1986)

Chapter 2
Dimensional Crossover in Liquids in Reduced Geometry

Alexander V. Chalyi

Abstract Problem of finding the pair correlation function G_2 and correlation length ξ of order parameter fluctuations in liquids in restricted geometry is discussed. Two types of dimensional crossover (DC) are studied. The 1st type (DC-1) corresponds to transition from $3D$ bulk to $3D$ bounded liquids. In this case the dependence of physical properties on thermodynamic variables (temperature, density, pressure, etc.) in bulk liquids with linear sizes $L \gg \xi$ may convert into dependence of these properties on linear sizes in bounded liquids with $L < \xi$. The 2nd type (DC-2) corresponds to the case when a further decreasing of linear sizes in confined liquids may be treated under certain conditions as the change of spatial dimensionality D (for example, $3D \Leftrightarrow 2D$ crossover in slit-like pores or $3D \Leftrightarrow 1D$ crossover in cylindrical pores). Smooth transition of effective critical exponents (say, from $\nu = 0.625$ for $D = 3$ to $\nu = 1$ for $D = 2$) as well as theoretical results versus experimental data are examined for liquids in reduced geometry.

2.1 Introduction

Phase transitions and critical phenomena in reduced geometry have been actively investigated in recent years. Many systems of experimental and theoretical interest are spatially bounded and have different forms of reduced geometry, such as thin surface layers, interfaces, porous media, biological membranes, vesicles, synaptic clefts, etc. This review paper is aimed at studying peculiarities of dimensional crossover in liquid systems in reduced geometry.

The problem going to be mainly discussed: how results of 3D systems transfer to results of 2D systems and vice versa. Obviously, such a 3D ⇔ 2D transition which may be called "dimensional crossover" or "dimensionality crossover" should be

A.V. Chalyi (✉)
Department of Medical and Biological Physics, Bogomolets National Medical University, 13 Shevchenko Blvd, Kyiv 01601, Ukraine
e-mail: avchal@nmu.kiev.ua

L. Bulavin and N. Lebovka (eds.), *Physics of Liquid Matter: Modern Problems*, Springer Proceedings in Physics 171, DOI 10.1007/978-3-319-20875-6_2

smooth and without discontinuities. While describing such a dimensional crossover (DC), one has to take into account the main theoretical ideas of bulk scaling and finite-size scaling (FSS) as well as experimental results (see e.g. [1–15]).

The topics under consideration are as follows: (a) pair correlation function and correlation length of order parameter fluctuations in liquids in reduced geometry; (b) effects of spatial limitation and two types of dimensional crossover; (c) comparison of theoretical results and experimental data in liquids in reduced geometry.

2.2 Pair Correlation Function and Correlation Length of Order Parameter Fluctuations

The major problem of the statistical physics approach is to find the pair correlation function G_2 and correlation length ξ of order parameter fluctuations in bounded systems undergoing phase transitions and critical phenomena. This problem is studied in liquid volumes in reduced geometry in the form of plane-parallel layers and cylindrical samples [4, 16].

To receive ***the pair correlation function*** of scalar order parameter fluctuations, namely density fluctuations for a single-component liquid near the critical point, one can use the well-known method of the statistical physics based on the Ornstein-Zernike (OZ) equation

$$G_2(r) = C(r) + <\rho> \int G_2(r - r')C(r')dr'. \tag{2.1}$$

Here $C(r)$ is the direct correlation function and $<\rho>$ is the average density. The direct correlation function $C(r)$ is usually short-range for real intermolecular potentials. It allows using the following differential equation instead of the integral (2.1):

$$(\Delta - \kappa^2)G_2(r) = -C(r)/C_2, \tag{2.2}$$

where $\kappa^2 = (1 - C_0)/C_2$ is the quantity related to the inverse isothermal compressibility, while $C_0 = <\rho> \int C(r)dr$, $C_2 = \frac{1}{6}<\rho> \int C(r)r^2 dr$ are the zero and the second spatial moments of the direct correlation function.

In order to obtain the pair correlation function G_2(r) one usually substitutes the short-range direct correlation function $C(r)$ in (2.2) with the delta function $\delta(r)$. Then G_2(r) can be found as the Green function for the Helmholtz operator $\hat{L} = \Delta - \kappa^2$, where Δ is the Laplacian.

Using this method, it is easy to obtain the OZ correlation function

$$G_2(r) = A \exp(-r/\xi)/r \tag{2.3}$$

for the 3-dimensional infinite system with zero boundary conditions. One has indeed for the pair correlation function $G_2(\mathrm{r}) \rightarrow 0$ if $r = |r_1 - r_2| \rightarrow \infty$. Here in (2.3) A is the constant coefficient, $\xi = \kappa^{-1}$ is the correlation length of order parameter fluctuations. It is important to mention that the anomalous growth of the correlation length ξ and the long-range behavior of the pair correlation function $G_2(r) \sim r^{-1}$ take place at the phase transition or critical points only for spatially infinite systems.

Let us consider the geometry of our problem in the form of a ***plane-parallel layer***: $-\infty < x, y < \infty$, $-L_0 < z < L_0$. It is possible to find the Green's function of the Helmholtz operator for such a layer with zero boundary conditions at the surfaces $z = \pm L_0$ in the following form [4]:

$$G_2(\rho, z) = \frac{1}{4\pi L_0} \sum_{n \geq 0} [1 - (-1)^n] * \\ K_0\left[\rho(\kappa^2 + n^2\pi^2/4L_0^2)^{3/2}\right] \cos(n\pi z/2L_0), \tag{2.4}$$

where $\rho = (x^2 + y^2)^{1/2}$ and $K_0(u)$ is the cylindrical Macdonald function.

The pair correlation function $G_2(\mathrm{r})$ may be obtained also by another method based on the result of acting the inverse Helmholtz operator $\hat{L}^{-1}$ on the delta function $\delta(r)$. It first requires the expression for the delta function constructed of orthonormal eigenfunctions of the operator $\hat{L}$ with zero boundary conditions and with eigenvalues λ_n satisfying the relation $\lambda_n = k_x^2 + k_y^2 + k_z^2 + \kappa^2$. Components of wave vector k_x, k_y change continuously ($-\infty < k_x$, $k_y < \infty$), while k_z is discrete ($k_z = n^2\pi^2/4L_0^2$, $n = 0, 1, 2, \ldots$). The following formula for $\delta(r)$ satisfies all these conditions:

$$\delta(r) = \frac{1}{8\pi L_0} \sum_{n \geq 0} \iint [1 - (-1)^n] \cos(n\pi z/2L_0) \exp\left[i(k_x x + k_y y)\right] dk_x dk_y. \tag{2.5}$$

With taking (5) into account, an expression for $G_2(\mathrm{r})$ can be obtained in the form

$$G_2(r) = \hat{L}^{-1}\delta(r) \\ = \frac{1}{8\pi^2 L_0^2} \sum_{n \geq 0} \iint \frac{\cos(n\pi z/2L_0) e^{i(k_x x + k_y y)}}{k_x^2 + k_y^2 + (n\pi/2L_0) + \kappa^2} dk_x dk_y. \tag{2.6}$$

Using polar coordinates $k_x = k_{xy}\cos\varphi$, $k_y = k_{xy}\sin\varphi$ and integrating (6) with such formulae for cylindrical functions

$$exp(ik\rho cos\varphi) = J_0(k\rho) + 2icos\varphi J_1(k\rho) + 2i^2 cos2\varphi J_2(k\rho) + \ldots,$$

$$\int_0^\infty \frac{kJ_0(k\rho)dk}{k^2 + a^2} = K_0(\rho a).$$

one can obtain the result given by formula (2.4) for the pair correlation function of a plane-parallel layer.

As in the previous case of a plane-parallel layer, let us find a solution of the differential OZ equation for a spatially restricted system in the form of a ***cylindrical sample*** ($0 \leq x, y \leq a, -Z_0 < z < Z_0, Z_0 \gg a$). The cylinder radius a is supposed to be much smaller than distance Z_0 along the cylinder axis. While solving the OZ equation in cylindrical coordinates

$$\frac{1}{r}\frac{\partial}{\partial r} r \frac{\partial G_2}{\partial r} + \frac{\partial G_2}{\partial z^2} - \kappa^2 G_2 = 0, \tag{2.7}$$

the nonsingular solution can be found as follows [16]:

$$G_2(r,z) = \sum_{n \geq 1} D_n J_0(\mu_n \frac{r}{a}) \exp\left[-(\kappa^2 + \mu_n^2/a^2)^{1/2}|z|\right]. \tag{2.8}$$

Here $J_0(u)$ is the Biessel function, μ_n are its nodes, i.e. the solutions of the equation $J_0(\mu_n) = 0$, and D_n are coefficients. The solution (2.8) is valid for all r except the nearest vicinity of the point $r = 0$.

The iterative procedure was proposed to find a non-singular solution of the OZ equation in [17]. The first iterations for the pair and direct correlation functions of scalar order parameter fluctuations near the critical point were found for bounded systems with geometry of a cylinder. These results, being valid even at $r = 0$, were used to study the shifts of the critical parameters.

The main contributions in both expressions (2.6) and (2.8) for the pair correlation function are given by the first terms with small n. Therefore, one has in the case of a plane-parallel layer

$$G_2(\rho, z) = \frac{1}{2\pi L_0} K_0\left[\rho(\kappa^2 + n^2\pi^2/4L_0^2)^{1/2}\right] \cos(\pi z/2L_0), \tag{2.9}$$

and in the case of a cylindrical sample

$$G_2(r,z) = D_1 J_0(\mu_1 r/a) \exp\left[-(\kappa^2 + \mu_1^2/a^2)^{1/2}|z|\right]. \tag{2.10}$$

Correlation functions (2.4) and (2.8) demonstrate an oscillatory behavior in the z direction for plane-parallel layers and in the r direction for cylindrical samples confirming the theoretical results and computer-simulation studies (see e.g. [18, 19]) for the radial distribution function $g(r)$ in liquids in restricted geometry).

The limiting case of the Ornstein-Zernike approximation (2.3) for an infinite system can be obtained from formulae (2.4) and (2.8) for G_2 with zero boundary conditions

$$G_2(\rho,z)=\frac{1}{4\pi L_0}\left(\frac{2L_0}{\pi}\right)^2\int_0^\infty K_0\left[\rho(\kappa^2+n^2\pi^2/4L_0^2)^{1/2}\right]\cos(k_z z)dk_z$$
$$=\frac{1}{4\pi}\exp\left[-\kappa(\rho^2+z^2)^{1/2}\right]/(\rho^2+z^2)^{1/2},$$
$$G_2(r,z)=\int_1^\infty e^{-x|z|}J_0(r\sqrt{x^2-\kappa^2})dx=\frac{\exp(-\kappa\sqrt{r^2+z^2})}{\sqrt{r^2+z^2}},$$

in geometries of plane-parallel layers and cylindrical samples, correspondingly.

As is seen from (2.4), (2.8–2.10), the pair correlation functions of liquids in reduced geometry have a non-exponential shape. Therefore, it is natural to determine ***the correlation length*** ξ of order parameter fluctuations in such bounded liquids according to the following relation:

$$\xi=\sqrt{M_2},M_2=\frac{\int G_2(r)r^2dr}{\int G_2(r)dr},\tag{2.11}$$

where M_2 is the second normalized spatial moment of the pair correlation function.

In the case of ***a plane-parallel layer***, taking into account the formula (2.9) for the pair correlation function $G_2(r)$ and expression for the cylindrical Macdonald function $K_\nu(u)$ and gamma-function $\Gamma(u)$ [20]

$$\int_0^\infty x^\mu K_\nu(ax)dx=2^{\mu-1}a^{-\mu-1}\Gamma\left(\frac{1+\mu+\nu}{2}\right)\Gamma\left(\frac{1+\mu-\nu}{2}\right),$$

one can derive the following formula for the correlation length ξ of order parameter fluctuations:

$$\xi=\xi_0\left[\frac{4}{\kappa^2+\pi^2/4L_0^2}+\frac{1}{4}(1-8/\pi^2)S^2\right]^{1/2}.\tag{2.12}$$

In the case of ***a cylindrical geometry*** in accordance with the main contribution for pair correlation function (2.10) and values of integrals [20]

$$\int_0^a J_0(a/\alpha)rdr=(a/\alpha)J_1(\alpha a),$$
$$\int_0^a J_0(\alpha r)r^3dr=\frac{a}{\alpha}(\alpha^2-\frac{1}{\alpha^2})J_1(\alpha a)+2\alpha^2a^2J_0(\alpha a),$$

the correlation length ξ can be written as

$$\xi = \xi_0 \left[\frac{2}{(\xi_0^{-1}\kappa)^2 + \mu_1^2/S^2} + (1 - \frac{4}{\mu_1^2})S^2 \right]^{1/2}. \quad (2.13)$$

Thecorrelation length ξ in (2.12) and (2.13) depends not only upon thermodynamic variables (because of relationships between κ and temperature, density, etc.) but also on the thickness $d = 2L_0$ or the radius a, i.e. on the geometric factors $S = d/\xi_0$ or $S = a/\xi_0$ related to the number of molecular monolayers along the direction of spatial limitation.

Let us summarize the approximations used to derive expressions for the pair correlation function G_2 and correlation length ξ:

(a) Here two methods have been used: (1) method of the OZ integral (2.1) which was transformed into the differential (2.2) with two spatial moments of the short-range direct correlation function. This method is quite equivalent to the well-known OZ approximation in the fluctuation theory of critical phenomena giving the critical exponent $\eta = 0$ in an scaling formula for the pair correlation function $G_2(r) = A \exp(-r/\xi)/r^{1+\eta}$ [21]; (2) method of the inverse Helmholtz operator $\hat{L} = \Delta - \kappa^2$ acting on the delta function and giving G_2 as the Green function for this operator.

(b) We used here only the main contributions to correlation function G_2 given by the first terms in (2.4) and (2.8). It was shown [22, 23] that next contributions, say, for a cylindrical sample are decreasing with the growth of number n of the Biessel function nodes μ_n ($\mu_1 = 2.4048$, $\mu_2 = 5.5201$, $\mu_3 = 8.8537$, $\mu_4 = 11.7915$, etc.) and due to reduction of cylindrical and exponential functions in (2.8) and (2.10) with increasing of their arguments. So, at $|z|/\kappa a \approx 1$ ratios of successive terms in (2.8) have such orders of magnitudes: $a_2/a_1 \approx 10^{-1} \sim 10^{-2}$, $a_3/a_1 \approx 10^{-3}$, $a_4/a_1 \approx 10^{-4}$.

(c) The pair correlation functions (2.4), (2.8–2.10) in liquids in reduced geometry are obtained for zero boundary conditions. The case of the arbitrary boundary conditions for the pair correlation function $G_2(r = a, z) = F(z)$ was studied for a cylindrical geometry in [22, 23]. Numerical values of the arbitrary function $F(z)$ depend on the concrete problems and change within the interval $0 \leq F \leq 1$. The case $F = 0$ corresponds to "hydrophobic" surfaces, the case $F = 1$ – to "hydrophilic" surfaces. The intermediate case $0 < F < 1$ describes obviously the situation of the so-called "incomplete wetting". Omitting immaterial details, one can receive finally the solution of the OZ differential (2.2) in the form (2.8) with the only important difference: values μ_n, being the nodes of the Biessel function and satisfying the equation $J_0(\mu_n) = 0$, have to be substituted in the formula (2.8) for the arbitrary boundary conditions by values ξ_n which are the nodes of the transcendental equation $J_0(\xi_n) = F(z) \exp[-(\kappa^2 + \xi_n^2/a^2)^{1/2}|z|]$.

2.3 Comparison with the Fisher Scaling Hypothesis, Anisotropic Effects in Finite-Size Scaling

Fisher [2] and other authors (see e.g. [8–10, 12, 13]) have proposed the ***FSS hypothesis for systems in reduced geometry***. According to this hypothesis (here we shall formulate it for classical liquids), the fluctuation part of the thermodynamic potential $\Delta\Phi_s$ and the correlation length ξ depend not only on the thermodynamic variables (the reduced temperature τ, etc.) and external fields h but on the linear size L of a system:

$$\Delta\Phi_s = L^{-d} f_\Phi(a\tau L^{1/v},\ bhL^{\beta\delta/v}),\ \xi = Lf\xi(a\tau L^{1/v},\ bhL^{\beta\delta/v}), \tag{2.14}$$

where a and b are nonuniversal constants (amplitudes).

The first scaling argument in (2.14) $x = a\tau L^{1/v}$ in both scaling functions f_Φ and f_R can be obtained from the following formulae:

$$\xi \sim \tau^{-n}, \tau \sim \xi^{-1/v},\ x \sim \frac{\tau}{\xi^{-1/v}} \sim \tau L^{1/v}\ \ (\xi \to L). \tag{2.15}$$

In analogous way one can easily obtain the expression for the second scaling argument y = $bhL^{\beta\delta/v}$ with taking such formulae into account

$$\begin{gathered}\xi \sim \tau^{-v} \sim \varphi^{-v/\beta} \sim \mathrm{h}^{-v/\beta\delta}, \varphi \sim \tau^{\beta}, \varphi \sim \mathrm{h}^{1/\delta}, \tau \sim \varphi^{1/\beta}, \\ \mathrm{h} \sim \xi^{-\beta\delta/v}, \mathrm{y} \sim \frac{h}{\xi^{-\beta\delta/v}} \sim \mathrm{hL}^{\beta\delta/v}(\xi \to \mathrm{L}).\end{gathered} \tag{2.16}$$

Here $\varphi = (\rho - \rho_c)/\rho_c$ is the order parameter for classical liquids, i.e. the reduced density.

The isothermal compressibility of liquids in reduced geometry in nonzero external fields may be written as

$$\begin{gathered}\beta_T = (\partial^2\Delta\Phi_s/\partial \mathrm{h}^2)_\tau = L^{-d+2\beta\delta/v} f_\beta(x,y), f_\beta = b(f_\Phi)''_h, \\ -d + 2\beta\delta/v = \gamma/v.\end{gathered} \tag{2.17}$$

The Fisher FSS hypothesis formulated for the susceptibility χ in [2]

$$\chi = \mathrm{S}^{\omega}\,\mathrm{F}\,(\mathrm{S}^{\theta}_{\tau}). \tag{2.18}$$

Equation (2.18) was stated for systems in reduced geometry in a zero external field. Here $S = L/a_0$ is the geometric factor and $F(x)$ is the scaling function of the susceptibility (the isothermal compressibility in liquids).

The most important consequences of (2.18) are as follows:

- the critical exponent $\omega = 2 - \eta$;
- the critical exponent $\theta = 1/\nu$;
- the scaling function $F(x)$ has such an asymptotic formula for large arguments $x = S^{\theta}\tau \gg 1 : F(x) \sim x^{-\gamma}$, where $\gamma = (2 - \eta)\nu$ is the critical exponent of the susceptibility for bulk systems.

Let us compare the Fisher FSS hypothesis for the susceptibility with results obtained above for liquids in reduced geometry of a plane-parallel layer. In accordance with the formula (2.12), one has such an expression for the susceptibility for a plane-parallel layer

$$\chi = \chi_0 S^2 \left\{ 4[S^2\tau^{2\nu} f_1(x) + \pi^2]^{-1} + (1-8/\pi^2)/4 \right\}, \tag{2.19}$$

where χ_0 is the amplitude of the susceptibility.

Equation (2.19) has the form analogous to the FSS hypothesis (2.18). Comparison between these formulae gives the following results:

(a) According to (2.18) and (2.19) the critical exponent $\omega = 2$. Taking into account $\omega = \gamma/\nu = 2 - \eta$ and the fact that the Helmholtz differential operator $\hat{L} = \Delta - \kappa^2$ corresponds to the OZ approximation with zero value of the critical exponent η, this result confirms the first consequence of the Fisher FSS hypothesis (2.18).

(b) Scaling function $F(x)$ in (2.19) depends on the argument $x = S^{2/\gamma}\tau$ for a zero external field, i.e. in the vicinity of the critical isochore. Therefore, for $\eta = 0$ one has $2/\gamma = 1/\nu = \theta$ in accordance with the second consequence of the scaling hypothesis (2.18).

(c) With increasing the thickness of a plane-parallel layer one has the following result for the scaling function $F(x)$ from (2.19) for large arguments $x = S^{2/\gamma}\tau \gg 1{:}F(x) \sim x^{-1} \sim x^{-\gamma}$ in accordance with the third consequence of the FSS hypothesis (2.18).

Idea of ***universality*** and ***isomorphism*** seems to be one of the most important features in the physics of critical phenomena and phase transitions. [5, 9–11, 24]. Within a certain universality class, critical behaviour of physical properties in bulk and confined systems is similar and allows generalizing on systems not only physical but also other nature.

There is another problem that is actively discussed in physics of the critical phenomena for bulk and confined systems [25–29]. Namely, ***influence of anisotropy*** on such quantities as ratios of amplitudes and scaling functions. Influence of anisotropy and universality on FSS was investigated in [25] on the basis of analysis of the critical Binder cumulant of a 2D Ising model. The Binder cumulant [26] characterizes the distribution of order parameter fluctuations and is defined by such a relation $U = 1 - \langle M^4 \rangle / (3 \langle M^2 \rangle^2)$, where M is the magnetization per spin, $\langle M^2 \rangle$ and $\langle M^4 \rangle$ are its second and fourth spatial moments. Results obtained in [25] support the validity of universal FSS for critical behavior of physical properties in the presence of a weak anisotropy. It is worthy to mention, that in our present approach using integral equations for the pair correlation function with the short-ranged direct correlation, the critical Binder cumulant $U = 1$ because of the OZ approximation in which $<M^4> = 0$. Correlative behavior of anisotropic binary liquid system were studied by the methods of integral equations in the three-moment approximation in [27] with taking into account the 3rd spatial moment of the direct correlation function.

Important results related to the problem of universality and effects of anisotropy were obtained in supercooled water [28] and in ionic fluids [29]. The authors [28] analyzed the critical behavior of supercooled water on the basis of two models: (1) the lattice-gas model used to describe the physical properties of liquid-vapor transitions, and (2) the lattice-liquid model, associated with an entropy-driven separation. The critical behavior of supercooled water appeared to be closer to the lattice-liquid model behavior, while the critical behavior of the lattice-gas model being equivalent to the critical behavior of Ising model for incompressible anisotropic ferromagnets.

The correlation length ξ in (2.12), (2.13) demonstrates an anisotropic behavior. Consider for definition the case of a plane-parallel layer, then correlation length ellipse related to (2.12) is determined by two contributions $\xi = (\xi_{xy}^2 + \xi_z^2)^{1/2}$ where $\xi_{xy} = 2/(\kappa^2 + \pi^2/4S_0^2)^{1/2}$ is the correlation length in the xy plane, and $\xi_z = (S_0/2)(1-8/\pi^2)^{1/2}$ is thecorrelation length in the z direction. In certain sense this result reminds the correlation length ellipse for a 2D Ising model with weak anisotropy caused by different amplitudes $\xi_{0,>}$ and $\xi_{0,<}$ above and below the critical point [25].

2.4 Effects of Spatial Limitation, Two Types of Dimensional Crossover

Spatial limitation of systems undergoing critical phenomena and 2nd order phase transitions causes the change of critical parameters and critical exponents. The most important feature of critical phenomena in bulk systems is the divergence of the bulk correlation length ξ in the critical point according to the formula

$$\xi = \xi_0|\tau|^{-\nu}, \tag{2.20}$$

where ξ_0 is the amplitude of correlation length which has the same order of magnitude as a_0, $\tau = (T - T_c(\infty))/T_c(\infty)$ is the temperature variable for a bulk liquid with the critical temperature $T_c(\infty)$.

However, the natural desire of investigators to study critical phenomena in systems in reduced geometry and perfect experimental technique allow one to realize a situation in which the correlation length becomes the same order of magnitude as the characteristic linear size of the sample. Thereby, theoretical and experimental results in such bounded systems yield not bulk critical parameters and bulk theoretical critical exponents but its effective values depending on linear sizes.

Here the problem of critical behavior of systems in reduced geometry is discussed for single-component liquids with scalar order parameters. Special attention is given to the dimensional crossover (DC) effects of two types as follows [30]:

1. ***The first type of dimensional crossover*** (***DC-1***) corresponds to transition from 3*D* bulk to 3*D* liquids in reduced geometry. In this case one has a situation in which dependence of physical properties on thermodynamic variables (temperature, density, pressure, etc.) in bulk liquids with linear sizes $L \gg \xi$ may convert into dependence of these properties on linear sizes in confined liquids with $L < \xi$.
2. ***The second type of dimensional crossover*** (***DC-2***) corresponds to the case when a further decreasing of linear sizes in confined liquids may be treated under certain conditions as the change of spatial dimensionality *D* (for example, 3*D* ⇔ 2*D* crossover in slitlike pores or 3*D* ⇔ 1*D* crossover in cylindrical pores).

2.4.1 Dimensional Crossover DC-1

Let us first consider the ***shifts of the critical parameters in liquids in reduced geometry***. Table 2.1 contains formulae for the temperature $T_c(S)$, density $\rho_c(S)$ and pressure $P_c(S)$ in bounded single-component liquids (3rd column) which are analogous to the critical temperature $T_c(\infty)$, critical density $\rho_c(\infty)$ and critical pressure $P_c(\infty)$ in bulk liquids (2nd column). An important difference between these parameters of bulk and confined liquids consist in as follows: while the critical parameters $T_c(\infty)$, $\rho_c(\infty)$, $P_c(\infty)$ of bulk liquids characterize coordinates of singularities for physical properties on the thermodynamic space, the analogous parameters $T_c(S)$, $\rho_c(S)$, $P_c(S)$ give coordinates of points in which these physical properties have only its rounding maxima or minima.

As an example, the shift of the critical temperature can be calculated for a slitlike pore with the geometric factor $S = 10$ and the pore's thickness $H \approx 3$ nm filled by water with the bulk critical temperature $T_c(\infty) \approx 647$ K. The result of calculation of $T_c(S)$ according to (21) for the critical exponent $\nu = 0.628$ gives $T_c(S) \approx 587$ K. In this case the shift of the critical temperature is negative and rather large: $\Delta T_c = T_c(S) - T_c(\infty) \approx -60K$ Analogues of other critical parameters—the critical

Table 2.1 Critical parameters for bulk and confined systems

Critical parameters	Bulk systems	Confined systems	
Critical temperature	$T_c = T_c(\infty)$	$T_c(S) = T_c(\infty)[1 + (G/S)^{1/\nu}]^{-1}$	(2.21)
Critical density	$\rho_c = \rho_c(\infty)$	$\rho_c(S) = \rho_c(\infty)[1 + (G/S)^{\beta/\nu}]^{-1}$	(2.22)
Critical density	$P_c = P_c(\infty)$	$P_c(S) = P_c(\infty)[1 + (G/S)^{\beta\delta/\nu}]^{-1}$	(2.23)

density and pressure—may be calculated for any liquid in reduced geometry with help of formulae (2.22) and (2.23).

Let us now derive relationships between thermodynamic variables in bulk and confined liquids. Namely, between the dimensionless deviation of temperature from its critical value $\tau = [T - T_c(\infty)]/T_c(\infty)$ in a bulk liquid system and the temperature variable $\tau(S) = [T - T_c(S)]/T_c(S)$ in a confined liquid system. For this purpose one has to substitute the formula (2.21) for the analogue of critical temperature $T_c(S)$ into $\tau(S)$ and obtains the following expression as a result of obvious algebraic transformations:

$$\tau(S,\xi) = (G/S)^{1/v} + [1 + (G/S)^{1/v}]|\tau| sign\tau. \tag{2.24}$$

Taking (2.20) into account, the formula (2.24) for the temperature variable in a confined system can be rewritten in a quite equivalent form

$$\tau(S,\xi) = (G/S)^{1/v} + [1 + (G/S)^{1/v}](\xi_0/\xi)^{1/v} sign\tau. \tag{2.25}$$

Here in (2.24) and (2.25) G is the geometrical factor depending on the low crossover dimensionality (geometrical form) of a liquid volume (thus, for plane-parallel layers or slit-like pores $G = \pi$, while for cylindrical samples $G = \mu_1 = 2.4048$ being the first zero of the Biessel function $J_0(z)$); $S = L/a_0$ is the number of molecular layers, where L is a linear size such as a layer's thickness or cylinder's radius in direction of system's spatial limitation, a_0 is an average diameter of molecule.

Quantity $\tau(S, \xi)$ in (2.24) and (2.25) can be considered as the ***temperature variable for confined systems***. For relatively large sizes $L \gg \xi$ it is easy to find from (2.25) that due to factor $S = L/a_0 \gg 1$ the correlation length ξ is approaching its bulk value $\xi = \xi_0\tau^{-v}$. In reduced geometry if an opposite inequality $L \ll \xi$ is valid, the first term in (2.25) becomes more important than the second one. It means that the correlation length ξ does not depend on thermodynamic variables and equals a system's linear size.

Similar to the temperature variable $\tau(S, \xi)$ in liquids in reduced geometry given by (2.25), the ***density*** $\Delta\rho(S, \xi)$ and ***pressure*** $\Delta p(S, \xi)$ ***variables*** may be introduced according to the following formulae:

$$\Delta\rho(S,\xi) = \left(\frac{G}{S}\right)^{\frac{\beta}{v}} + \left(1 + \left(\frac{G}{S}\right)^{\frac{\beta}{v}}\right)\left(\frac{\xi_0}{\xi}\right)^{\frac{\beta}{v}} sign\Delta\rho, \tag{2.26}$$

$$\Delta p(S,\xi) = \left(\frac{G}{S}\right)^{\frac{\beta\delta}{v}} + \left(1 + \left(\frac{G}{S}\right)^{\frac{\beta\delta}{v}}\right)\left(\frac{\xi_0}{\xi}\right)^{\frac{\beta\delta}{v}} sign\Delta p. \tag{2.27}$$

Equations (2.26) and (2.27) give correct asymptotical expressions for the correlation length in bulk liquids with $L \gg \xi$, namely $\xi = \xi_0'\Delta\rho^{-v/\beta}$, $\xi = \xi_0''\Delta p^{-v/\beta\delta}$.

Here the amplitudes ξ_0', ξ_0'' of the correlation length have such relationships with the amplitudeξ_0 : $\xi_0' = \xi_0 B_0^{\nu/\beta}$, $\xi_0'' = \xi_0 (B_0/D_0)^{\nu/\beta\delta}$. The coefficients B_0 and D_0 characterize the shapes of the coexistence curve $\Delta\rho = B_0|\tau|^{\beta}$ and the critical isotherm $\Delta\rho = D_0 \Delta p^{1/\delta}$ in the close vicinity of the bulk critical point.

Let us illustrate effects of spatial limitation and dimensional crossover of the 1st type (DC-1) on the temperature dependence of ***diffusion coefficient*** described by the following formula in confined liquids:

$$D = \{a_R + a_S^0 [\tau(S,\xi)]^{-\nu}\} (\partial\mu/\partial\rho)_T^0 [\tau(S,\xi)]^{\gamma}. \quad (2.28)$$

Here a_R is the regular part and a_s^0 is the amplitude of the singular part of the Onsager coefficient, $(\partial\mu/\partial\rho)_T^0$ is the amplitude of the inverse isothermal susceptibility. For small volumes ($\xi \gg L$) the term $(G/S)^{1/\nu}$ for $\tau(S, \xi)$ in (2.25) will prevail. This is a reason why the diffusion coefficient D in formula (2.28) is decreasing at the fixed temperature with increasing linear sizes S of liquid volumes. For large volumes ($L \gg \xi$) the term $(1/\xi)^{1/\nu}$ for $\tau(S, \xi)$ in (2.25) has a greater role with increasing linear sizes $S = L/a_0$. That is why the diffusion coefficient D in (2.28) will increase and asymptotically approach its value D_0 for spatially infinite liquid volumes.

Another important consequences of finite-size effects on the critical behavior of the diffusion coefficient are as follows: the temperature variable $\tau_M(S) = [T_M - T_c(\infty)]/T_c(\infty)$ corresponding to the minimum value of the diffusion coefficient, i.e. the shift of the critical temperature in liquids in restricted geometry, (i) has a negative value in agreement with the FSS theories [2, 8–10, 12, 13]; (ii) goes to zero with increasing the geometric factor S (linear size L) of a system; (iii) increases with transition from plane-parallel to cylindrical geometry, in other words, while the lower crossover dimensionality d_{LCD} decreases.

Table 2.2 explains the notion of the ***lower crossover dimensionality*** which has to be introduced to characterize the type of restricted geometry [30, 31]. For real confined 3D systems (1st column) the lower crossover dimensionality d_{LCD} (2nd column) determines the spatial dimensionality of limiting geometrical objects (3rd column) for cases when system's linear sizes in directions of spatial limitation are approaching its minimum possible value.

The ***effects of spatial dispersion*** (nonlocality) being neglected gives unreal consequences for physical properties with approaching the critical and 2nd order phase transitions points in bulk systems as well as its analogous points in confined systems.

Table 2.2 Lower crossover dimensionality

Real confined 3D systems	Lower crossover dimensionality	Limiting geometric objects
Plane-parallel layer, slit-like pore, membrane, synaptic cleft	2	Molecular plane
Cylindrical pore, bar, ionic channel	1	Molecular line
Sphere, cube, vesicle	0	Point (one molecule)

Namely, in bulk systems the isothermal compressibility, magnetic susceptibility, isobaric and isochoric heat capacities have infinite values, while the diffusion coefficient, thermal conductivity, and sound velocity approach zero values. To take effects of spatial dispersion into account, such an idea has to be used [32]: spatial dispersion contributions must be added to the quantities which become zero in the bulk critical point (e.g. added to the diffusion coefficient or to the reverse isothermal compressibility, etc.). Thus, the self-diffusion coefficient has a minimum nonzero value at the bulk critical temperature in accordance with the following formula [30]:

$$D_{\min}(L_{\max}) = D_0 \frac{(G/S_{\max})^{\gamma/\nu} + 4\pi B/L^2_{\max}}{(G/S_{\max})^{\gamma/\nu} + 4\pi b/L^2_{\max}}. \quad (2.29)$$

Here $S_{\max} = L_{\max}/a_0$ is the number of molecular layers in the direction of a maximum system's linear size $L_{\max}$, B and b are the coefficients describing the effects of spatial dispersion in the Ornstein-Zernike approximation.

Figure 2.1 illustrates this result. Here, the solid curve corresponds to bulk liquids. The data are also presented for a plane-parallel (dotted curve) and cylindrical (curve with points) reduced geometry.

As is seen from Fig. 2.1, the self-diffusion coefficient $D(\tau, S, k)$ demonstrates an asymmetrical behavior with change of sign of the temperature variable $\tau = [T - T_c(\infty)]/T_c(\infty)$ in accordance with the inequalities $D(\tau, S, k) > D(-|\tau|, S, k)$, $D_{0,\tau>0} > D_{0,\tau<0}$. Such an asymmetry of the self-diffusion coefficient is confirmed by theoretical studies in [33].

2.4.2 Dimensional Crossover DC-2

While describing the dimensional crossover DC-2, the fact of different numerical values of the critical exponents in 3D and 2D liquids [11] (see Table 2.3) and the results of computer experiment [15] are taken into account.

For definition let us consider confined liquids in slitlike pores. While reducing linear sizes L (for plane-parallel geometry—the thickness H of pores or number of its molecular layers $S = H/a_0$, a liquid system will transfer from 3D to 2D geometry. This transition should be resulting in change of the critical exponents of classical liquids belonging to the Ising-model universality class. For example, the critical index ν describing the temperature dependence of the correlation length ξ has to shift its value from 0.638 to 1.0 (see Table 2.3).

The following expression for any effective critical exponents n_{eff} is proposed to study a smooth transition between its 3D value n_3 and 2D value n_2 [30]:

$$n_{eff} = n_3 + \left[\frac{2}{\pi}\arctan(ax - b) - 1\right]\frac{n_3 - n_2}{2}. \quad (2.30)$$

Fig. 2.1 Relative self-diffusion coefficient in bulk D/D_0 versus the temperature variable $\tau = (T - T_c(\infty))/T_c(\infty)$ in bulk and confined liquids

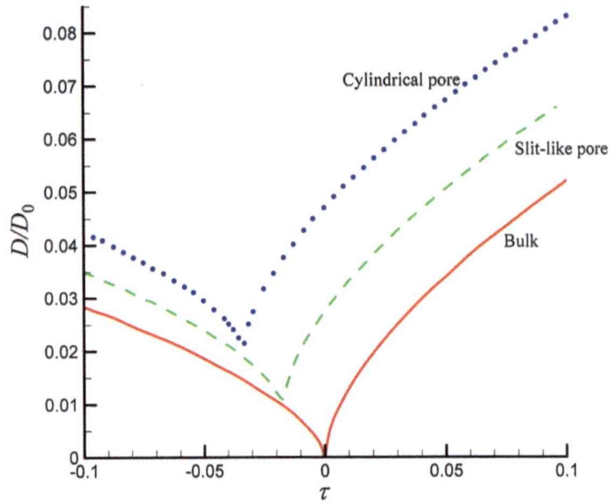

Table 2.3 Numerical values of critical exponents in 3D and 2D Ising models [11]

Physical property	Critical exponents	3D Ising model	2D Ising model
Heat capacity	α	0.125[a]	$0(C_V \sim \ln\lvert\tau\rvert)$
		0.110[b]	
Coexistence curve	β	0.3125[a]	1/8
		0.325[b]	
Isothermal compressibility	γ	1.250[a]	7/4
		1.241[b]	
Critical isotherm	δ	5.0[a]	15
		4.8[b]	
Correlation length	υ	0.638[a]	1
		0.630[b]	
Correlation function	η	0.041[a]	1/4
		0.031[b]	

[a] sum of series; [b] RG-approach

Here in (2.30) $x = H/H_{cr}$ is the dimensionless size (in fact—thickness H) of a plane-parallel layer; H_{cr} is the layer's thickness at which the 3D $\Leftrightarrow$ 2D crossover occurs (authors [15] consider $H_{cr} \approx 2, 4$ nm for a slitlike pore filled by water molecules).

As an example, formula (2.30) illustrates the $3D \Leftrightarrow 2D$ crossover for the critical exponent ν of the correlation length ξ in a slitlike pore. Figure 2.2 shows the theoretical dependence $\nu(S)$ on number of monolayers S in accordance with (2.30).

As is mentioned before, there are always factors which distort the idealized bulk critical behavior and give the effective values of critical parameters and exponents. Among these factors: effects of spatial limitation, crossover effects, external fields, temperature gradients, impurities, multiple-scattering effects, etc. [34–39].

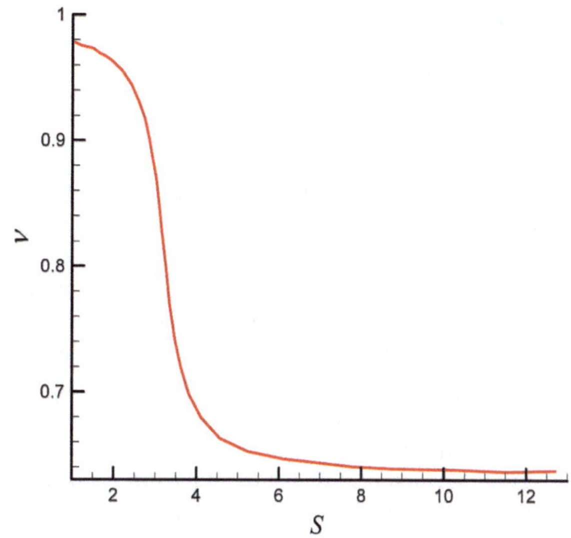

Fig. 2.2 Dimensional 3*D* ⇔ 2*D* crossover in the critical exponent ν versus number of molecular layers *S*

Table 2.4 Dependence of the effective critical exponent ν_{eff} on *S*

S	1.0	2.4	3.1	3.5	5.6	8.0
H, (nm)	0.3	0.7	0.9	1.1	1.7	2.4
ν_{eff}	1.0	0.95	0.85	0.75	0.65	0.64

The problem of the effective critical parameters and critical exponents is examined in detail in a number of papers (see e.g. references in [14]).

Table 2.4 demonstrates the dependence of the effective critical exponent ν_{eff} on number of molecular layers *S* (dimensionless thickness *H* of a slitlike pore) calculated in accordance with (2.30) for water molecules.

Parameters *a* and *b* characterizing slope and position of the 3D ⇔ 2D crossover were chosen to fit the following condition: 2D value of the critical exponent $\nu = 1$ corresponds to a system's thickness consisting of almost one molecular layer.

Such an interpolation (2.30) looks similar in a certain sense to approach used in [40] to receive the crossover between hydrodynamic and fluctuation regions for the central Rayleigh component of the light-scattering spectrum in mode-mode coupling version of the dynamical theory of critical phenomena.

2.5 Theoretical Results Versus Experimental Data in Confined Liquids

In this section the relationship between theoretical and experimental data will be analyzed with taking into account effects of spatial limitation and dimensional crossover in liquids in reduced geometry.

Table 2.5 Diffusion coefficient of water molecules in cylindrical pores [41, 42]

R (nm)	62.5	88.0	117.0	130.0	150.0
D_{exp} 10^{-9} (m^2/s)	1.59	1.95	2.13	2.19	2.30
D_{theor} 10^{-9} (m^2/s)	1.79	2.00	2.11	2.14	2.17

Experimental studies (see e.g. references in [41]) of ***the size dependence of the diffusion coefficient D*** of water molecules in a porous glass allow considering the dimensional crossover of the 1st type (DC-1) in cylindrical pores. These results give also an opportunity to investigate the asymptotical transition of the diffusion coefficient D from its values in liquid in restricted geometry to the bulk value $D_0 = 2.3 \cdot 10^{-9}$ m^2/s for water molecules. Table 2.5 demonstrates the size experimental dependence of $D_{exp}(R)$ on the pore's radius R together with results of the theoretical calculation $D_{theor}(R)$ for $G = 2.4048$ and $S = R/\xi_0$ in the case of a large pore's radius ($R \gg \xi$) [42].

Thus, asymptotical transition of the diffusion coefficient D of water molecules to its bulk value takes place in cylindrical pores if a pore's radius R ≥ 150 nm. It is possible to conclude that experimental and theoretical results of study the dependence of water diffusion coefficient on size of cylindrical pores demonstrate not only qualitive, but also quantitive agreement with average deviation about 4,8 %.

The computer experiment [15] allows verifying the interpolation formula (2.30) with taking into account ***the size dependence of the critical temperature of a fluid in a pore***. Figure 2.3 (see black circles) demonstrates computer experiment results for dependence of the critical temperature $T_c(H)/T_c(\infty)$on the pore's size, i.e. on the thickness H of a slitlike pore filled by water.

To compare the interpolation formula (2.30) with results of computer experiment [15], we substitute the size dependence of the critical exponent $\nu(H)$ into formula (2.21) for the critical temperature $T_c(H)$ of a fluid in a pore. The agreement between computer experiment data and theoretical calculations seems to be quite good. The lowest experimental point corresponds to the critical temperature of almost 2D water in slitlike pore with its thickness $H = 0.5$ nm. This value of thickness H refers to almost one monolayer plane with taking into account that the diameter of water molecule equals $d \approx 0.3$ nm. It is interesting to stress that the beginning of the dimensional crossover DC-2 from 3D to 2D critical behavior takes place at the slitlike pore's thickness $H_{cr} \approx 2.4$ nm. This value of H_{cr} was mentioned in [15] and corresponds to approximately 8 molecular layers of water molecules in slitlike pores.

Finally we would like to compare theoretical and experimental ***shifts of the critical temperature in confined liquids.*** Table 2.6 (2nd line) contains experimental results for shifts of the water critical temperature $\Delta T_c = T_c(\infty) - T_c(R)$ in cylindrical pores of different radius R obtained by different experimental methods (see references in review [15]). Results of theoretical caculation $\Delta T_c = T_c(\infty) - T_c(R) \sim (\mu_1 \xi_0/R)^{1/\nu}$ in accordance with the formula (2.21) are presented in the 3rd line of Table 2.6 [23, 43].

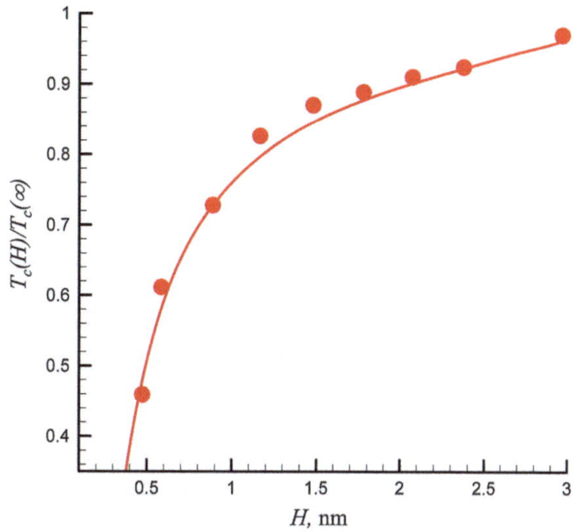

Fig. 2.3 Related critical temperature of a fluid in a pore $T_c(H)/T_c(\infty)$ versus the thickness H of a slitlike pore (*black circles* correspond to the data computer experiment [15]; solid curve was obtained using the formulae (2.21) for $T_c(H)$ and (2.30) for ν [30])

Table 2.6 Experimental and theoretical shifts of the critical temperature

Pore radius R (nm)	3.9	5.0	11.0	12.1	15.7
$(\Delta T_c/T_c)_{exp}$	0.047	0.044	0.033	0.029	0.0015
$(DeltaT_c/T_c)_{theor}$	0.065	0.044	0.013	0.011	0.0071

In spite of that is qualitative agreement between experimental and theoretical data, these and other results demonstrate a considerable enough distinction of the data received by different methods for the same liquid in pores with approximately equal radius.

2.6 Conclusion

In this paper we studied critical phenomena in liquid in restricted geometry. Thermodynamic, kinetic and correlation properties of confined liquids are essentially different from those properties of bulk systems. The influence of spatial limitation effects (among them—effects of dimensional crossover) is more pronounced in cases when linear sizes L of systems are less or the same order of magnitude as the correlation length ξ of order parameter fluctuations.

The study of critical phenomena and 2nd order (continuous) phase transitions in systems in reduced geometry allows one to formulate important conditions of universality classes for confined soft matter. Namely, to the following well-known basic conditions of universality classes in bulk systems with $L \gg \xi$:

(a) the same space dimensionality;
(b) the same dimensionality of order parameter;

(c) the same type of intermolecular interaction (short- or long-range);
(d) the same symmetry of Hamiltonian (fluctuation part of thermodynamic potential), it is necessary to add the following conditions of universality classes in confined systems with $L \leq \xi$;
(e) the same type of system's geometry (lower crossover dimensionality);
(f) the same type of boundary conditions;
(g) the same type of physical properties under consideration.

The last factor appears to be important because coordinates of maxima or minima in systems in restricted geometry, being analogues of bulk critical parameters, are different (nonuniversal) for various physical properties.

The verification of theoretical results for critical phenomena in confined systems needs experimental studies which are much more complicated than those experimental studies in bulk systems. Nevertheless, we hope that the attractive FSS hypothesis of finite-size scaling for the diverse critical phenomena and phase transitions in the nano- and mesoscale world around and inside us will prove experimentally to be correct.

Acknowledgments The author wishes to thank Prof. M.A. Anisimov, Prof. L.A. Bulavin, Prof. K.A. Chalyy, Prof. N.I. Lebovka, Prof. H.-J. Moegel, Prof. W. Schroer, Prof. V.M. Sysoev, Prof. A.N. Vasilev, Dr. V.G. Bojko, Dr. L.M. Chernenko, Dr. G.V. Hrapijchuk, Dr. E.L. Lakoza, Dr. A.V. Oleinikova, Dr. E.V. Zaitseva for helpful advices, remarks and collaboration.

References

1. L.P. Kadanoff, Phys. **2**, 263 (1966)
2. M.E. Fisher, ed. by M.S. Green (Academic Press, New York, 1971)
3. H.E. Stanley, *Introduction to Phase Transitions and Critical Phenomena* (Glarendon Press, Oxford, 1971)
4. E.L. Lakoza, V.M. Sysoev, A.V. Chalyi, JETP **65**, 605 (1973) Sov.Phys.-JETF,**38**, 298 (1974)
5. M.E. Fisher, Rev. Mod. Phys. **46**, 597 (1974)
6. K.G. Wilson, J. Kogut, Phys. Rep. **12**, 75 (1974)
7. A.Z. Patashinskii, V.L. Pokrovskii, *The Fluctuation Theory of Phase Transitions* (Pergamon Press, Oxford, 1979)
8. M.N. Barber,ed. by C.Domb and J.L.Lebowitz (Academic Press, 1983), p.145
9. V. Privman, M.E. Fisher, Phys. Rev. B **30**, 322 (1984)
10. V. Privman, ed. by V. Privman (World Scientific, Singapore, 1990), p.1
11. M.A. Anisimov, *Critical Phenomena in Liquids and Liquid Crystals* (Gordon & Breach, Philadelphia, 1991)
12. V. Privman, P.C. Hohenberg, A. Aharony, ed. by C. Domb and J.L. Lebowitz (Academic, New York, 1991)
13. K. Binder, Annu. Rev. Phys. Chem. **43**, 33 (1992)
14. A.V. Chalyi, A.V. Lebed, *Non-Homogeneous Liquids Near the Critical Point and the Boundary of Stability and Theory of Percolation in Ceramics* (Harwood Academic Publishers, London, 1993)
15. I. Brovchenko, A. Oleinikova, Handb. Theor. Comput. Nanotechnol. **62**, 1 (2005)
16. A.V. Chalyi, J. Mol. Liquids **58**, 179 (1993)
17. A.V. Chalyi, A.N. Vasilev, J. Mol. Liquids **84**, 203 (2000)

18. I.Z. Fisher, *Statistical Theory of Liquids* (Fizmatgiz, Moscow, 1961)
19. V.Y. Antonchenko, A.S. Davydov, V.V. Ilyin, *Basic Principles of Physics of Water* (Naukova Dumka, Kiev, 1991)
20. I.S. Gradshtein, I.M. Ryzhyk, *Tables of Integrals, Sums, Series and Products* (Nauka, Moscow, 1971)
21. M.E. Fisher, J. Math. Phys. **5**, 944 (1964)
22. L.A. Bulavin, A.V. Chalyi, K.A. Chalyy, L.M. Chernenko, L.G. Grechko. Preprint of the Bogolyubov Institute for Theoretical Physics, ITP-93-15E, Kiev (1993)
23. A.V. Chalyi, Y.V. Tsekhmister, K.A. Chalyy, *Processes of Ordering and Self-organization in Fluctuation Models of Open Systems* (Vipol, Kiev, 2001)
24. M.A. Anisimov, A.V. Voronel, and E.E. Gorodetskii, Sov.Phys.-JETF,**33**, 605 (1971)
25. B. Kastening, Phys. Rev. E **86**, 041105 (2012); arXiv: 1209.0105v2 [cond-mat.stat-mech] (2013)
26. K. Binder, Phys. Rev. Lett. **47**, 693 (1981)
27. A.V. Chalyi, A.N. Vasilev, Ukr. Phys. J. **45**, 118 (2000)
28. C.E. Bertrand, M.A. Anisimov, J. Phys. Chem. B **115**(48), 14099 (2011)
29. H. Weingartner, W.Schroer, ed. by I.Prigogine, S.A.Rice (Interscience Publications, Wiley, 2001)
30. A.V. Chalyi, L.A. Bulavin, V.F. Chekhun et al., Condens. Matter Phys. **16**, 23008 (2013)
31. M.O. Kimball, K.P. Mooney, F.M. Gasparini, Phys. Rev. Lett. **92**, 15301 (2004)
32. V.M. Sysoev, A.V. Chalyi, Theor. Math. Phys. **26**, 126 (1976)
33. W. Koch, V. Dohm, Phys. Rev. E **58**, R1179 (1998)
34. A.V. Chalyi, L.M. Chernenko, JETP **87**, 187 (1984)
35. E.K. Riedel, F.J. Wegner, Phys. Rev. Lett. **29**, 349 (1972)
36. S.D. Landau, V.G. Mukhailovsky, A.V. Chalyi, Ukr. Phys. J. **15**, 423 (1971)
37. M.A. Anisimov and J.V. Sengers, ed. by E.Kivan, P.G. Debenedetti, C.J. Peters (Kluver, Dordrecht, 2000), 89121
38. Y.C. Kim, M.A. Anisimov, J.V. Sengers, E. Luijten, J. Stat. Phys. **110**, 591 (2003)
39. M.A. Anisimov, Condens. Matter Phys. **16**, 1 (2013)
40. K. Kawasaki, ed. by C. Domb, M.S. Green (Academic Press, New York, 1976)
41. L.A. Bulavin, T.V. Karmazina, V.V. Klepko, V.I. Slisenko, *Neutron Spectroscopy of Condensed Systems* (Academperiodyka, Kiev, 2005)
42. L.A. Bulavin and K.A. Chalyy, Bulletin of the Kiev University (Phys.&Math.), №1, 328 (2006)
43. L.A. Bulavin, K.A. Chalyy, *Neutron Optics of Mesoscale Liquids* (Naukova Dymka, Kiev, 2006)

Part II
Phase Transitions, Supercritical Liquids & Glasses

Chapter 3
Global Isomorphism Approach: Main Results and Perspectives

Leonid Bulavin, Vadim Cheplak and Vladimir L. Kulinskii

Abstract In this chapter we review main results of the Global Isomorphism approach and discuss possible routes for further studies. The approach is based on the minimal geometric reformulation of the (approximate) linearities of the binodal diameter and the unit compressibility line (Batschinsky law). Explicit relations between the thermodynamic functions of the Lattice Gas model and the fluid within the framework of the approach proposed earlier in [V.L. Kulinskii, J. Phys. Chem. B **114** 2852 (2010)] are discussed. On this basis we show that the critical compressibility factor of molecular fluids can be related with that of the lattice gas. We show how the associative properties of a fluid can be taken into account via the structure of the isomorphic lattice. Also we derive the relation between the entropies of a fluid and its lattice analog. The entropy of the fluid is decomposed into symmetrical and asymmetrical parts. We demonstrate that such decomposition is consistent with the basic Clausius-Clapeyron relation and the binodal asymmetry represented by the law of the rectilinear diameter.

3.1 Introduction

The idea of similarity for the physical properties of the systems is widely used in physics. It is based on the possibility to introduce dimensionless quantities which are invariant with respect to the scaling transformations. Since the seminal work of

L. Bulavin
Department of Molecular Physics, Faculty of Physics, Taras Shevchenko National University of Kyiv, 2, Prosp. Academician Glushkov, Kyiv 03022, Ukraine
e-mail: bulavin221@gmail.com

V. Cheplak · V.L. Kulinskii (✉)
Department of Theoretical Physics, Odessa National University, Dvoryanskaya 2, 65082 Odessa, Ukraine
e-mail: kulinskij@onu.edu.ua

V. Cheplak
e-mail: vadimcheplak@yandex.ru

L. Bulavin and N. Lebovka (eds.), *Physics of Liquid Matter: Modern Problems*, Springer Proceedings in Physics 171, DOI 10.1007/978-3-319-20875-6_3

van der Waals [1] his phenomenological equation plays important role in the theory of Soft Matter. In the theory of Liquid Matter the classical result of van der Waals (vdW)—the Principle of Corresponding States (PCS) is one of its cornerstones. Despite pure phenomenological and rather crude approximation the vdW equation has many remarkable features. This made it the basic model of the equation of state (EoS) for different fluids.

Two competitive effects: the attraction due to long ranged polarizational forces and the repulsion because of hardness of the electronic molecular shells bring the critical state where liquid and vapor become indistinguishable. In accordance with the PCS the critical compressibility factor $Z_c = \frac{P_c}{n_c T_c}$ is the same for all substances with similar interaction potentials. For vdW EoS $Z_c = 3/8$ while for real substance Z_c varies in the interval $0.1 < Z_c < 0.5$ [2]. Therefore additional parameters should be added to characterize the classes of thermodynamic universality. In fact, the PCS in rigorous sense is valid for the systems with the potentials which can be related via simple scaling transformation of the parameters [3].

Due to interaction of long range fluctuations the critical universality emerges in the fluctuational region. Functional form of the critical asymptotics of thermodynamic quantities is universal and determined by the general symmetries of the system. Critical behavior of the molecular fluids belongs to the universality class of the Ising model [4]. Thus there are two notions of thermodynamic universality. The similarity of the EoS in a sense of the PCS and the isomorphism of the critical behavior. It is clear that the first one is much stronger and implies the similarity of the critical behavior. In particular, the critical amplitudes for the systems of the same class of thermodynamic similarity can be related by some factors which depend on the parameters of the interaction.

Despite the fact that vdW EoS fails to predict the critical behavior for the system with realistic potentials some relations can be derived on its basis. Long time these relations were considered as mere curious facts restricted to simplified structure of the vdW EoS. But as the body of thermophysical data expanded it appeared that these relations could be applied to much wider class of substances than the vdW EoS itself.

One of the well-known relation is the law of the rectilinear diameter (LRD) [5, 6]. It states that the diameter of the coexistence curve in terms of density-temperature is the straight line:

$$\tilde{n}_d = \frac{n_l + n_g}{2n_c} = 1 + A\,\frac{T_c - T}{T_c}\,, \quad A > 0, \tag{3.1}$$

where n_i, $i = l, g$ are the densities of the liquid and the gas phases correspondingly, n_c is the critical density, T is the temperature. Further we put the Boltzmann constant to unit: $k_B = 1$. Actually, for the vdW EoS the diameter is not straight but the curvature is indeed very small.

Another simple linear relation is the Batschinski law [7] for the vdW EoS:

$$P = \frac{nT}{1 - nb} - an^2, \tag{3.2}$$

where P—is the pressure. It states that the set of points of unit compressibility factor $Z = P/(nT)$, i.e. $Z = 1$, is the straight line:

$$\frac{T}{T_B} + \frac{n}{n_B} = 1, \tag{3.3}$$

where $T_B = a/b$ and $n_B = 1/b$. In general, the line on on $n - T$ plane where $Z = 1$ is called the Zeno line [8]. The parameter T_B is the Boyle temperature:

$$B_2(T_B) = 0 \tag{3.4}$$

and the characteristic density n_B is determined by the relation [8]:

$$n_B = \frac{T_B}{B_3(T_B)} \frac{dB_2}{dT}\bigg|_{T=T_B}. \tag{3.5}$$

Here B_n is the virial coefficient of nth order [9]. Additionally, it is known that the Zeno line is tangent to the extension of the binodal into the low temperature region so that the binodal appears to be inscribed into the triangle formed by the T-axis and the Zeno line [2, 10]. Apfelbaum and Vorob'ev also revealed a number of correlations between the critical point parameters T_c, n_c and those of Zeno line T_B, n_B [11–13].

Recently, simple geometrical formulation of the results on the Zeno line, the LRD and the triangle of liquid-gas states was proposed in [14] and developed in series of publications [15–17]. It is based on the fact that the lines of the phase equilibria determine the partitions for the space of thermodynamic states. These partitions are topologically equivalent (see Fig. 3.1). The linear character of (3.1) and (3.3) allows to search the isomorphism in a form of the correspondence between the phase diagram of the projective 1–1 mapping the LG and the liquid-vapor states of the Lennard-Jones fluid and the lattice gas (Ising model). This approach provides the explanation for some the phenomenological findings of

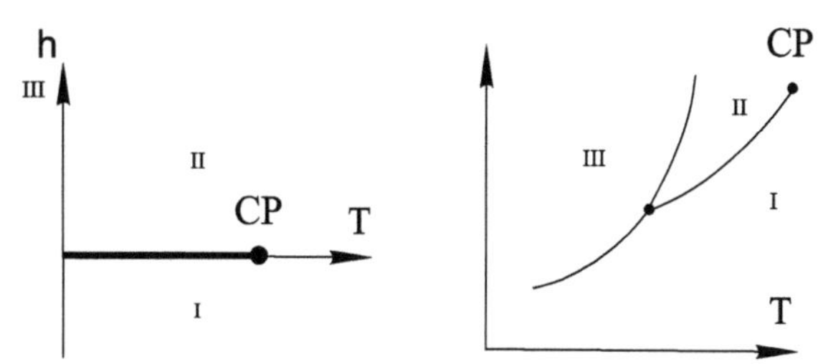

Fig. 3.1 Phase diagrams of the Lattice Gas (Ising model) (*left*) simple fluid (*right*)

Apfelbaum and Vorob'ev concerning the triangle of liquid-gas states and the line of the critical points [18]. The generalizations of this approach of Global Isomorphism have been proposed recently in [19–21].

In this contribution we review main results of the Global Isomorphism approach. They include the coordinates of the Lennard-Jones (LJ) fluids in different dimensions and the relation with thermodynamical potential of the lattice gas (Ising model). Discussion of physical basis of the linearities (3.1) and (3.3) on the liquid-vapor part of the phase diagram of the fluids follows the results of [14, 16]. We expand some arguments of [16], and give corrected interpretation of the Batchinski law in a way consistent with the van der Waals approximation for the EoS. Also we derive the relation between the thermodynamic potentials for the continuum and the lattice models of fluids. On this basis we obtain the relation between the entropies of these systems and show that this relation is consistent with the basic thermodynamic Clausius-Clapeyron relation. Also we present recent results on the critical compressibility factor values of molecular fluids [22] and discuss routes for further research. The structure of the paper is as follows. In Sect. 3.2 we give the basic grounds for the Global Isomorphism approach and relation of the parameters of the transformation with the microscopic (model) potential. Section 3.3 is devoted to the relation between thermodynamic potentials and its consequences for the calculation of the critical point loci for the fluids with Mie-class potentials. In particular we derive the relation between the critical values of compressibility factor for fluid and lattice gas. In Sect. 3.4 we derive the relation between the entropies of the LG and LJ fluid and consider its consequences from the point of view of symmetrization of the liquid-gas binodal. The results and perspectives are summarized in Discussion section.

3.2 Global Isomorphism

The idea of Global Isomorphism between liquid system which demonstrate the linearities like LRD and the Zeno line and the lattice gas model is very simple. It starts with the fact that all this linearities trivially hold in simple Ising model due to its symmetry spin-flip symmetry:

$$H = \frac{1}{2}\sum_{i,j} J_{ij}\sigma_i\sigma_j + h\sum_i \sigma_i . \tag{3.6}$$

where $\sigma_i = \pm 1$ as the random variables, h is the "magnetic" field and J_{ij} is the interaction strength between cites i, j. The lattice gas (LG) representation is given in terms of the occupation number variables: $x_i = \frac{1+\sigma_i}{2} = 0.1$. Correspondingly the order parameters magnetization (in terms of the Ising model) $m = \langle \sigma_i \rangle$ and the density $x = \langle x_i \rangle$ (for the LG representation) are related as: $m = 2x - 1$. Because of the spin flip symmetry $\sigma_i \to -\sigma_i$ the phase diagram of the LG is symmetrical with respect to the line $x = 1/2$ which is the diameter of the binodal. Other obvious linear elements are the lines $x = 0$ and $x = 1$ which are the tangents to the binodal in corresponding points. The lattice models are widely used in statistical physics. For the continuous systems they serve as useful approximations suitable for numerical simulations. Yet in the thermodynamic limit the continuum and discrete systems are assumed to demonstrate the same behavior. This idea is the basis of the modern theory of the phase transitions and critical phenomena (see [23, 24]). The critical behavior of many liquids has the same asymptotics as the Ising model. The latter is equivalent to the φ^4 field theory [23, 25].

Apart from the universal critical characteristics which are determined by the long range fluctuations there are nonuniversal quantities, e.g. the locus of the CP, the critical amplitudes. Such quantities depend on the details of the interparticle interactions and are governed by the noncritical fluctuations of short scales. Similarly the linearities of the Zeno-line and binodal diameter lie beyond the critical region and obviously are not connected with the critical universality.

The analog of the Zeno-line for the LG obviously is the line $x = 1$ where the "holes" are absent. This is consistent with the basic expression for the compressibility factor [9]:

$$Z = \frac{P}{nT} = 1 - \frac{2\pi n}{3T} \int r^3 \frac{\partial \Phi(r)}{\partial r} g_2(r; n, T)\, dr, \tag{3.7}$$

If $x = 1$ then the perfect configurational order takes place and the site-site correlation function for (3.6) vanishes:

$$\langle\langle \sigma_i \sigma_j \rangle\rangle = \langle \sigma_i \sigma_j \rangle - \langle \sigma_i \rangle \langle \sigma_j \rangle = 0\,.$$

The critical isochore $x_c = 1/2$ coincides with the diameter. The Zeno-line is the tangent to the binodal which in this case expands up into the region $t \to 0$.

The empirical linearities of the Zeno-line and the binodal diameter allow to state that:

$$n = n_* \frac{x}{1 + z\mathrm{t}}, \qquad T = T_* \frac{z\mathrm{t}}{1 + z\mathrm{t}}, \tag{3.8}$$

where

$$z = \frac{T_c}{T_* - T_c}\,. \tag{3.9}$$

The inverse transformations has the form:

$$x = \frac{n/n_*}{1 - T/T_*}, \qquad \mathrm{t} = \frac{1}{z} \frac{T}{T_* - T}\,. \tag{3.10}$$

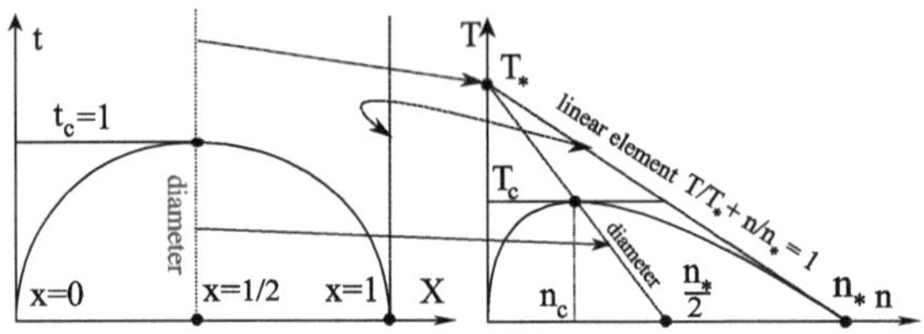

Fig. 3.2 Correspondence between the linear elements of the phase diagrams. Zeno-line, generally, is not linear and we introduce the linear element $T/T_* + n/n_* = 1$ [14]

Here n, T are the density and the temperature of the LJ fluid, $\mathrm{t} = t/t_c$ is the temperature variable of the LG reduced to its critical value t_c, n_* and T_* are the parameters of the Zeno element [15].

The transformation (3.8) is uniquely determined by the correspondence between the characteristic *linear* elements on the phase diagrams of the fluid and the LG (see Fig. 3.2). In particular, the line $x = 1$ of the lattice gas corresponds to the Zeno-element:

$$\frac{n}{n_*} + \frac{T}{T_*} = 1\,. \tag{3.11}$$

Here T_* is the Boyle temperature in the van der Waals approximation [26]:

$$T_* = \frac{a}{b}\,, \tag{3.12}$$

where

$$a = \; - 2\pi \int\limits_{\sigma}^{+\infty} \Phi_{attr}(r)\, r^2\, dr \tag{3.13}$$

and $\Phi_{attr}(r)$ is the attractive part of the full potential $\Phi(r)$, σ is the effective diameter of the particle so that $b = \frac{2\pi}{3}\sigma^3$. The definition for the density parameter n_* is analogous to (3.5):

$$n_* = \frac{T_*}{B_3(T_*)}\left.\frac{dB_2}{dT}\right|_{T=T_*}. \tag{3.14}$$

In general, T_* and n_* are different from commonly used Zeno-line parameters T_B and n_B and therefore (3.11) is not the classical Zeno-line determined by (3.3). They are the same only in a case of the vdW EoS. The Mie-potentials are introduced as follows:

$$Mie(r;m,q) = \varepsilon\Phi_0(q/m)\left(\left(\frac{\sigma}{r}\right)^q - \left(\frac{\sigma}{r}\right)^m\right), \quad q > m. \tag{3.15}$$

$$\Phi_0(q/m) = \frac{m}{q-m}\left(\frac{q}{m}\right)^{q/(q-m)}.$$

We consider the case with $m = 6$ and $7 \le q \le 18$. The value of T_* is as follows:

$$T_*/\varepsilon = \frac{\left(\frac{q}{m}\right)^{\frac{q}{q-m}}}{(m/d-1)(q/m-1)}. \tag{3.16}$$

Note that at $q \to \infty$ we get the Sutherland potential:

$$\Phi_S(r;m) = \begin{cases} \infty & r \le \sigma, \\ -\frac{\varepsilon}{r^m} & r > \sigma. \end{cases} \tag{3.17}$$

with $T_B \approx 1.17$, $n_B \approx 1.79$ [27]. Further we use common dimensionless units for the temperature $T \to T/\varepsilon$ and the density $n \to n\sigma^d$, where d is the dimension. In general d-dimensional case the following dimensionless representation of the second virial coefficient is useful:

$$B_2(T)/v_0 = d2^{d-1}\int_0^\infty \left(1 - e^{-\Phi(r)/T}\right) r^{d-1}\, dr, \tag{3.18}$$

where $S_d = \pi^{d/2}/\Gamma\left(\frac{d}{2}\right)$ is the area of unit sphere in d-dimensional space, $v_0 = S_d/d(\sigma/2)^d$ is the d-dimensional volume of the particle with the diameter σ. Of course, the spatial scale defined by σ is determined by the repulsive part of the interaction and in the absence of hard wall has some arbitrariness. Only for the Sutherland potential (3.17) σ is determined unambiguously. In this particular case ($q = \infty$) the integration in (3.18) can be performed analytically. E.g. in 3D:

$$B_2(T)/v_0 = \frac{8}{3}\left(\sqrt{\frac{\pi}{T}}\operatorname{erfi}\left(\frac{1}{\sqrt{T}}\right) - e^{\frac{1}{T}}\right). \tag{3.19}$$

If the potential $\Phi(r)$ can be splitted into repulsive and attractive part via introduction of some short scale then the high-temperature expansion of (3.18) has the form [26]:

$$B_2(T)/v_0 = d2^{d-1}(1 - T_*/T) + o(1/T). \tag{3.20}$$

The binodals of the systems with the Mie-potentials obey the LRD (3.1) as it follows from the simulations [28, 29]. The natural restriction here is the stability of whole liquid-gas region which is fulfilled is the attractive part is not too short ranged [30–33], i.e. if m is not very big ($m < 8$ in 3D) [34].

In view of the "fluid–lattice model" isomorphism the usage of T_* instead of T_Z can be supported by the following argument. In lattice gas model (Ising model) there is only attractive part of the interaction. The repulsive part of the potential is substituted by the strict filling rule—"one site–one particle" which contribute to the configurational entropy. From this point of view the usage of T_* which depend only on the attractive part of the potential appears to be consistent. Also this option can be tested via the calculation of the CP locus and the binodal of the LJ fluid using the corresponding information about lattice gas (Ising model). This was done in [16, 17]. In following section we review these results.

3.3 The Relation Between the Thermodynamic Potentials of the Fluid and the Lattice Gas

The transformation (3.8) states the 1–1 correspondence between the thermodynamic states within the liquid-gas triangle of the LJ fluid and the corresponding region $t > 0,\ 0 < x < 1$ of the lattice gas. Naturally, this relation is equivalent to the statement about the relation between the thermodynamic potentials of these systems.

Let us consider the thermodynamic potentials of the grand canonical ensembles for the LG and the fluid:

$$\mathfrak{G}(h, t, \mathcal{N}) = \mathcal{N}\, \mathfrak{g}(h, t), \quad \text{and} \quad J(T, \mu, V) = P(T, \mu)\, V, \tag{3.21}$$

Here $\mathcal{N}$ is the number of sites in a lattice. First it is natural to state the following relation $\mathcal{N} = n_* V$ between the extensive variables of these ensembles. The results of [35] allow to interpret $1/n_*$ as the volume per particle in the ideal crystal state at $T \to 0$. Such a state defines the lattice which may serve as the basis for the determination of the corresponding Lattice Gas model.

Using the standard definitions:

$$n = \frac{1}{V} \frac{\partial J}{\partial \mu}\bigg|_T, \quad x = \frac{1}{\mathcal{N}} \frac{\partial \mathfrak{G}}{\partial h}\bigg|_t \tag{3.22}$$

and taking into account that in accordance with (3.8) T depends on t only, we get the following relation between the potentials:

$$J(\mu, T, V) = \mathfrak{G}(h(\mu, T), t(T), \mathcal{N}) \Rightarrow P(\mu, T) = n_* \mathfrak{g}(h(\mu, T), t(T)). \tag{3.23}$$

In terms of the potentials it is as follows:

$$P(T, \mu)\, V = \mathfrak{G}(t(T), h(\mu, T)), \tag{3.24}$$

Further we will use the dimensionless quantities $\tilde{n} = n/n_*$, $\tilde{T} = T/T_*$, $P = P/P_*$, $P_* = n_* T_*$. Since the pressure of the lattice gas $\mathfrak{P}(h,t)$ is related with the grand canonic potential (see e.g. [36]):

$$\mathfrak{G} = \mathfrak{P}(h,t)\,V \tag{3.25}$$

therefore:

$$P(T,\mu) = \mathfrak{P}(t(T), h(\mu,T)). \tag{3.26}$$

From (3.8) the relation between the density of the fluid and the density of the LG can be written as following:

$$n(\mu,T)/n_* = x(h(\mu,T), t(T))\,(1 - T/T_*). \tag{3.27}$$

Taking into account that:

$$\left.\frac{\partial}{\partial\mu}\right|_T = \left.\frac{\partial h}{\partial\mu}\right|_T \left.\frac{\partial}{\partial h}\right|_t,$$

from (3.22), (3.23) and (3.27) we get the following relation:

$$h(\mu,T) = \left(1 - \tilde{T}\right)\left(\tilde{\mu} - \tilde{\mu}_0(T)\right), \tag{3.28}$$

or:

$$\tilde{\mu} - \tilde{\mu}_0(T) = h\,(1 + zt)\,. \tag{3.29}$$

Note that this relation is consistent with the basic relation between chemical potential and pressure:

$$\tilde{\mu}(P,T) - \tilde{\mu}_0(T) = \int\limits_{P_0(T)}^{P} \frac{dp}{n(p,T)}, \tag{3.30}$$

where P_0 is the saturation pressure.

We remind that $h = 0$ below CP is the coexistence line for the LG and is mapped onto the saturation curve of the continuum fluid. Therefore $\mu_0(T)$ should be identified with the chemical potential along the saturation curve below the critical point $T < T_c$.

Let us use the obtained relation (3.23) for the calculation of the critical properties of the fluids basing on the information about lattice model. According to (3.8) it is possible to calculate the state the LJ fluids based on the information about the lattice

Table 3.1 The results of computation of CP locus according to (3.31) for LJ potential $Mie(6,12)$ for different dimensions with the parameters T_*, n_* of the Zeno-element

LJ "6–12" fluid	2D	3D	4D	5D
T_*	2	4	8	20
n_*	0.94	0.967	1.36	2.87
T_c	0.5	1.33	3.2	9.1
n_c	0.353	0.322	0.41	0.78
$T_c^{(num)}$	0.515, [38]	1.312, [39]	3.404, [40]	8.8, [40]
$n_c^{(num)}$	0.355, [38]	0.316, [39]	0.34, [40]	?
T_Z	1.56	3.418	9.01	40.8
n_Z	1.19	1.14	1.21	1.67

The parameters T_B, n_B are given for comparison

model [17, 22, 37] (see Fig. 3.2). In particular, as it directly follows from (3.8) the locus of the CP of the LJ fluid is:

$$n_c = \frac{n_*}{2(1+z)}, \qquad T_c = T_* \frac{z}{1+z}. \tag{3.31}$$

As was shown in [17, 37] the parameter z can be related with the scaling property of the attractive potential $\Phi_{\mathrm{attr}}(r) \propto -1/r^m$:

$$z = d/m \tag{3.32}$$

where d is the dimension [15]. So that for LJ in 2D and 3D we have $z = 1/3$ and $z = 1/2$ correspondingly. This gives the possibility to apply and check the global isomorphism approach for different dimensions $d < 6$ where T_* is finite for the LJ potential. The results of calculations for the CP locus for the fluid with "6–12" LJ interaction potential in different dimensions are presented in Table 3.1.

Overall agreement with the numerical result are good and gives the grounds to apply the approach to potentials of type $Mie(6,q)$. It can be also obtained using their simple relation with the characteristics of the LJ potential. Indeed, according to (3.31) the CP locus is determined by the Boyle temperature T_* in the vdW approximation. From (3.15) it follows that attractive part of the potentials $Mie(6,q)$ differs only by the amplitude factor and therefore the parameters T_* for such potentials are related via simple scaling:

$$\frac{T_*(6,q_1)}{T_*(6,q_2)} = \sqrt{\frac{q_2-6}{q_1-6}} \left(\frac{q_1}{6}\right)^{\frac{q_1}{2(q_1-6)}} \left(\frac{q_2}{6}\right)^{-\frac{q_2}{2(n_2-6)}} \tag{3.33}$$

We choose the standard LJ potential $\Phi(r;6,12)$ as the reference case. Then from simple rescaling following from (3.33) we get the critical temperatures for $\Phi(r;6,n)$ potential in 3D:

$$T_*(q) = 2\sqrt{\frac{6}{q-6}}\left(\frac{q}{6}\right)^{\frac{q}{2(q-6)}} \quad \Rightarrow \quad T_c(q) = \frac{2}{3}\sqrt{\frac{6}{q-6}}\left(\frac{q}{6}\right)^{\frac{q}{2(q-6)}}. \tag{3.34}$$

The analogous results for other potentials in dimensions $D \le 5$ are in [16, 41].

3.3.1 The Relation Between Critical Compressibility Factors

The critical compressibility factor (CCF) $Z_c = P_c/(n_c T_c)$ is the simplest scaling invariant widely used in the formulation of the PCS. For the vdW EoS $Z_c = 3/8$. Simple fluids, e.g. noble gases Ar, Kr, Xe, have Z_c ranged between 0.28 and 0.3. Using (3.23) and the results about the loci of the CP the relation between the critical compressibility factors of the LJ-fluid and the lattice gas can be derived [22].

With this (3.24) allows to derive the relation between the critical compressibility factors Z_c of a fluid and the corresponding lattice model. Indeed, from (3.8) and (3.26) we get:

$$Z_c^{(fl)} = \frac{P_c}{n_c T_c} = \frac{(1+z)^2}{z}\frac{t_c}{T_*} Z_c^{(LG)}, \tag{3.35}$$

where t_c is the critical temperature of the lattice gas model and $Z_c^{(LG)}$ is its critical compressibility factor value. In accordance with [36] it is related with the partition function per spin $G^{1/N}$ of the Ising model $Z_c^{(LG)} = 2 \ln G_c^{1/N}$. The value $G_c^{1/N}$ can be obtained for lattice models using (high) low-temperature expansions [42, 43]. E.g. in 3D case of cubic lattice $t_c \approx 4.51J$ [44] and $Z_c^{(LG)} = 0.221$ [36].

To consider the case of the LJ-fluid we take into account that for the LJ potential:

$$V_{LJ}(r) = 4\varepsilon\left(\left(\frac{\sigma}{r}\right)^{12} - \left(\frac{\sigma}{r}\right)^{6}\right), \tag{3.36}$$

$T_* = 4\varepsilon$ and the relation between the LJ parameter ε and the spin-spin interaction J of the Ising model is $\varepsilon = 4J$ [45]. We assume that the LJ parameter ε can be identified with the interaction constant of the lattice gas. This is based on the fact that ε is the energy of the LJ interaction at equilibrium distance. The latter can be identified with the period of the cubic lattice. Further we use conventional dimensionless units for the temperature $T \to T/\varepsilon$ density $n \to n\sigma^3$ and pressure $P \to P\sigma^3/\varepsilon$.

Calculation of the factors in (3.35) results in:

$$Z_c^{(fl)} \approx 1.27 Z_c^{(LG)} = 0.281, \tag{3.37}$$

which correlates well with the CCF value for simple fluids. This also clarifies the nature of the difference between value of Z_c for simple fluids and that of classical vdW EoS. The possibility to include the Pitzer's acentric factor into consideration was considered in [20, 22].

The estimate for the value of critical pressure is:

$$P_c = 0.121\,, \tag{3.38}$$

and agrees well with the result of simulations [46] $P_c = 0.126$ for the 3D LJ fluid.

Our main result (3.35) can be also applied in 2D case. From the Onzager's solution we know the basic quantities exactly:

$$Z_c^{(LG)} \approx 0.097\,, \quad t_c \approx 2.27J\,,$$

and we get the estimate:

$$Z_c^{(fl)} \approx 1.513 Z_c^{(LG)} = 0.146\,. \tag{3.39}$$

The value of the critical pressure of the 2D LJ fluid is:

$$P_c = Z_c^{(fl)} n_c T_c \approx 0.026\,, \quad n_c = 0.353\,, \quad T_c = 0.5\,. \tag{3.40}$$

The estimates for critical density and temperature are consistent with known results for two dimensional LJ fluid [38]. The values of Z_c and P_c lower than those obtained within perturbation theory $P_c = 0.046$ and in simulations $P_c = 0.037$ [47]. Possible explanation of this fact is that these approaches underestimate the flatness of the binodal of the 2D LJ fluid. This is due to the global power law behavior with the critical exponent $\beta = 1/8$ because of the global isomorphism with the 2D Ising model [17]. For convenience all the parameters and the results obtained are summarized in Table 3.2.

As has been shown recently in [18] the relation (3.35) can be easily modified so that to include the associative properties of the fluids in near critical region. From the point of view of the global isomorphism approach it is natural to relate the associative property of a fluid and the structure of the lattice. Namely, the prevalence of q-mers (dimers, trimers etc.) contribution into the density can be interpreted in terms of the isomorphic lattice with q particles per unit cell. The formula (3.35) is modified as follows:

$$Z_c^{(fl)} = \frac{1}{q}\frac{(1+z)^2}{z}\frac{t_c}{T_*} Z_c^{(LG)}\,. \tag{3.41}$$

Table 3.2 The results for the critical point parameters of *LJ* fluid in 2D ($z = 1/3$) and 3D ($z = 1/2$) cases [15]

LJ "6–12" fluid	P_c	$Z_c^{(fl)}$	$Z_c^{(LG)}$
2D	0.026	0.147	0.097
3D	0.122	0.281	0.221

Table 3.3 Critical parameters (t_c and $Z_c^{(LG)}$) of the isotropic Ising model on different 3D lattices taken from [42, 48] and corresponding values of $Z_c^{(fl)}$ for the LJ fluid obtained using (3.41)

	q	t_c	$Z_c^{(LG)}$	$Z_c^{(fl)}$
sc	1	4.51	0.221	0.281
bcc	2	6.35	0.239	0.214
fcc	4	9.80	0.246	0.17

Table 3.4 Experimental values of Z_c for some molecular fluids and liquid metals [51]

	Li	Na	K	Rb	Cs	CH_3OH [54]	H_2O [54]	NH_3 [54]
Z_c	0.21	0.23	0.21	0.22	0.21	0.19	0.23	0.25

The results for the lattices are given in Table 3.3. The comparison of theoretical estimates (3.41) for Z_c shows good correlation with the data (see Table 3.4). Because of high temperature CP and indirect measurements the uncertainty in value Z_c is quite high especially for light alkali metals like lithium and sodium (see e.g. [49]). The extrapolation methods for the low temperature data are used to obtain the parameters of the critical state. Commonly, such extrapolation exploits the extrapolation based on (3.1) (see e.g. [50]). For light alkali metals we choose the values for Z_c from [51] basing on the similarity reasonings and identity of their electronic structure with more heavier alkali metals. For high temperature metals like Zn, Pb and Au such analysis was performed in [52]. As the result the values $Z_c \approx 0.22$ for Au and $Z_c \approx 0.29$ for Zn and Pb were obtained. In view of the difference of the electronic structure of these metals it could be explained easily. The electronic configuration of Au is $6s^1$ which is qualitatively identical with alkali metals, while electronic structure of Zn and Pb is give by $4s^2$ and $6p^2$ correspondingly. This makes the non universal critical properties of these metals close to those of monoatomic molecular fluids leading to the value Z_c similar to that of noble gases.

From the results obtained above we can conclude that water, ammonia and alkali liquid metals fall into the same class of the ILM with $q = 2$ which reflects their dimerization near the critical point. Methanol is known for the abundance of the tetramers in saturated vapor [53] and identified with $q = 4$. Thus low values of Z_c for the associative LJ fluids can be explained by the significant contribution of the associates in the critical density.

As a summary we note the main result (3.26) of the Global Isomorphism approach. It allows to get the well known critical compressibility factor $Z_c \approx 0.28$ of classical LJ fluids like noble gases basing on the critical parameters of the Ising model.

3.3.2 *Mapping the Binodals*

Additionally to the calculations of the critical points (3.8) gives the possibility to obtain the binodal of the LJ fluid as the image of the binodal of the lattice model [17].

Indeed, the transformation (3.8) allows to map the binodal $T_{bin}(n)$ of the Lennard-Jones fluid onto the binodal $t_{bin}(x)$ of the corresponding lattice gas model. In parametric form it is as follows:

$$n(x) = n_* \frac{x}{1 + z t_{bin}(x)}, \quad T_{bin}(x) = T_* \frac{z t_{bin}(x)}{1 + z t_{bin}(x)}. \tag{3.42}$$

Due to the particle-hole symmetry of the Hamiltonian the lattice gas binodal has explicitly symmetric shape with respect to the critical isochore $x_c = 1/2$: $t_{bin}(x) = t_{bin}(1 - x)$. For 2D Ising model the binodal is given by the relations:

$$t_{bin}(x)/t_c = \frac{\mathrm{ArcSinh}1}{\mathrm{ArcSin}\left(\frac{1}{\left(1-(2x-1)^8\right)^{1/4}}\right)}. \tag{3.43}$$

where $t_c = 2/\mathrm{ArcSinh}1 \approx 2.269$ is the critical temperature of the 2D Ising model. Applying the transformation (3.42) according to (3.8) with $T_* = 2$, $n_* = 0.94$ and $z = 1/3$ we get the binodal of the 2D LJ fluid (see Fig. 3.3). To check the approach the available numerical data [55, 56] on the liquid-vapor equilibrium in 2D LJ fluid were processed using (3.42) with T_* and n_* as fitting parameters [17] (see Fig. 3.4). The agreement between theoretical values and numerical data is quite good.

As has been noted above the projective form of the transformation (3.8) is due to the linear character of the density diameter of the binodal and the Zeno-line. Both these linearities seem to be rather approximations than the strict laws. From the general point of view (3.8) can be viewed as the restoration of the the binodal symmetry in proper variables. Thus one can treat (3.8) as the procedure of the symmetrization of the phase diagram. Indeed, suppose that $n_d(T)$ is the dependence of the density diameter. Then the variable $n/n_d(T)$ is symmetrical over the binodal. Thus the symmetrization of the fluid binodal is given by:

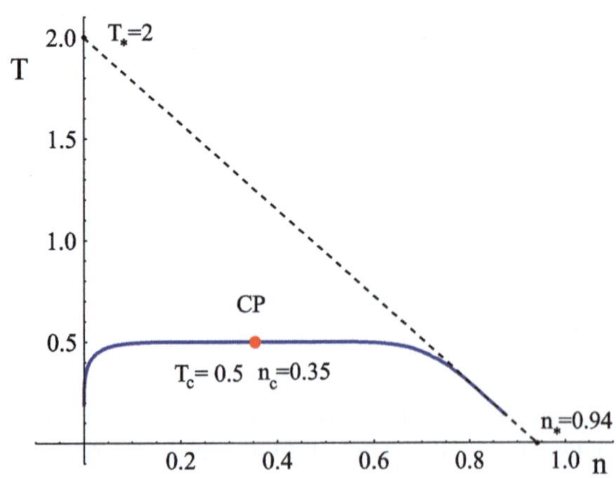

Fig. 3.3 The result of transformation (3.42) the binodal of 2D lattice gas (3.43) with $T_* = 2$, $n_* = 0.94$ and $z = 1/3$

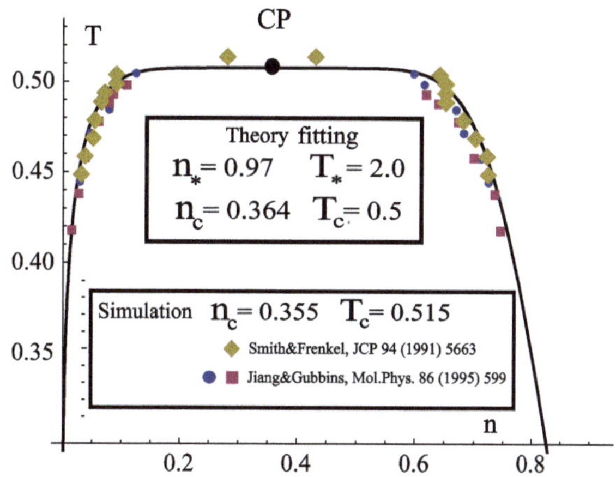

Fig. 3.4 Fitting the binodal of two dimensional LJ fluid (3.42) with $z = 1/3$ to the results of the simulations [55, 56]. Best fitting gives $T_* = 2.0$ and $n_* = 0.971$. Also the locus of the CP is shown

$$x = \frac{n}{2n_d(T)}, \tag{3.44}$$

with

$$\tilde{T} = 1 - f(x), \tag{3.45}$$

being the equation of the binodal. Here the parametrization function f is chosen so that to fulfil the conditions of the thermodynamic equilibrium:

$$P(n_g(T), T) = P(n_l(T), T). \tag{3.46}$$

The equality of the chemical potentials is fulfilled due to symmetry of the binodal of the "lattice" variables x, t. Clearly, (3.44) assumes that the equation for the Zeno-element which is the image of the line $x = 1$ is:

$$n/n_* - 2n_d(T) = 0, \tag{3.47}$$

but in general it is different from the Zeno-line $Z = 1$.

To illustrate this procedure let us consider the classical vdW EoS (3.2) for which the diameter is slightly nonlinear. In view of this and the symmetry with respect to the diameter $x_c = 1/2$ it is convenient to represent the densities of the coexisting phases as following:

$$\tilde{n}_g = xf(x), \quad \tilde{n}_l = (1 - x)f(x). \tag{3.48}$$

Substituting these relations into (3.46) and taking into account that the pressure P is given by (3.2) we get simple algebraic equation for the function f:

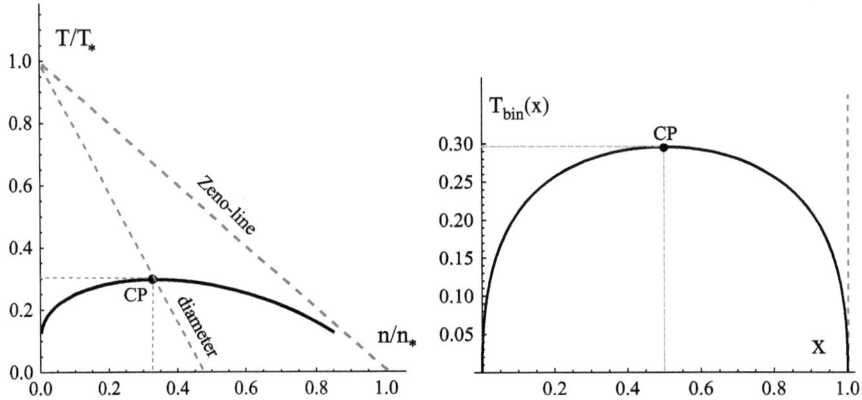

Fig. 3.5 The symmetrization of the vdW binodal using the parametrization (3.48) with (3.49) corresponding to the linear approximation for the diameter

$$1-2f+f^2+f^3(x-1)x=0\,. \tag{3.49}$$

The result is shown in Fig. 3.5. The corresponding coordinates of the critical point are:

$$\tilde{n}=\frac{1}{2}f(1/2)=0.352\,,\ \tilde{T}_c=1-f(1/2)\approx 0.296\,,\ \tilde{P}_c\approx 0.037\,. \tag{3.50}$$

The difference between the exact values:

$$\tilde{n}_c=1/3\,,\ \tilde{T}_c=8/27\,,\ \tilde{P}_c=1/27\,,$$

is caused by the deviation of the diameter for the vdW EoS form the linear behavior. As we see the differences are rather small. Of course, the example of EoS considered is not adequate for real fluids and therefore the symmetrical EoS in coordinates (x,t) may not correspond to any realistic lattice model. Nevertheless, this approach allows to avoid ambiguity in extrapolation of the binodal into the region $T\to 0$ [10] because $0\le x\le 1$.

3.4 Relation Between the Entropies of the LG and LJ Fluid

In this section we derive the relation between the entropies and check it using the available data for argon as the classical example of the LJ simple fluid.

According to the definition, from (3.23) we get:

$$\begin{aligned}\tilde{\mathfrak{s}}(\mu, T) &= \left.\frac{\partial \tilde{P}}{\partial \tilde{T}}\right|_{\mu} = \left.\frac{dt}{dT}\frac{\partial \mathrm{g}}{\partial t}\right|_{h} + \left.\frac{\partial h}{\partial T}\right|_{\mu}\left.\frac{\partial \mathrm{g}}{\partial h}\right|_{t} \\ &= \frac{(1+zt)^2}{z}\,\mathfrak{s}_h + \left[\tilde{\mu}_0 - \tilde{\mu} + \left(1-\tilde{T}\right)\left(-\frac{d\tilde{\mu}_0}{d\tilde{T}}\right)\right]x.\end{aligned} \tag{3.51}$$

Here $\tilde{\mathfrak{s}}$ is the dimensionless entropy of a fluid per unit volume, $\mathfrak{s}_h$ is the entropy of the lattice gas (per site), $\tilde{\mu} = \mu/T_*$ is the dimensionless chemical potential. Let us consider (3.51) along the coexistence curve, where $\mu = \mu_0(T)$:

$$\tilde{\mathfrak{s}}(\mu_0, T) = \frac{(1+zt)^2}{z}\,\mathfrak{s}_{bin} + \left(1-\tilde{T}\right)\left(-\frac{d\tilde{\mu}_0(T)}{d\tilde{T}}\right)x(t). \tag{3.52}$$

Here $\mathfrak{s}_{bin}$ is the entropy of the LG along the binodal which according to the particle-hole symmetry is the same for the coexisting phases. From (3.52) one can obtain the latent heat for the transition liquid-gas of the fluid:

$$L_{1\to 2} = T(S_{gas} - S_{liq}) = T\left(1-\tilde{T}\right)\left(-\frac{d\tilde{\mu}_0(T)}{d\tilde{T}}\right)\mathcal{N}(x_-(t) - x_+(t))\,. \tag{3.53}$$

From this we get:

$$(x_-(t) - x_+(t)) = \frac{(s_{gas} - s_{liq})}{\left(1-\tilde{T}\right)\left(-\frac{d\tilde{\mu}_0(T)}{d\tilde{T}}\right)}, \tag{3.54}$$

where $s = S/\mathcal{N}$. Taking into account (3.27) it is easy to see that (3.54) is nothing but the Clausius-Clapeyron relation:

$$\frac{d\mu_0(T)}{dT} = \frac{s_{liq} - s_{gas}}{\tilde{n}_{gas} - \tilde{n}_{liq}}\,. \tag{3.55}$$

This shows the consistency of the approach based on the relations (3.23)–(3.28) with the basic thermodynamic relations of phase equilibrium. As is clear from the consideration above the relation (3.54) is reduced to the relation (3.27). In its turn (3.27) is based on the LRD. The LRD is valid for wide range of molecular fluid and liquid metals except for the fluctuational region where the most strong $|\tau|^{2\beta}$ and $|\tau|^{1-\alpha}$ anomalies of the binodal diameter emerge [57]. To exclude such singularities from the consideration we use the EoS for the lattice gas (Ising model) of the form:

$$x(t) = \frac{1}{2}\left(1 \pm \tau^{\beta}\left(b_0 + b_1\tau + b_\Delta \tau^{\Delta}\right)\right), \tag{3.56}$$

and check the relation

$$(x_-(t) - x_+(t)) = \frac{n_{gas} - n_{liq}}{1 - \tilde{T}}\,, \tag{3.57}$$

substituting the available equation for the densities $n_{liq,gas}$ from NIST [54]. The result of such fitting is on Fig. 3.6. We use T_c/T_* as the fitting parameter and match the value of critical amplitude b_0 for argon to the value $b_0 \approx 1.7$ for $3D$ Ising model obtained in [58] with $\beta \approx 0.327$. The gives $T_c/T_* = 0.315$ (see Fig. 3.7) which is close to the theoretical estimation $T_c/T_* = 1/3$ for $3D$ LJ fluid. Also this justifies the estimate for the amplitude of the 3D LJ fluid obtained in [37]. The results for the amplitude of the isomorphic lattice gas model b_0 obtained from (3.57) are shown in Fig. 3.8 as the linear correlation with the value of Z_c:

$$b_0 = 5.5\,(1 - 2.3Z_c)\,. \tag{3.58}$$

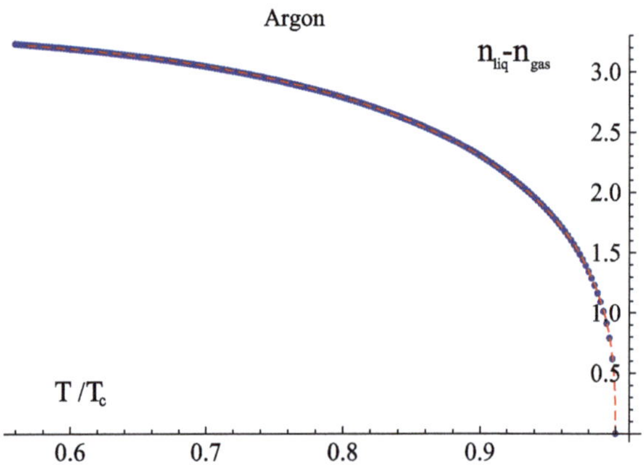

Fig. 3.6 Fitting [54] using (3.57) and (3.56) (*dashed line*) for the density difference along the binodal (*points*) for argon (NIST)

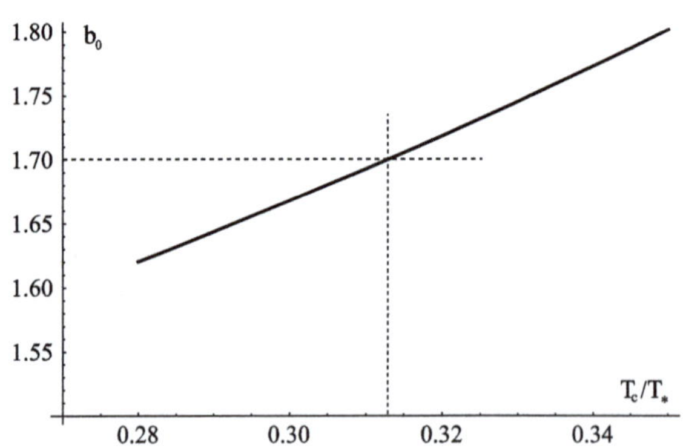

Fig. 3.7 Dependence of the amplitude b_0 on T_c/T_* for argon

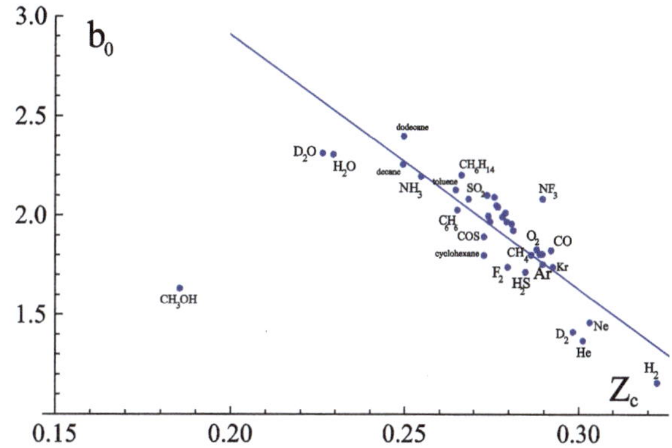

Fig. 3.8 The values of b_0 and Z_c for different molecular liquids

One can note that methanol stays far from the main body of the correlation (3.58). Numerically this happened because of anomalously low values of both of $Z_c \approx 0.19$ and $B_0 \approx 1.6$ obtained using NIST database. E.g. according to the data of direct measurements of Abdulagatov et al. [59] $B_0 \approx 1.86$, $Z_c \approx 0.22$. But it still does not allow to put methanol to the correlation (3.58). We attribute this to the strong self-associative properties of this fluid [53] noted above (see Sect. 3.3.1).

Since the critical amplitudes of the binodal for the fluid and the isomorphic lattice model differ by the scaling factor we can compare this result with that for fluid binodal amplitude B_0 in [60]:

$$B_0 = 4.2\left(1 - 2.0 Z_c\right). \tag{3.59}$$

It was obtained by processing the data in close vicinity of the CP $(10^{-4} < \tau < 10^{-2})$ with the exponent $\beta = 0.338$. As we see, despite the difference between (3.58) and (3.59) in absolute values, caused by the different methodology and the data range, the slopes for the amplitude Z_c-dependence of normalized values are close to each other (2.0 and 2.3 correspondingly). This reflects the fact that the slope of the normalized value of the amplitude should be universal.

3.5 Discussion

The idea that the difference between irregularity configuration for continuum fluids and regularity of configurations of lattice models is unimportant for consideration of the order-disorder transitions in the fluctuational region belongs to K.S. Pitzer (see [61]). However, beyond the fluctuational region the shape of the holes in real or continuum liquid and that in the lattice gas causes the main difference between the configurations of these systems. From this point of view the global character of the transformation (3.8) shows that the particle-hole simplified picture still can be useful. Though, it is not the particle density which reflects such symmetry. Rather the combination of the density of the particles and the density of the holes (i.e. the entropy) is the symmetrical variable [62]. This statement generalizes the idea of N. Mermin and J. Rehr about the symmetrization of the description of the critical EoS (see also complete scaling approach of Fisher [63]). This rehabilitates the hole theory for expanded liquids [36, 64]. Such caricature picture of the liquid state gives the possibility to relate the thermodynamic functions of these systems using (3.23).

The main results of the proposed concept of the global isomorphism obtained up to now include the calculation of the nonuniversal critical parameters like the critical point locus and the compressibility factor Z_c for the LJ fluids and their binodals in dimensions $D = 2, 3$ using the corresponding information for Ising model. The applications to the surface tension and the spinodal are also possible [65]. It is interesting to elaborate the approach to other potentials like Yukawa or square well which also known to lead to the linearities of the diameter and the Zeno-line for corresponding range of parameters (see [27] and references therein). Here the invariant z should be related with the parameters of these potentials function. One of the obvious options is to use the second virial coefficient as the general characteristic because it does not depend on fine details of the potential. All these results signify that classical vdW corresponding states similarity can be included into broader context of the Global Isomorphism between continuum fluid and discrete lattice model.

It should be noted that the fluid-lattice gas isomorphism formalized by the transformation (3.8) refers only to the states inside the liquid-gas triangle. Solid phase is excluded from the consideration because the lattice is considered as incompressible. This is because the oscillatory degrees of freedom were not taken into account. Their contribution becomes important in the vicinity of the triple point where the stability of the liquid phase terminates. It is interesting option to incorporate the compressibility of the lattice into consideration in order to expand the isomorphism ideology over the liquid—solid coexistence where lattice models can be used extensively. For solid state thermodynamic potential the contribution from the phonon part is essential and should be taken into account along with the particle-hole configurational part.

In addition to obvious problem of microscopic derivation of the transformation (3.8) based on some model Hamiltonians there is the route of its phenomenological generalization for those substances where the deviations from the linearity of both the Zeno line and the binodal diameter are essential. Note that for the diameter of the density binodal these deviations should be distinguished from the nonlinear fluctuational corrections to the mean-field behavior in fluctuational region. The approach developed in [66, 67] for the construction of Ising like variables through the canonical form of the Hamiltonian can be considered as the nonlinear variant of (3.8) in the fluctuational region. From such point of view (3.8) can be viewed as the linearized mean-field version of the nonlinear canonical transformation of the density order parameter [67].

Acknowledgments Vladimir Kulinskii thanks Konstantin Yun for the financial support of the research.

References

1. J.D. van der Waals, De constinuitet van den gas—en floeistoftoestand. Ph.D. thesis, Leiden University, Leiden (1873)
2. L.P. Filippov, *Metody Rascheta i Prognozirovaniya Svoistv Veshchestv (Methods for Calculating and Predicting the Properties of Substances)* (Moscow State University, Moscow, 1988)
3. I. Prigogine, *The Molecular Theory of Solutions* (Interscience Publihers, NY, 1957)
4. A.Z. Patashinskii, V.L. Pokrovsky, *Fluctuation Theory of Critical Phenomena* (Pergamon, Oxford, 1979)
5. L. Cailletet, E. Mathias, Seances Acad. Sci. **102**, 1202 (1886). doi:http://dx.doi.org/10.1051/jphystap:018860050054900
6. E.A. Guggenheim, J. Chem. Phys. **13**(7), 253 (1945). doi:10.1063/1.1724033. http://link.aip.org/link/?JCP/13/253/1
7. A. Batschinski, Ann. Phys. **324**, 307 (1906). doi:10.1002/andp.19063240205. http://onlinelibrary.wiley.com/doi/10.1002/andp.19063240205/abstract
8. D. Ben-Amotz, D.R. Herschbach, Isr. J. Chem. **30**, 59 (1990). doi:10.1002/ijch.199000007. http://onlinelibrary.wiley.com/doi/10.1002/ijch.199000007/abstract
9. J.P. Hansen, I.R. Mcdonald, *Theory of Simple Liquids,* 3rd edn. (Academic Press, San Diego, 2006)
10. E.M. Apfelbaum, V.S. Vorob'ev, G.A. Martynov, J. Phys. Chem. B **110**, 8474 (2006). doi:10.1021/jp057327c
11. E.M. Apfelbaum, V.S. Vorob'ev, J. Phys. Chem. B **113**(0), 3521 (2009). doi:10.1021/jp808817p. http://pubs.acs.org/doi/abs/10.1021/jp808817p
12. E. Apfelbaum, V. Vorob'ev, Chem. Phys. Lett. **591**, 212–215 (2013). doi:http://dx.doi.org/10.1016/j.cplett.2013.11.041. http://www.sciencedirect.com/science/article/pii/S0009261413014541
13. E.M. Apfelbaum, V.S. Vorob'ev, J. Phys. Chem. B **117**(0), 7750–7755 (2013). doi:10.1021/jp404146h. http://pubs.acs.org/doi/abs/10.1021/jp404146h
14. V.L. Kulinskii, J. Phys. Chem. B **114**(8), 2852 (2010). doi:10.1021/jp911897k. http://pubs.acs.org/doi/full/10.1021/jp911897k
15. V.L. Kulinskii, J. Chem. Phys. **133**(3), 034121 (2010). doi:10.1063/1.3457943. http://link.aip.org/link/?JCP/133/034121/1
16. L.A. Bulavin, V.L. Kulinskii, J. Chem. Phys. **133**(13), 134101 (2010). doi:10.1063/1.3496468. http://link.aip.org/link/JCPSA6/v133/i13/p134101/s1
17. V.L. Kulinskii, J. Chem. Phys. **133**(13), 131102 (2010). doi:10.1063/1.3499857. http://link.aip.org/link/?JCP/133/131102/1
18. V.L. Kulinskii, J. Chem. Phys. **141**(5), 054503 (2014). doi:http://dx.doi.org/10.1063/1.4891806. http://scitation.aip.org/content/aip/journal/jcp/141/5/10.1063/1.4891806
19. E.M. Apfelbaum, V.S. Vorob'ev, J. Phys. Chem. B 114(30), 9820 (2010). doi:10.1021/jp1022899. http://dx.doi.org/10.1021/jp1022899
20. Q. Wei, D.R. Herschbach, J. Phys. Chem. C **117**, 22438 (2013). doi:10.1021/jp403307g. http://pubs.acs.org/doi/abs/10.1021/jp403307g
21. V. Vorob'ev, Chem. Phys. Lett. **605–606**(0), 47 (2014). doi:http://dx.doi.org/10.1016/j.cplett.2014.05.013. http://www.sciencedirect.com/science/article/pii/S0009261414003790
22. V.L. Kulinskii, J. Chem. Phys. **139**, 184119 (2013). http://scitation.aip.org/content/aip/journal/jcp/139/18/10.1063/1.4829837
23. K. Wilson, J. Kogut, The renormalization group and the epsilon expansion. Phys. Lett. (North-Holland, 1974). http://books.google.com.ua/books?id=hanYHAAACAAJ
24. S. Ma, *Modern theory of critical phenomena* (W.A. Benlamin Inc, London, 1976)
25. M. Fisher, Rev. Mod. Phys. **70**, 653 (1998)
26. L.D. Landau, E.M. Lifshitz, *Statistical Physics (Part 1)*, 3rd edn. (Pergamon Press, Oxford, 1980)

27. E.M. Apfelbaum, V.S. Vorob'ev, J. Chem. Phys. **130**(21), 214111 (2009). doi:10.1063/1.3151982. http://link.aip.org/link/?JCP/130/214111/1
28. H. Okumura, F. Yonezawa, J. Chem. Phys. **113**(20), 9162 (2000). doi:10.1063/1.1320828. http://link.aip.org/link/?JCP/113/9162/1
29. P. Orea, Y. Reyes-Mercado, Y. Duda, Phys. Lett. A **372**(47), 7024 (2008). doi:http://dx.doi.org/10.1016/j.physleta.2008.10.047. http://www.sciencedirect.com/science/article/pii/S0375960108015417
30. M.H.J. Hagen, D. Frenkel, J. Chem. Phys. **101**(5), 4093 (1994). doi:10.1063/1.467526. http://link.aip.org/link/?JCP/101/4093/1
31. C.F. Tejero, A. Daanoun, H.N.W. Lekkerkerker, M. Baus, Phys. Rev. Lett. **73**(5), 752 (1994). doi:10.1103/PhysRevLett.73.752
32. M. Hasegawa, K. Ohno, J. Phys.: Condens. Matter **9**(16), 3361 (1997). http://stacks.iop.org/0953-8984/9/i=16/a=008
33. M. Hasegawa, J. Chem. Phys. **108**(1), 208 (1998). doi:10.1063/1.475392. http://link.aip.org/link/?JCP/108/208/1
34. P.J. Camp, Phys. Rev. E: Stat., Nonlinear, Soft Matter Phys. **67**(1), 011503 (2003). doi:10.1103/PhysRevE.67.011503
35. E.M. Apfelbaum, V.S. Vorob'ev, G.A. Martynov, J. Phys. Chem. A **112**, 6042 (2008). doi:10.1021/jp802999z
36. O. Rice, *Statistical Mechanics, Thermodynamic and Kinetics* (W.H. Freeman and Co, San Francisco, 1967)
37. L.A. Bulavin, V.L. Kulinskii, J. Phys. Chem. B **115**(19), 6061 (2011). doi:10.1021/jp201872f. http://pubs.acs.org/doi/abs/10.1021/jp201872f
38. B. Smit, D. Frenkel, J. Chem. Phys. **94**(8), 5663 (1991). doi:10.1063/1.460477. http://link.aip.org/link/?JCP/94/5663/1
39. J.J. Potoff, A.Z. Panagiotopoulos, J. Chem. Phys. **109**(24), 10914 (1998). doi:10.1063/1.477787. http://link.aip.org/link/?JCP/109/10914/1
40. M. Hloucha, S.I. Sandler, J. Chem. Phys. **111**(17), 8043 (1999). doi:10.1063/1.480138. http://link.aip.org/link/?JCP/111/8043/1
41. V.L. Kulinskii, J. Chem. Phys. **134**(14), 144111 (2011). doi:10.1063/1.3578469. http://link.aip.org/link/?JCP/134/144111/1
42. C. Domb, M.F. Sykes, Proc. R. Soc. Lond. A. Math. Phys. Sci. **235**(1201), 247 (1956). doi:10.1098/rspa.1956.0080. http://rspa.royalsocietypublishing.org/content/235/1201/247.abstract
43. E. Montroll, in *Statistical Physics, Phase Transitions and Superfluidity (Brandeis University Summer Institute in Theoretical Physics (1966))*, vol. 2, ed. by M. Chretien, S. Deser, E. P. Gross (Gordon and Breach, New York, 1968), pp. 195–267
44. A.J. Liu, M.E. Fisher, Physica A **156**(1), 35 (1989). doi:http://dx.doi.org/10.1016/0378-4371(89)90109-X. http://www.sciencedirect.com/science/article/pii/037843718990109X
45. R.J. Baxter, *Exactly Solved Models in Statistical Mechanics* (Dover, New York, 2007)
46. A. Lotfi, J. Vrabec, J. Fischer, Mol. Phys. 76(6), 1319 (1992). doi:10.1080/00268979200102111. http://dx.doi.org/10.1080/00268979200102111
47. F.F. Abraham, Phys. Rep. **80**(5), 340 (1981). doi:http://dx.doi.org/10.1016/0370-1573(81)90099-5. http://www.sciencedirect.com/science/article/pii/0370157381900995
48. R. Abe, Prog. Theor. Phys. **81**, 990 (1989). http://ptp.oxfordjournals.org/content/81/5/990.full.pdf
49. P.I. Bystrov, D.N. Kagan, G.A. Krechetova, E.E. Shpilrain, *Liquid Metal Coolants for Heat Pipes and Power Plants* (Hemisphere, New York, 1990)
50. D. Frenkel, B. Smit, *Understanding Molecular Simulation: From Algorithms to Applications*, 2nd edn. Computational Science Series, vol 1. (Academic Press, San Diego, 2001)
51. N.B. Vargaftik, *Tables on the Thermophysical Properties of Liquids and Gases in Normal and Dissociated States* (Hemisphere, Washington, 1975)
52. W. Schröer, G. Pottlacher, High Temp. -High Pressures **43**, 201 (2014)
53. A.N. Fletcher, J. Phys. Chem. 75(12), 1808 (1971). doi:10.1021/j100681a009. http://dx.doi.org/10.1021/j100681a009

54. P. Linstrom, W. Mallard (eds.), *NIST Chemistry WebBook, NIST Standard Reference Database 69* (National Institute of Standards and Technology, Gaithersburg MD, 2004). http://webbook.nist.gov
55. R.R. Singh, K.S. Pitzer, J.J. de Pablo, J.M. Prausnitz, J. Chem. Phys. **92**(9), 5463 (1990). doi:10.1063/1.458524. http://link.aip.org/link/?JCP/92/5463/1
56. S. Jiang, K.E. Gubbins, Mol. Phys. **86**(4), 599 (1995). doi:10.1080/00268979500102221. http://dx.doi.org/10.1080/00268979500102221
57. N.D. Mermin, J.J. Rehr, Phys. Rev. Lett. **26**(19), 1155 (1971). doi:10.1103/PhysRevLett.26.1155
58. A.L. Talapov, H.W.J. Blote, J. Phys. A: Math. Gen. **29**(17), 5727 (1996). doi:10.1088/0305-4470/29/17/042. http://dx.doi.org/10.1088/0305-4470/29/17/042
59. N.G. Polikhronidi, I.M. Abdulagatov, G.V. Stepanov, R.G. Batyrova, Int. J. Thermophys. **28**, 163 (2007)
60. A. Alekhin, B. Abdikarimov, L. Bulavin, Y. Ostapchuk, E. Rudnikov, Y. Shimanskaya, Ukrainian J. Phys. **55**, 897 (2010)
61. K.S. Pitzer, Pure Appl. Chem. **61**, 979 (1989). doi:10.1351/pac198961060979
62. V. Kulinskii, N. Malomuzh, J. Mol. Liq. **158**, 166 (2011). doi:10.1016/j.molliq.2010.11.013. http://dx.doi.org/10.1016/j.molliq.2010.11.013
63. Y.C. Kim, M.E. Fisher, J. Phys. Chem. B **108**, 6750–6759 (2004). doi:10.1021/jp037071f
64. J.A. Barker, D. Henderson, Rev. Mod. Phys. **48**(4), 587 (1976). doi:10.1103/RevModPhys.48.587
65. V.L. Kulinskii, Liquid-gas spinodal and the interfacial properties from the lattice gas-fluid isomorphism approach, in *6th International Workshop on Nonequilibrium Thermodynamics and 3rd Lars Onsager Symposium IWNET 2012 at Røros, Norway*, http://www.complexfluids.ethz.ch/CONF/g/SLIDES/Kulinskii.pdf. Accessed 19–24 Aug 2012
66. V.L. Kulinskii, J. Mol. Liq. **105**(2–3), 273 (2003). doi:10.1016/S0167-7322(03)00067-9. http://dx.doi.org/10.1016/S0167-7322(03)00067-9
67. V. Kulinskii, N. Malomuzh, Physica A **388**(5), 621 (2009). doi:10.1016/j.physa.2008.11.014. http://dx.doi.org/10.1016/j.physa.2008.11.014

Chapter 4
Collective Excitations in Supercritical Fluids

Taras Bryk, Federico Gorelli, Giancarlo Ruocco, Mario Santoro and Tullio Scopigno

Abstract Recent progress in theoretical and simulation studies of collective excitations in supercritical fluids is reviewed. We discuss a methodology of fit-free estimation of dispersion of longitudinal and transverse excitations in simple fluids. The issue of vanishing positive sound dispersion—a viscoelastic increase of the speed of sound from adiabatic one to its high-frequency (elastic) value—with reduction of density, as it was observed in inelastic X-ray scattering experiments on supercritical Ar, is discussed from the point of view of finding distinctions in collective dynamics of low—and high-density supercritical fluids. On the basis of several theoretical models within the extended hydrodynamic description of density-density correlations in liquids analytical expressions for positive sound dispersion are obtained and applied for analysis of time correlation functions obtained from molecular dynamics simulations. A location of a crossover from the liquid-like to gas-like types of collective dynamics is discussed based on general findings for spectra of collective excitations in supercritical Ar and soft sphere fluids.

T. Bryk (✉)
Institute for Condensed Matter Physics of the National Academy of Sciences of Ukraine, 1, Svientsitskii Street, Lviv 79011, Ukraine
e-mail: bryk@icmp.lviv.ua

F. Gorelli · M. Santoro
Istituto Nazionale di Ottica, CNR-INO, and LENS, 50019 Sesto Fiorentino, Italy

G. Ruocco · T. Scopigno
Dipartimento di Fisica, Universita di Roma La Sapienza, 00185 Rome, Italy

G. Ruocco
Center for Life Nano Science @Sapienza, Istituto Italiano di Tecnologia, 295 Viale Regina Elena, 00161 Rome, Italy

L. Bulavin and N. Lebovka (eds.), *Physics of Liquid Matter: Modern Problems*, Springer Proceedings in Physics 171, DOI 10.1007/978-3-319-20875-6_4

4.1 Introduction

Collective excitations in disordered systems being one of the most sophisticated and fascinating problems of modern condensed matter physics attracted focus of many theoretical, experimental and simulation groups. An interplay of different spatial and temporal scales makes the probes of collective excitations in liquids and glasses by scattering experiments and computer simulations very important for unveiling details of their dispersion at different thermodynamic conditions. In particular, the change in collective dynamics of fluids above the liquid-gas critical point can reveal fundamental dissimilarities in collective behaviour of matter in different states—from a liquid-like type of dynamics for dense fluids to a gas-like one for rarified fluids.

In his Nobel lecture Johannes van der Waals told that "I conceived the idea that there is no essential difference between the gaseous and the liquid state of matter" [1]. Until recently the supercritical fluids have been considered as a unique state with intermediate properties of the liquid and gas phases. There exists a traditional point of view on liquid and gas as phases of the same symmetry, that implies these phases cannot be distinguished above the critical point where they do not coexist. Structural studies of supercritical fluids completely support this point of view. However, there are examples, like the liquid-glass transition when the structure of both liquid and glass phases is of the same symmetry while the dynamic properties and in particular the behaviour of density-density time-dependent correlations reveal fundamental difference connected with the non-ergodicity of the glass state [2]. According to [2] below the glass transition temperature the density-density time correlation functions have non-vanishing with time tail, which is a measure of the non-ergodicity of the system. This effect corresponds to the dynamical arrest of particles in contrast to the liquid state for which the density-density correlations decay with time to zero. This example of the dissimilarity of collective dynamics for liquids and glasses reflects a sensitivity of dynamic quantities to the states of matter and implies a possibility to find some dynamic dissimilarities in the collective dynamics of liquids and gases.

Recent inelastic X-ray scattering (IXS) experiments performed for supercritical Oxygen [3] and Argon [4, 5] and experimentally obtained dispersions of collective excitations in these supercritical fluids revealed a strong reduction of a typical for dense fluids viscoelastic effect known as positive sound dispersion (PSD) with the decrease of density of the fluids. The viscoelasticity of liquids [6] is manifested in their collective dynamics mainly via several effects: (i) existence of a deviation from the linear hydrodynamic dispersion law for acoustic excitations towards higher frequencies-PSD, (ii) emergence of the short-wavelength shear waves (SW) and (iii) a crossover between thermal and structural relaxations as the main contributions to the relaxing behaviour of density-density correlations in long- and short-wavelength regions. It is known that liquids do not sustain macroscopic shear and therefore macroscopic transverse sound waves cannot propagate in liquids [7], however on molecular length scales the microscopic elasticity of fluids can cause emergence of the short-wavelength transverse collective excitations-SW, which however are quite overdamped. The dynamic features of fluids due to their viscoelasticity-PSD and SW-can be the candidates to find dissimilarities in liquid-like and gas-like types of dynamics in supercritical region.

On the theoretical side the description of collective excitations in fluids must account for different space and time scales in dynamics of the studied systems. On macroscopic scales the analytical results must recover hydrodynamic expressions [8–12] for collective modes. The hydrodynamic theory [7, 13, 14] being a collection of fundamental local conservation laws treats the fluid as continuum without any atomistic structure. Therefore any extension of hydrodynamic theory must account for dynamic processes occuring on molecular space and time scales. Namely the generalized hydrodynamic models [7, 14] should be applied for correct understanding the origin of the viscoelastic effects. It is important to stress that the theoretical description of collective dynamics in fluids will be performed for the weakly nonequilibrium states of fluids and all the fluctuations are quite small that allows to neglect non-linear hydrodynamic fluctuations.

The paper is organized as follows: in the next section we will discuss the extended models of collective dynamics of liquids which enable calculations of dispersion of collective excitations and description of the viscoelastic effects such as positive sound dispersion and shear waves. The third section gives information on the computer simulations for supercritical Ar and application of the developed theoretical schemes to analysis of time correlation functions obtained in molecular dynamics simulations. A crossover in collective dynamics of supercritical fluids is discussed in the fourth section, and the last section contains conclusions of this study.

4.2 Theoretical Models of Collective Dynamics in Fluids

4.2.1 Hydrodynamic Approach

Perhaps historically first analytical treatment of collective modes in liquids was performed by Landau and Placzek [15] who published (without any details—just mentioning that the hydrodynamic equations were used) a very short communication on their famous Landau-Placzek ratio for the integral intensities of the scattered light in liquids that reads

$$\frac{I_R}{2I_B} = \gamma - 1.$$

Here $\gamma = C_P/C_V$ is the ratio of specific heats at constant pressure and volume, I_R and I_B are the integral intensities of the Rayleigh and Brillouin peaks of dynamic structure factor $S(k, \omega)$ with k and ω being wave number and frequency, respectively. The Landau-Placzek ratio is applicable only for essentially small wave numbers. The dynamic structure factor as a function of wave number and frequency contains all the information about the spectral distribution of the scattered light by fluid, and for essentially small wave numbers the $S(k,\omega)$, as it was observed in the light-scattering experiments, has a three-peak shape. The Rayleigh peak centered at zero frequency is due to thermal relaxation while two side (Brillouin) peaks of $S(k,\omega)$ at frequencies

$$\pm\omega_B(k) = \pm c_s k, \tag{4.1}$$

are coming from acoustic collective excitations that propagate with the adiabatic speed of sound

$$c_s = [\frac{\gamma}{\rho\kappa_T}]^{1/2},$$

where ρ is mass density and κ_T is isothermal compressibility. The Landau-Placzek ratio is a measure of the visibility of the Brillouin peaks in $S(k,\omega)$ for small wave numbers. Since the dynamic structure factor is connected with the density-density time correlation function $F_{nn}(k,t) = \langle n(k,t)n^*(k,t=0)\rangle$ via time-Fourier transformation

$$S(k,\omega) = \int_0^\infty F_{nn}(k,t)e^{-i\omega t}dt \tag{4.2}$$

the main focus in theoretical studies is on obtaining analytical expressions for the density-density time correlation functions. In the definition of $F_{nn}(k,t)$ the $n(k,t)$ denotes the spatial-Fourier component of fluctuations of number density.

Active theoretical studies of hydrodynamic collective modes and their contributions to the density-density time correlation functions for simple and binary fluids have been performed during the period 1960–1970s [8–12]. The hydrodynamic approach for collective excitations in simple one-component fluids consists in solving a system of three equations—continuity equation, longitudinal part of Navier-Stokes equation and balance equation for energy—in terms of dynamic eigenmode. These three equations represent balance equations for three conserved quantities: total number of particles, longitudinal component of total momentum and total energy. Within the linearized hydrodynamics the transverse component of total momentum is decoupled from the other equations—hence, transverse collective modes in the long-wavelength can be treated separately. For transverse dynamics there exists only one balance equation for conserved quantity, which is the transverse component of total momentum.

According to [9] three balance equations for the longitudinal dynamics in hydrodynamic treatment can be represented in the Laplace-transformed matrix form as follows

$$[z\mathbf{I} + \tilde{\mathbf{M}}(k, z = 0)]\tilde{\mathbf{F}}(k, z) = \mathbf{F}(k, t = 0), \tag{4.3}$$

where the 3×3 matrices of time correlation functions $\mathbf{F}(k, t)$, its Laplace transform $\tilde{\mathbf{F}}(k, z)$ and others are defined on the set of three dynamic variables

$$\mathbf{A}^{(hyd)}(k, t) = \{n(k, t), J^L(k, t), e(k, t)\}, \tag{4.4}$$

where $J^L(k, t)$ and $e(k, t)$ correspond to spatial Fourier components of longitudinal component of density of total momentum and energy density, respectively. The structure of the matrix equation (4.3) exactly corresponds to the Laplace-transformed form of the generalized Langevin equation [7, 14] with the $\tilde{\mathbf{M}}(k, z)$ being the matrix of Laplace-transformed memory functions [16].

Among the eigenmodes of the hydrodynamic model (4.3), (4.4) there are: a single pure real eigenvalue

$$d_1(k) = D_T k^2, \quad Re[z_1](k) = d_1(k),$$

where D_T is thermal diffusivity, and a pair of complex-conjugated eigenvalues

$$z_\pm(k) = \Gamma k^2 \pm i c_s k,$$

where the imaginary part corresponds to the hydrodynamic dispersion law

$$\omega(k) = c_s k$$

and real part describes damping of collective excitations with the damping coefficient

$$\Gamma = \frac{1}{2}[D_L + (\gamma - 1)D_T],$$

which depends on thermal diffusivity D_T and longitudinal kinematic viscosity D_L. The eigenvectors associated with the eigenvalues result in amplitudes of mode contributions to the density-density time correlation functions

$$F_{nn}(k, t) = A_{nn} e^{-D_T k^2 t} + [B_{nn} \cos c_s kt + D_{nn}(k) \sin c_s kt] e^{-\Gamma k^2 t}, \tag{4.5}$$

where $A_{nn}(k \to 0) = 1 - \gamma^{-1}$, $B_{nn}(k \to 0) = \gamma^{-1}$ and $D_{nn}(k) = \frac{3\Gamma - D_L}{\gamma c_s} k$.

An important insight for the propagation of macroscopic sound in fluids can be obtained from application of a perturbation approach [17, 18] for cross-correlations in the hydrodynamic model. In particular, if the cross-correlations between thermal and viscous processes are small, i.e. for small ratio of specific heats γ, it is easily to show that the macroscopic sound propagation in fluids is caused by coupled fluctuations of $n(k, t)$ and $J^L(k, t)$ while the cross-correlation with thermal processes renormalizes its propagation speed from the isothermal c_T to the adiabatic c_s value.

4.2.2 Extended Hydrodynamic Approach

In order to account for the short-time dynamic processes on molecular spatial and temporal scales in addition to hydrodynamic ones (which correspond to very small wave numbers and are quite slow) one has to apply generalized hydrodynamics for analysis of time-dependent correlations. A straightforward way is to introduce in addition to the hydrodynamic set of three variables some new ones (non-hydrodynamic variables of fluctuations of non-conserved quantities) which would describe more short-time processes. Suppose that the non-hydrodynamic variables are chosen to be "orthogonal" to the hydrodynamic ones in the sense of statistically independent correlation of corresponding fluctuations, i.e. their cross-correlators are zero

$$<A^{hyd}(-k)A^{non-hyd}(k)> \; \equiv 0,$$

that allows them to describe dynamic processes in liquids beyond the hydrodynamic regime [7, 14]. Henceforth we will denote by angle brackets an average over the equilibrium canonic ensemble. The most obvious choice for the non-hydrodynamic variables is to sample the first time derivatives of the hydrodynamic ones, because [7]

$$<A(-k)\frac{\partial}{\partial t}A(k)> \; = \; <A(-k)\dot{A}(k)> \; \equiv 0.$$

In [19, 20] it was proposed to generate an extended set of N_v dynamic variables by taking the first and next time derivatives of the hydrodynamic variables, generate the $N_v \times N_v$ generalized hydrodynamic matrix, find its N_v eigenvalues and associated eigenvectors and construct from them the theoretical density-density time correlation function, which can be compared either with the simulation-derived $F_{nn}(k,t)$, or via the time-Fourier transform compared with the experimental dynamic structure factor $S(k,\omega)$. Such an approach is known in the literature as the approach of generalized collective modes (GCM). While the approach of deSchepper and Cohen [19] makes use of fitting parameters connected with unknown wavenumber-dependent transport coefficients, the methodology proposed by Mryglod, Omelyan and Tokarchuk [20] allows to have the unknown for the theory wavenumber dependences of different correlation times $\tau_{ij}(k)$ via their direct estimation in molecular dynamics (MD) simulations. Extensive studies of generalized collective modes [21, 22] showed a convergence of the GCM eigenvalues with systematic extension of the set of dynamic variables up to the third time derivatives of the hydrodynamic ones. The GCM methodology is very close to the generalized hydrodynamic theory by Kivelson and Keyes [23, 24] developed for molecular fluids. The microscopic theory of extended collective modes enables correct description of the hydrodynamic modes as well as of the well-defined short-wavelength collective excitations for large wave numbers [25, 26].

In general within the GCM approach the extended set of N_v dynamic variables is used for estimation of the matrix elements of the generalized hydrodynamic matrix $\mathbf{T}(k)$ obtained from the matrix of time correlation functions $\mathbf{F}(k,t)$ and its Laplace transform $\tilde{\mathbf{F}}(k,z)$ taken in Markovian approximation $(z=0)$

$$\mathbf{T}(k) = \mathbf{F}(k,t=0)\tilde{\mathbf{F}}^{-1}(k,z=0). \tag{4.6}$$

The N_v eigenvalues and associated eigenvectors yield GCM expression for any time correlation function between two dynamic variables of the given set of N_v ones via a separable sum of contributions from N_v collective modes

$$F_{ij}^{GCM}(k,t) = \sum_{\alpha}^{N_v} G_{ij}^{\alpha}(k)e^{-z_\alpha(k)t}. \tag{4.7}$$

Here the sum is over N_v mode contributions, the $z_\alpha(k)$ and weight coefficients $G_{nn}^{\alpha}(k)$ are in general case complex functions of wave number k. The mode contributions can be transformed to purely real weight coefficients as it was proposed in [27].

Very important is the issue of exact sum rules which should satisfy the theoretical time correlation functions $F_{ij}(k,t)$. Any time correlation function between dynamic variables of the classical system has the following short-time expansion [7, 14]

$$F_{ij}(k,t) = F_{ij}(k,t=0)[1 - \frac{1}{2!}\langle\omega^2\rangle_{ij}(k)t^2 + \frac{1}{4!}\langle\omega^4\rangle_{ij}(k)t^4 - \frac{1}{6!}\langle\omega^6\rangle_{ij}(k)t^6 + \ldots]. \tag{4.8}$$

Here the wavenumber-dependent coefficients $\langle\omega^{2n}\rangle_{ij}(k)$, $n=0,1,2,\ldots$ are the normalized $2n$th frequency moments of corresponding spectral function (dynamic structure factor for the case of density-density fluctuations)

$$\langle\omega^{2n}\rangle_{ij}(k) = \frac{1}{2\pi F_{ij}(k,t=0)} \int_{-\infty}^{\infty} \omega^{2n} S_{ij}(k,\omega)d\omega. \tag{4.9}$$

Since the dynamic structure factors are even functions of frequency for classical systems—all odd frequency moments are equal to zero

$$\langle\omega^{2n+1}\rangle_{ij}(k) \equiv 0 \quad n=0,1,2,\ldots$$

For the density-density time correlation functions there exist exact analytical expressions for first eight frequency moments [7]. The zeroth and second frequency moments in the expansion of the density-density time correlation function have very simple form

$$\langle \omega^0 \rangle_{nn}(k) \equiv 1, \quad \langle \omega^2 \rangle_{nn}(k) = \frac{k_B T}{mS(k)} k^2.$$

The higher even frequency moments of dynamic structure factor are expressed via spatial integrals involving effective interactions and their spatial derivatives [7, 14]. The extended sets of dynamic variables enable derivation of theoretical expressions for time correlation functions (4.7) with different level of the sum rules fulfilled. As it was shown in [20] if the model included in the extended set of variables all time derivatives of a hydrodynamic variable $A(k,t)$ up to the sth order, then the corresponding GCM time correlation function $F_{AA}^{GCM}(k,t)$ would exactly fulfill first $2s+1$ sum rules for corresponding frequency moments.

The eigenvalues and corresponding eigenvectors of the generalized hydrodynamic matrix represent the dynamic eigenmodes that can exist in fluids on the spatial scale $\sim 2\pi/k$ for the given wave number k. The eigenvalues of $\mathbf{T}(k)$ are either real numbers $d_\alpha(k)$ or pairs of complex-conjugated numbers $z_\alpha(k) = \sigma_\alpha(k) \pm i\omega_\alpha(k)$. The formers correspond to relaxing mode with the relaxation time $d_\alpha^{-1}(k)$, while the latters—to propagating modes with dispersion $\omega_\alpha(k)$ and damping $\sigma_\alpha(k)$. The GCM methodology was developed in spirit of the traditional for solids eigenvalue problem for dynamic matrix, however it is free from an assumption of local potential energy minima for atoms and additionally allows to study relaxing modes and their effects on the dispersion of collective excitations. Analysis of the dispersion $\omega(k)$ of acoustic modes should clarify the issue of the existence and behaviour of positive sound dispersion and non-hydrodynamic modes in supercritical fluids in very wide range of densities.

In the long-wavelength limit one knows the exact asymptotes of the correlation times and different correlators, that allows analytical studies of non-hydrodynamic collective modes in fluids [28]. Below we will show how analytical results can be obtained within the GCM approach for the positive sound dispersion in supercritical fluids. The GCM approach was successfully applied in analytical theories of non-hydrodynamic optic-like excitations in binary liquids [29, 30], "fast sound" excitations in binary liquids with disparate masses [31] and propagating charge modes in molten salts [32], and all the analytical results were supported by the MD simulations.

4.2.3 Viscoelastic Model

A simple dynamical model for longitudinal dynamics of fluids, which consists of three dynamical variables

$$\mathbf{A}^{(ve)}(k,t) = \left\{ n(k,t), J^L(k,t), \dot{J}^L(k,t) \right\}, \tag{4.10}$$

is well known as the viscoelastic one, because in addition to the hydrodynamic variables of particle density and longitudinal component of mass-current density, a non-hydrodynamic variable connected with the elastic properties of the liquid is taken into account. Since the extended variable $\dot{\mathbf{J}}(k,t)$ is connected to the stress tensor $\sigma_{\alpha\beta}(k,t)$ via [33]:

$$\frac{d}{dt}\mathbf{J}(k,t) = i\mathbf{k}\hat{\sigma}(k,t),$$

there appear in the viscoelastic approach the elastic quantities connected with microscopic forces acting on particles. From the point of view of exact sum rules the density-density time correlation function obtained from this model will have exact first five sum rules satisfied, that is two more than within the hydrodynamic description.

In this set of dynamic variables $\mathbf{A}^{(ve)}(k,t)$ one does not take the energy fluctuations into account, hence the density-density correlation time can be obtained from the hydrodynamic one $\tau_{nn}(k)$ defined as

$$\tau_{nn}(k) = \frac{1}{S(k)}\int_0^\infty F_{nn}(k,t)dt$$

by setting $\gamma = 1$ in the analytical hydrodynamic expression for $F_{nn}(k,t)$ [9]:

$$\tau_{nn}(k) = \frac{D_L}{c_s^2 + 4D_L^2k^2}.$$

Now the generalized hydrodynamic matrix generated on the basis set (4.10) can be written down as follows:

$$\mathbf{T}^{(ve)}(k) = \begin{pmatrix} 0 & -i\frac{k}{m} & 0 \\ 0 & 0 & -1 \\ -imkc_T^2\frac{c_\infty^2-c_T^2}{D_L} & k^2c_\infty^2 & \frac{c_\infty^2-c_T^2}{D_L} \end{pmatrix}, \tag{4.11}$$

where the following shortcut was introduced $k^2c_\infty^2 = \langle \dot{J}^L\dot{J}^L\rangle/\langle J^LJ^L\rangle$ with c_∞ being the high-frequency (elastic) speed of sound. The quantities D_L and c_T in (4.11) are the kinematic viscosity and isothermal speed of sound. One can easily find eigenvalues of the matrix $\mathbf{T}^{(ve)}(k)$ within the precision of $O(k^2)$, which are a pair propagating modes

$$z_\pm(k) = \frac{D_L}{2}k^2 \pm ic_Tk \equiv \sigma(k) \pm i\omega(k), \tag{4.12}$$

and a non-hydrodynamic relaxing collective mode

$$d_{ve}(k) = \frac{c_\infty^2 - c_T^2}{D_L} - D_L k^2 \equiv d^0 - D_L k^2, \tag{4.13}$$

which tends to a nonzero constant d^0 in the long-wavelength limit. The factor $(c_\infty^2 - c_T^2)$ is usually called as the "strength" of the structural relaxation process. Besides, the constant d^0 goes to zero when the kinematic viscosity tends to infinity, that means an almost infinite relaxation time of the mode $d_{ve}(k)$ at the glass transition. All this implies, that the non-hydrodynamic mode $d_{ve}(k)$ is connected to the structural relaxation.

The expression for the sound dispersion (imaginary part of eigenvalues $z_\pm(k)$) does not show any effect due to coupling of acoustic excitations with the non-hydrodynamic mode of structural relaxation $d_{ve}(k)$, that can appear only in the $O(k^3)$ order in the sound dispersion. In order to estimate the imaginary part of the complex eigenvalues corresponding sound dispersion within the precision of $O(k^3)$ one can derive from (4.11) an effective equation for sound eigenmodes by eliminating the known real eigenvalue $d_{ve}(k)$ (4.13). The effective equation reads

$$z^2 - D_L k^2 z + \frac{c_T^2 d^0}{d_{ve}(k)} k^2 = 0, \tag{4.14}$$

and now the sound dispersion can be obtained as

$$\omega_{ve}(k) = c_T k \sqrt{1 + \frac{D_L}{d^0} k^2 - \frac{D_L^2}{4 c_T^2} k^2 + O(k^4)} = c_T k + \beta_{ve} k^3 + \ldots \tag{4.15}$$

In the expression under the square root there are two contributions proportional to k^2: the first one is positive and comes from the coupling of acoustic excitations with structural relaxation, and the second contribution, the negative one, is the standard renormalization down of the dispersion law due to the damping effects. One can obtain the first correction to the linear viscoelastic dispersion law, proportional to the k^3 having the following coefficient:

$$\beta_{ve} \approx c_T \frac{D_L^2}{8} \frac{5 - (c_\infty / c_T)^2}{c_\infty^2 - c_T^2}. \tag{4.16}$$

The most interesting consequence is, that in general the sign of the $O(k^3)$ correction to the linear dispersion law can be different depending on the ratio between the high-frequency speed of sound and the isothermal one. Note, that the high-frequency speed of sound is always higher than the adiabatic one c_s [34, 35], which in its turn is $\sqrt{\gamma}$ times higher than c_T.

For the case of transverse dynamics the viscoelastic model contains just two dynamic variables

$$\mathbf{A}^{(ve,T)}(k,t) = \{J^T(k,t), \dot{J}^T(k,t)\}. \tag{4.17}$$

These two dynamic variables can be used for estimation of the 2 × 2 generalized hydrodynamic matrix for the transverse case. The two eigenmodes for this model easily can be obtained [28, 36] in the whole range of wave numbers. In the long-wavelength region they are purely real eigenvalues corresponding to two (one hydrodynamic and one non-hydrodynamic) relaxing modes, while starting from a nonzero wave number

$$k_s = \frac{\sqrt{\rho G}}{2\eta}, \tag{4.18}$$

where G and η are the high-frequency shear modulus and shear viscosity, respectively, in the fluid emerge short-wavelength shear waves (a pair of complex-conjugated eigenvalues of the generalized hydrodynamic matrix for the transverse dynamics). The long-wavelength region $k < k_s$ in which the transverse collective excitations are not supported by the fluid is called a propagation gap for shear waves. The dispersion of the shear waves is

$$\omega^T(k) = \left[\langle\omega^2\rangle_{JJ}(k) - \sigma^2(k)\right]^{1/2}, \tag{4.19}$$

where the second frequency moment of the transverse current-current spectral function

$$\langle\omega^2\rangle_{JJ}(k) = k^2 G(k)/\rho \equiv \frac{\langle \dot{J}^T(-k)\dot{J}^T(k)\rangle}{\langle J^T(-k)J^T(k)\rangle}$$

should be larger than the square of the damping coefficient $\sigma(k)$ in order the shear waves to exist. The $G(k)$ in the expression above is the wavenumber-dependent shear modulus which tends to its macroscopic value G in the long-wavelength limit. It is seen that for small wave numbers the second frequency moment of the transverse current-current spectral function tends to zero resulting in the negative expression under the square root in (4.19) that corresponds to the region where no transverse propagating modes can exist in the fluid.

4.2.4 Thermoviscoelastic Approach

Now we will take into account the coupling of density fluctuations to thermal processes and find out how this effect contributes to the deviation of the dispersion curve from the linear hydrodynamic dispersion law. We want to obtain the analytical expressions for the eigenmodes in a pure fluid, within a five-variable thermo-viscoelastic model in the long-wavelength limit [37]. The dynamical model $\mathbf{A}^{(5)}(k,t)$

$$\mathbf{A}^{(5)}(k,t) = \{n(k,t), J^L(k,t), e(k,t), \dot{J}^L(k,t), \dot{e}(k,t)\} \tag{4.20}$$

contains additionally to the viscoelastic model the energy density $e(k,t)$ and the corresponding extended variable $\dot{e}(k,t)$. The five eigenmodes within the precision of $O(k^2)$ for the thermo-viscoelastic model were reported in [37] and contained three hydrodynamic modes:

$$\begin{aligned} d_1(k) &= D_T k^2, \\ z_\pm(k) &= \Gamma k^2 \pm i[c_s k + O(k^3)], \end{aligned} \tag{4.21}$$

exactly as they appear in the hydrodynamic approach, and two non-hydrodynamic relaxing modes:

$$d_2(k) = d_2^0 - D_L k^2 + (\gamma - 1)\Delta k^2, \tag{4.22}$$

and

$$d_3(k) = d_3^0 - \gamma D_T k^2 - (\gamma - 1)\Delta k^2, \tag{4.23}$$

where the following shortcuts were introduced:

$$d_2^0 = \frac{c_\infty^2 - c_s^2}{D_L},$$

and

$$d_3^0 = \frac{C_V}{m\lambda}[G^h - \frac{(\gamma - 1)}{\kappa_T}],$$

and

$$\Delta = \frac{d_2^0 d_3^0}{d_3^0 - d_2^0} \frac{D_T}{D_L c_s^2} (D_T - D_L)^2.$$

The G^h corresponds to the heat rigidity modulus and λ is the thermal conductivity. The last terms in right hand sides of (4.22) and (4.23) appear only due to coupling between the heat and density fluctuations. When this coupling is neglected, i.e. $\gamma = 1$, one obtains for the non-hydrodynamic modes $d_2(k)$ and $d_3(k)$ the same expressions as within the separated treatment of two-variable heat—and three-variable viscoelastic dynamical models (at $\gamma = 1$ the non-hydrodynamic mode $d_2(k)$ is reduced to (4.13)).

The complex eigenvalues of the thermo-viscoelastic dynamic model within the precision of $O(k^3)$ are

$$z_s(k) = \Gamma k^2 \pm i(c_s k + \beta k^3), \tag{4.24}$$

where the coefficient at k^3 reads as follows:

$$\beta = -\frac{\Gamma^2}{2c_s} - (\gamma - 1)D_T \frac{D_L - D_T}{2c_s} + \frac{c_s D_L}{2d_2^0} + (\gamma - 1)\frac{c_s(\gamma - 1)D_T}{2d_3^0}. \tag{4.25}$$

In the case when $\gamma \approx 1$ the estimate for the sound dispersion at the boundary of hydrodynamic regime within the thermo-viscoelastic model reads:

$$\omega(k) \approx c_s k + \frac{c_s D_L^2}{8} \frac{5 - (c_\infty / c_s)^2}{c_\infty^2 - c_s^2} k^3, \tag{4.26}$$

which is identical to (4.15) and (4.16) for the case of $\gamma = 1$, with c_s being equal to c_T.

4.3 Computer Simulations of Time-Dependent Correlations

Computer molecular dynamics simulations represent a powerfull tool which being based on laws of microscopic statistical mechanics is in fact a computer experiment for realistic systems, that allows direct comparison with the output of real experiments. The time correlation functions directly obtained in MD simulations can be compared via the time-Fourier transform with the experimentally measured dynamic structure factor $S(k, \omega)$. The dispersion of collective excitations can be obained either directly from observation of the locations of the visible Brillouin peaks (4.1) in dynamic structure factor, or from a fitting procedure based either on a simple model of damped harmonic oscillator (DHO) [38, 39] or expressions for dynamic structure factor obtained within second order memory function approach [7, 14]. The direct observation of the Brillouin peaks is not really precise methodology because they are well-pronounced only for essentially small wave numbers. Therefore the main approach of the direct estimation from MD of the dispersion of collective excitations is from the peak position in the longitudinal current spectral function $C^L(k, \omega)$, which according to the continuity equation has the following relation with the dynamic structure factor [7, 14]

$$C^L(k, \omega) = \frac{\omega^2}{k^2} S(k, \omega). \tag{4.27}$$

The methodology based on identification of the dispersion of collective excitations with the peak locations of $C^L(k, \omega)$ also has its drawbacks connected with the neglect of contributions from relaxation processes to $C^L(k, \omega)$. The strong contributions from relaxation processes can essentially shift the contribution from collective excitations and therefore the apparent peak position of $C^L(k,\omega)$ can essentially differ from the real frequency of the collective excitation.

Therefore more advanced methods of estimation of dispersion of collective excitations in fluids are based on theoretical expressions for the time correlation functions or $S(k,\omega)$ and $C^L(k,\omega)$ obtained within generalized hydrodynamic approach. The theoretical expressions are either fitted (memory function approach) or directly compared without any fit (GCM approach) with the simulation data. Several fitting schemes were proposed for the density-density time correlation functions or dynamic structure factors within different approximations for memory functions [40–47] with the purpose to estimate dispersion of collective excitations in liquids while parameter-free calculation of the dispersion was developed only within the GCM approach.

In order to perform analysis of collective dynamics from MD simulations one has to calculate hydrodynamic time correlation functions. For this purpose one has to sample in MD simulations the spatial Fourier components of density of hydrodynamic quantities: number density of particles

$$n(k,t) = \frac{1}{\sqrt{N}} \sum_{j=1}^{N} e^{i\mathbf{k}\mathbf{r}_j(t)}, \tag{4.28}$$

where $\mathbf{k}$ is the sampled wave vector and $\mathbf{r}_j(t)$ is the trajectory of the jth particle; longitudinal and transverse components of momentum density

$$\begin{aligned} J^L(k,t) &= \frac{m}{k\sqrt{N}} \sum_{i=1}^{N} \mathbf{k}\mathbf{v}_j(t) e^{i\mathbf{k}\mathbf{r}_j(t)}, \\ \mathbf{J}^T(k,t) &= \frac{m}{k\sqrt{2}\sqrt{N}} \sum_{j=1}^{N} [\mathbf{k} \times \mathbf{v}_j(t)] e^{i\mathbf{k}\mathbf{r}_j(t)}, \end{aligned} \tag{4.29}$$

and energy density

$$e(k,t) = \frac{1}{\sqrt{N}} \sum_{j=1}^{N} \varepsilon_j(t) e^{i\mathbf{k}\mathbf{r}_j(t)}. \tag{4.30}$$

In (4.29) and (4.30) $\mathbf{v}_j$ and ε_j are the velocity and single-particle energy of the jth particle. Note that the dynamic variables of number density and longitudinal component of momentum density are connected by the fundamental continuity equation

$$\frac{dn(k,t)}{dt} = \frac{ik}{m} J^L(k,t). \tag{4.31}$$

The extended dynamic variables are obtained from (4.29) and (4.30) as their first time derivatives:

$$\dot{j}^L(k,t) = \frac{m}{k\sqrt{N}} \sum_{j=1}^{N} [\mathbf{k}\mathbf{a}_j(t) + i[\mathbf{k}\mathbf{v}_j(t)]^2] e^{i\mathbf{k}\mathbf{r}_i(t)},$$

$$\dot{\mathbf{j}}^T(k,t) = \frac{m}{k\sqrt{2}\sqrt{N}} \sum_{j=1}^{N} [\mathbf{k} \times \mathbf{a}_j(t) + i[\mathbf{k} \times \mathbf{v}_j(t)]\mathbf{k}\mathbf{v}_j(t)] e^{i\mathbf{k}\mathbf{r}_j(t)},$$

$$\dot{e}(k,t) = \frac{1}{\sqrt{N}} \sum_{j=1}^{N} [\dot{\varepsilon}_j(t) + i\varepsilon_j(t)\mathbf{k}\mathbf{v}_j(t)] e^{i\mathbf{k}\mathbf{r}_j(t)},$$

where the overdot means the time derivative and $\mathbf{a}_j(t)$ is the acceleration of the jth particle.

The density-density time correlation functions $F_{nn}(k,t)$ in different regions of wave numbers are shown for supercritical Ar at $T = 280\,\mathrm{K}$ and density $921.885\,\mathrm{kg/m^3}$ in Fig. 4.1. The MD simulations [48] were performed for 13 densities along the isothermal line $T = 280\,\mathrm{K}$ for supercritical Ar using systems of 2000 particles interacting via ab initio Woon potentials [49]. Parameters of the potentials were taken from [50] and cut-off radius was 12 Å. These potentials were the same as used in the experimental and MD study of supercritical Ar at 573 K [5]. The time step in simulations was 2 fs. All the simulations were performed in microcanonical ensemble. Energy conservation was on very good level: the energy drift was less than 0.02 percent over the production runs of 480,000 time steps. Every sixth configuration was used for sampling of dynamic variables. Dynamic variables of particle density, momentum density and energy density as well as their time derivatives needed for GCM analysis were sampled for thirty different wave numbers directly in MD simulations. The averages of static and time correlation functions over all possible directions of different wave vectors with the same magnitude were performed.

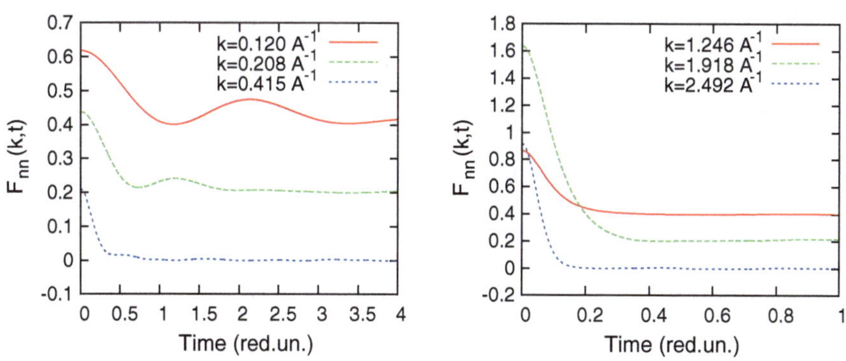

Fig. 4.1 Density-density time correlation functions for supercritical Ar at $T = 280\,\mathrm{K}$ and density $921.885\,\mathrm{kg/m^3}$ at six wave numbers as directly obtained in MD simulations. The time scale for reduction of units is 3.45494276 ps. A progressive vertical shift of 0.2 was applied with decreasing wave numbers for eye convenience

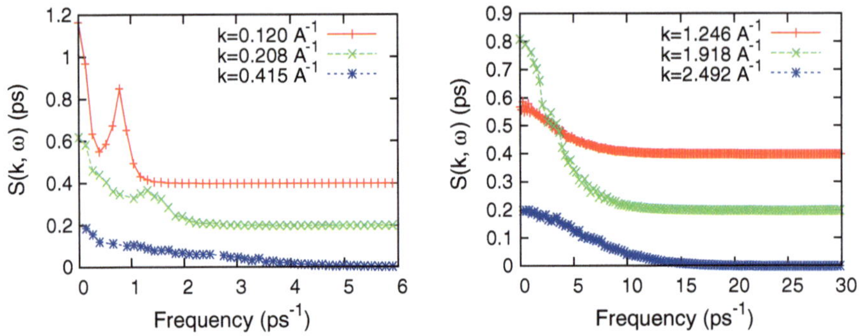

Fig. 4.2 Dynamic structure factors for supercritical Ar at 280 K at six wave numbers obtained from time-Fourier transformation of corresponding density-density time correlation functions shown in Fig. 4.1. A progressive vertical shift of 0.2 was applied with decreasing wave numbers for eye convenience

In Fig. 4.1 one can see that only for very small wave numbers the density-density time correlation functions show oscillating behaviour due to propagating collective excitations, while with increasing wave numbers the oscillations got hidden under the relaxation shape of the $F_{nn}(k,t)$. It is obvious that the purely relaxation shape of $F_{nn}(k,t)$ as is observed for $k > 0.5$ Å^{-1} for this thermodynamic point of supercritical Ar does not mean the absence of collective excitations. The corresponding dynamic structure factors are shown in Fig. 4.2. According to the Landau-Placzek ratio the visibility of the side Brillouin peaks depends on the ratio of specific heats γ, which was obtained in [48] from MD simulations for this thermodynamic point $\gamma = 1.98$ in good agreement with the NIST database [51]. It is seen from Fig. 4.2 that is is impossible to observe the exact location of the side peak of $S(k,\omega)$ for wave numbers $k > 0.3$ Å^{-1} that excludes the possibility to estimate dispersion of collective excitations for these wave numbers directly from the obtained dynamic structure factors.

Since the fluctuations of density are connected with the fluctuations of the longitudinal mass-current density via the time derivative of the former, the time correlation functions $F^L_{JJ}(k,t)$ show well pronounced oscillations (Fig. 4.3) in the region of small wave numbers. However even far outside the hydrodynamic region the longitudinal current-current time correlation functions show negative minimum at small times that is an evidence of the presence of collective excitations in contrast to what was observed in the shape of the density-density time correlation functions at large wave numbers.

The time-Fourier transform of $F^L_{JJ}(k,t)$, which is denoted as $C^L(k,\omega)$, shows the well-pronounced maximum as a function of frequency for all wavenumbers as it is seen from Fig. 4.4. The maxima positions of $C^L(k,\omega)$ give the direct, though being exact only in the long-wavelength limit and therefore—not really precise for large wave numbers, way of estimation of the dispersion $\omega(k)$ of longitudinal collective excitations in fluids.

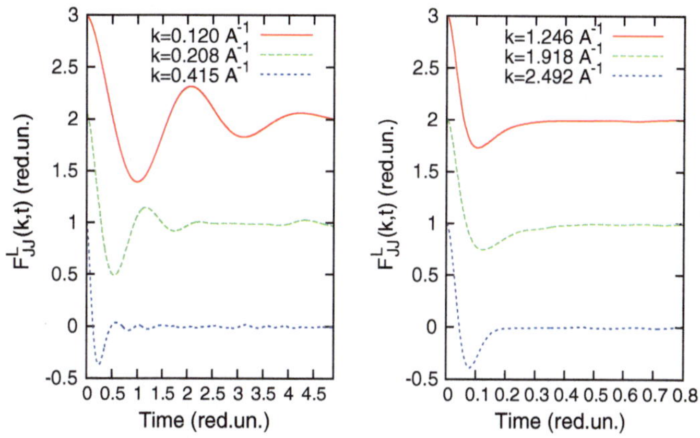

Fig. 4.3 Longitudinal current-current time correlation functions for supercritical Ar at $T = 280\,\mathrm{K}$ and density $921.885\,\mathrm{kg/m^3}$ at six wave numbers as directly obtained in MD simulations. The time scale for reduction of units is 3.45494276 ps. A progressive vertical shift of 1.0 was applied with decreasing wave numbers for eye convenience

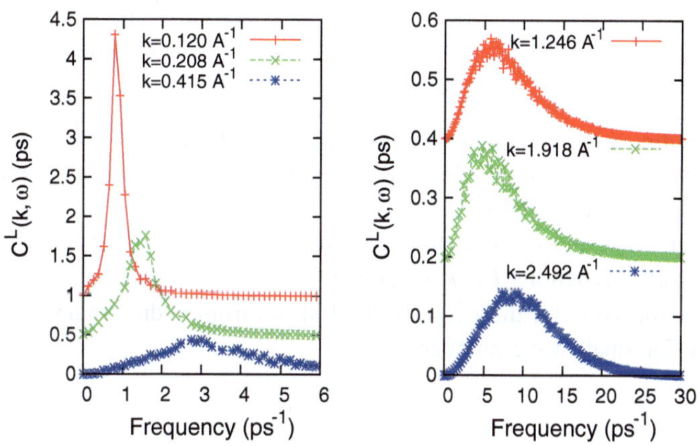

Fig. 4.4 Longitudinal current spectral functions $C^L(k,\omega)$ for supercritical Ar at $T = 280$ K and density 921.885 kg/m^3 at six wave numbers as directly obtained in MD simulations. A progressive vertical shifts of 0.5 and 0.2 were applied in the left and right frames with decreasing wave numbers for eye convenience

The ability of the generalized hydrodynamics to describe correctly the time-dependent correlations are shown in Fig. 4.5, where density-density, density-energy and energy-energy time correlation functions obtained from MD simulations are compared without any fit with the corresponding functions obtained within the five-variable thermo-viscoelastic GCM approach (4.20). The theoretical curves (4.7) almost perfectly recover the damped oscillations in the shape of the MD-derived time correlation functions and nicely reproduce their short-time

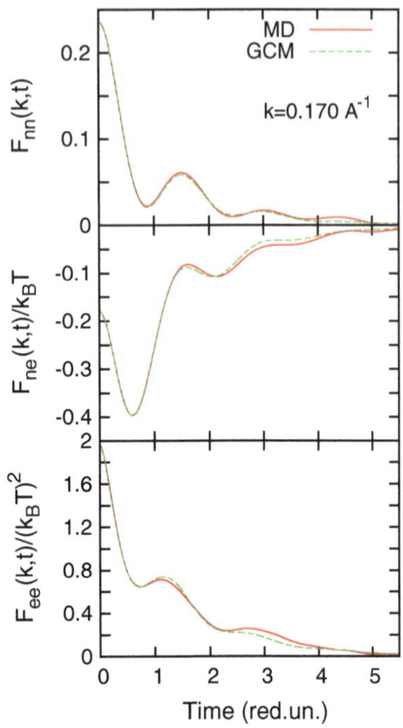

Fig. 4.5 Density-density, density-energy and energy-energy time correlation functions obtained from MD simulations (*solid line*) and from fit-free thermo-viscoelastic (20) generalized hydrodynamic GCM approach (*dashed line*) for supercritical Ar at $T = 280\,\mathrm{K}$ and density $1621.2\,\mathrm{kg/m^3}$

behaviour due to high number of the sum rules fulfilled. This means that the five-variable thermo-viscoelastic model is able to yield correct dispersion of the collective excitations in fluids.

Transverse dynamics can be studied via analysis of the transverse current-current time correlation functions $F^T_{JJ}(k,t)$, shown for six wave numbers in Fig. 4.6. An analytical expression for the $F^T_{JJ}(k,t)$ that follows from hydrodynamic theory [7] is valid only for small wave numbers:

$$F^{T,hyd}_{JJ}(k,t) = \frac{k_B T}{m} e^{-\frac{\eta k^2}{\rho}t}, \tag{4.32}$$

where η is shear viscosity. In Fig. 4.6 the time correlation function $F^T_{JJ}(k,t)$ with the smallest wave number corresponds well to the single-exponential hydrodynamic form (4.32). However outside the hydrodynamic region a deviation from the hydrodynamic form is increasing. For very dense fluids outside the hydrodynamic region a similar as in longitudinal case (Fig. 4.3) negative minimum at small times is observed, which is an evidence of emerging in the liquid short-wavelength shear waves [29, 36].

The time-Fourier transformed transverse functions $F^T_{JJ}(k,t)$ give transverse current spectral functions $C^T(k,\omega)$ shown in Fig. 4.7. In contrast to the longitudinal case $C^T(k,0) \neq 0$, hence for the transverse spectral functions $C^T(k,\omega)$ the contributions from transverse collective excitations can be hidden under the relaxing

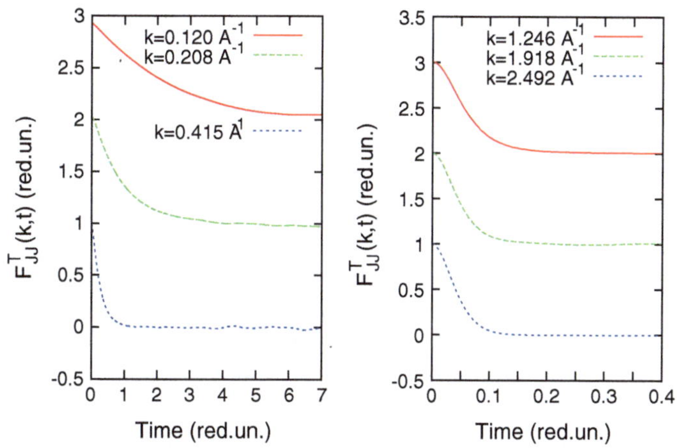

Fig. 4.6 Transverse current-current time correlation functions for supercritical Ar at $T = 280\,\mathrm{K}$ and density $921.885\,\mathrm{kg/m^3}$ at six wave numbers as directly obtained in MD simulations. The time scale for reduction of units is 3.45494276 ps. A progressive vertical shift of 1.0 was applied with decreasing wave numbers for eye convenience

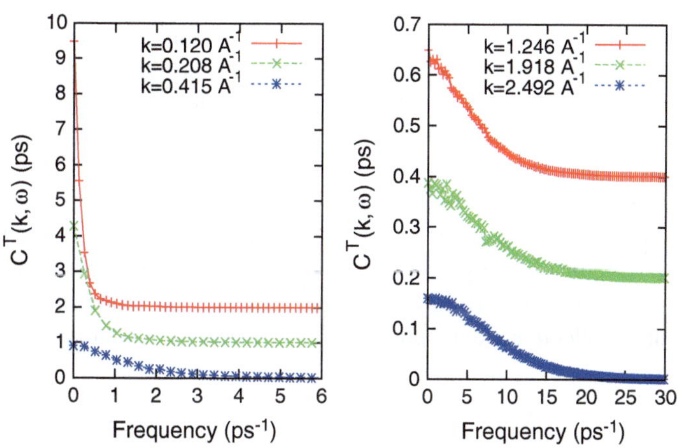

Fig. 4.7 Transverse current spectral functions $C^T(k,\omega)$ for supercritical Ar at $T = 280\,\mathrm{K}$ and density $921.885\,\mathrm{kg/m^3}$ at six wave numbers as directly obtained in MD simulations. A progressive vertical shifts of 1.0 and 0.2 were applied in the left and right frames with decreasing wave numbers for eye convenience

part of $C^T(k,\omega)$. Indeed, the zero-frequency limit of the transverse spectral functions reads $C^T(k,0) \sim \rho/(k^2\eta(k))$, where $\eta(k)$ is the wavenumber-dependent shear viscosity, which tends in the long-wavelength limit to its macroscopic value η. The issue of the visibility of transverse collective excitations in $C^T(k,\omega)$ is not really well elaborated in the literature, but it is obvious that the absence of the well-defined peak in $C^T(k,\omega)$ does not mean the complete absence of shear waves propagating on nanoscales $L \sim 2\pi/k$. Similarly, the absence of a side peak in dynamic structure factors $S(k,\omega)$ for large wave numbers does not mean the absence of short-wavelength longitudinal collective excitations. In fact their contribution simply is too weak in comparison with the one from relaxation processes. Therefore in the case of transverse dynamics only the proper analysis based on dynamic eigenmode calculations can reveal the existence of transverse excitations, their dispersion and contribution to $C^T(k,\omega)$ in a wide range of densities.

The two-variable viscoelastic model of transverse dynamics (4.17) is able to fulfill only first three frequency moments of the transverse current spectral function, therefore the theoretical curves recover the MD-derived transverse time correlation functions not so perfectly as it is for the longitudinal case. One should note that the short-time behaviour of the transverse time correlation functions is exactly the same as for the regular time correlation function (4.8) and depends on the first few k-dependent frequency moments. The long-time behaviour though is different in the small-k and large-k limits. While in the hydrodynamic regime the transverse current functions have typical single-exponential decay, in the limit of large wave numbers these function should have Gaussian-type tail of the time dependence [7]. In Fig. 4.8 theoretical and MD-derived transverse time correlation functions are compared for three wave numbers. One can see that close to the hydrodynamic region the quality of theory is very good, while in the region where exist shear waves the theoretical curves correctly recover the frequency of damped oscillation, but underestimate their damping. In general one can expect that further extension of the two-variable set of transverse dynamics would allow better description of the damping of shear waves. Note that for the highest wave number shown in Fig. 4.8 the Gaussian regime for the long-time behaviour has not been reached.

Having the eigenvalues obtained from the generalized hydrodynamic matrix in the longitudinal and transverse cases one can compare the dispersion of the eigenmodes with the dispersion curves estimated from the peak positions of current spectral functions $C^{L/T}(k,\omega)$. In Fig. 4.9 one observes deviation of the dispersion of longitudinal collective excitations from the linear hydrodynamic dispersion law

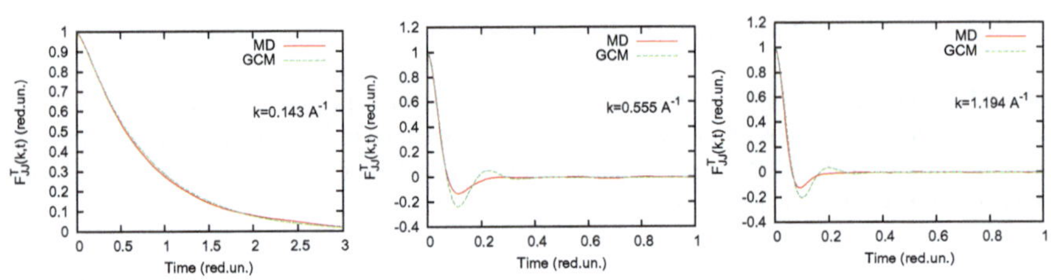

Fig. 4.8 Transverse current-current time correlation functions for three wave numbers as obtained from MD simulations (*solid line*) and from the fit-free viscoelastic (4.17) generalized hydrodynamic GCM approach (*dashed line*) for supercritical Ar at $T = 280\,\mathrm{K}$ and density $1621.2\,\mathrm{kg/m^3}$

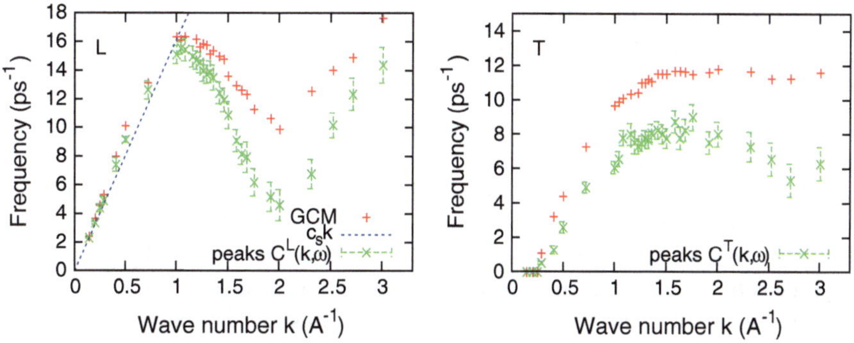

Fig. 4.9 Dispersion of longitudinal (L) and transverse (T) collective excitations for supercritical Ar at $T = 280$ K and density 1621.2 kg/m^{3}obtained from the peak positions of the L-and T-current spectral functions (cross symbols with error bars) and from the theoretical thermo-viscoelastic (4.20) and viscoelastic (for T) models of collective dynamics (plus symbols)

in perfect agreement of theory and MD simulations, that is an evidence of correct description of the positive sound dispersion within the thermo-viscoelastic dynamic model. The macroscopic adiabatic speed of sound c_s is not an easy task to estimate from molecular dynamics simulations. We used one of the most reliable approaches to calculate c_s via the long-wavelength extrapolation of a smooth dependence $\sqrt{\gamma(k)/S(k)}$ multiplied by the thermal velocity. Here $\gamma(k)$ is the wavenumber-dependent ratio of specific heats which easily can be expressed via correlators used in the GCM approach [20, 52]. To date this is the most precise methodology of calculations of adiabatic speed of sound from classical [35, 48] and ab initio [53–55] simulations. A comparison of the calculated adiabatic speed of sound for supercritical Ar with the NIST database showed almost perfect agreement in the whole density range, see [48].

For the transverse case both the GCM theory and MD simulations are in nice agreement for the existence and the width of the propagation gap for shear waves. In the region of wave numbers where the density-density time correlation functions show relaxing behaviour with no oscillations the peak positions of $C^L(k,\omega)$ are located at smaller frequencies with respect to the dispersion obtained from GCM eigenmodes, that is explained by the neglect of the shift of the frequency to the apparent one by contributions from relaxing modes. Similar tendency is observed also for the transverse case. This means that the dispersion of collective excitations should be estimated with the full account for different contributions to $C^{L/T}(k,\omega)$ spectral functions from all relaxing and propagating modes, that would allow separation of actual contribution from L/T collective excitations to the $C^{L/T}(k,\omega)$. Namely the GCM approach allows such a separation of different contributions to the spectral functions of interest. In any case the region, where the positive sound dispersion and emergence of shear waves are observed can be studied by different methodologies, because they lead to consistent results for dispersion of collective excitations.

4.4 Crossover in Collective Dynamics of Supercritical Fluids

Attempts to identify liquid-like and gas-like features in supercritical fluids have been undertaken since the derivation of the van der Waals equation of state. More recent approaches targeted on static density fluctuations [56] or suggested a continuous phase transformation in the supercritical region [57, 58]. In particular, in [58] a mesophase in a finite density range separating the low-density gaseous and high-density liquid phases in supercritical region was studied.

Here we will focus on dynamic manifestations of the possible crossover. A crossover in collective dynamics from gas-like to liquid-like behaviour was suggested in [5] on the basis of experimental study of positive dispersion of collective excitations in supercritical Ar. The idea was based on observation that the positive sound dispersion practically vanished at some low density of the supercritical fluid, that led to a suggestion that the dense fluids have liquid-like type of dynamics with nonzero PSD. This was supported by analytical treatment of the positive sound dispersion caused by coupling of acoustic excitations to structural relaxation [34]. Later on another specific non-zero positive dispersion was found in MD simulations of supercritical Argon for low-density states [59]. In was found in [59] that a weak positive deviation from the hydrodynamic dispersion remains even for the low-density supercritical fluids, however it is not anymore connected with the structural relaxation but with coupling with non-hydrodynamic heat waves which can emerge in fluids on short wave lengths. The coupling to heat waves causes the "residual PSD" far outside the hydrodynamic region. If the PSD was solely caused by the structural relaxation it would vanish as it was shown in [34, 35] within the viscoelastic (no coupling with thermal processes) analysis of PSD. In Fig. 4.10 one can observe how the PSD changes with reduction of the density for the case of supercritical Ar. The apparent speed of sound estimated as $\omega_{sound}(k)/k$ shows rapid increase in the long-wavelength region for the high-density fluid, while decreasing density causes much wider region of small wave numbers in which the dispersion remains linear with k. Simulations performed in [34] showed that further

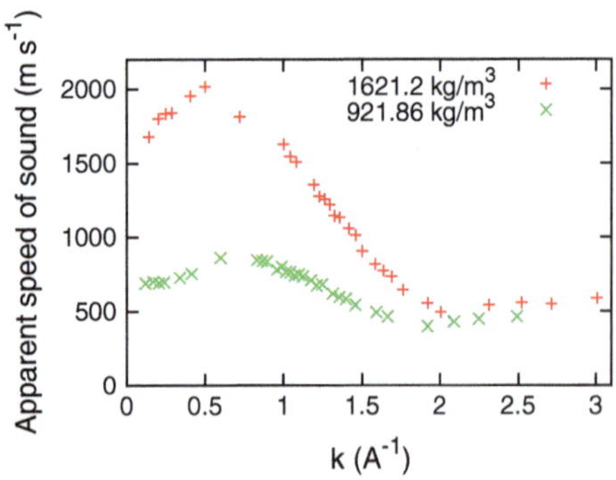

Fig. 4.10 Apparent speed of sound $c_{app}(k) = \omega_{sound}(k)/k$ for two densities of supercritical Ar at $T = 280\,\text{K}$

reduction of the density can cause even a region with "negative" sound dispersion, possibility of which is also supported by expressions (4.15) and (4.26).

Interestingly, that for model fluids of soft spheres, which do not have liquid-gas coexistence, the density dependence of collective dynamics was found very similar as in the supercritical Ar [35]. The apparent speed of sound calculated for soft sphere fluids [35] gives evidence of a strong positive dispersion for the dense fluids and its reduction with the decrease of density reaching almost zero-PSD state at some density. Further decrease of density for soft sphere fluids resulted again in non-zero PSD, that gives evidence of very similar scenario of behaviour of PSD as it was observed in the case of Lennard-Jones fluids [5, 59]. This means that for the systems with and without the gas-liquid binodal and critical point the behaviour of the PSD is very similar.

In order to make a link between the dynamics of systems with and without the coexistence gas-liquid binodal in [35] were calculated density dependences of the thermal diffusivity D_T and of the longitudinal kinematic viscosity D_L. Their density dependences were obtained qualitatively similar with a minimum right at the density where the vanishing PSD was observed. For the soft sphere fluids was obtained similar density dependence of D_T as it was known for real supercritical fluids [51] and were reported in studies of critical behaviour of thermal diffusivity D_T of CO_2, C_2H_6 [60] and H_2O [61]. Both quantities D_T and D_L define the behaviour of the PSD as it follows from the analytical expression reported in 4.25. On the other hand, in more realistic Lennard-Jones supercritical fluids (see [59]) minima in D_T and D_L have been found to correspond to maxima in the specific heat C_P. Note that in the soft sphere fluids the Widom line is no longer defined (no gas-liquid coexistence line and its continuation into supercritical regime) however a link between high frequency dynamics and macroscopic transport/thermodynamic quantities is retained. Put in different words, when soft sphere potential is turned into the Lennard-Jones one by adding the attractive part, the correspondence between the dynamic crossover in PSD and the Widom line is reproduced. In this sense, the soft sphere fluids support such a relationship. For realistic fluids the critical behaviour of the thermal diffusivity

$$D_T = \frac{\lambda}{nC_P}$$

depends on the divergence of thermal conductivity λ and specific heat at constant pressure C_P at the critical point [62]. The experiments [60, 61] give evidence of a rapid decay of D_T on approaching the critical point, i.e. leading contribution from the C_P, and consequently, the line of the minimum of $D_T(n)$ for realistic fluids should be very close to their Widom line. This makes a strong argument in connecting the dynamic crossover in system with and without coexistence gas-liquid binodal. It was suggested [35] that the dynamic crossover in soft sphere fluids takes place at the line of minima of $D_T(n)$, which almost coincides with the line of minima of $D_L(n)$. This is in agreement with the previous observations for supercritical Ar [59] as well as the very first suggestions on the role of the Widom line in the observed dynamic crossover [5]. Both quantities, thermal diffusivity and longitudinal kinematic viscosity, define the relaxation behaviour of fluids, because they define the hydrodynamic correlation times and damping of the long-wavelength collective excitations [7, 14] as well as they define the width of hydrodynamic regime [34]. It seems that namely these two quantities are responsible for the dynamic crossover for all fluids: supercritical ones and soft-sphere systems. It is necessary to stress, that the experimental studies [60, 61] were performed for fluids near the critical point. Therefore for temperatures far away from the critical region new simulation studies for Lennard-Jones fluids are required in order to check the behaviour of positive sound dispersion in a wide range of temperatures: from critical region up to very high temperatures. So far the only study [59] was performed in this direction on supercritical Ar. Its results were in agreement with the findings for soft sphere fluids [35] on the connection of thermal diffusivity D_T and kinematic viscosity D_L with the non-monotonic behaviour of positive sound dispersion. Furthermore, the NIST database [51] allows to follow the non-monotonic behaviour of the density dependence of D_T and D_L in very wide temperature and pressure ranges, that will definitely help in establishing their connection to the positive sound dispersion far away from the critical region in realistic liquids.

4.5 Conclusions

The positive sound dispersion—a viscoelastic increase of the speed of sound from the adiabatic one to its high-frequency (elastic) value—changes drastically with density and practically disappears for gas-like fluids. This was observed in IXS experiments on supercritical Ar [3, 5] and in MD simulations on many Lennard-Jones fluids. Such a change of the pressure (density) dependence of PSD was observed to take place in the supercritical fluids in the region of the Widom line being a natural extension of the gas-liquid coexistence curve. The only unclear issue of a possible separation of the liquid-like and gas-like fluids remained the case of fluids without interparticle attraction which do not have the gas-liquid coexistence and the Widom line. Recently this problem was in the focus of a molecular dynamics study [35]. That study gave evidence that the PSD behaves in the same way in soft-sphere fluids as in the Lennard-Jones ones, the crossover in PSD takes place on the line of minima of thermal diffusivity and kinematic viscosity—the fact observed for supercritical fluids too [59]. The theory of PSD, developed within the GCM approach, allowed one to connect the behaviour of PSD with the location of the Widom line in supercritical fluids above the coexistence curve. The theory is a general one—it does not depend on the particular interaction potentials. The results of [35, 59] generalize the behaviour of supercritical and soft-sphere (without the gas-liquid coexistence) fluids and extend our understanding of dynamic crossover on all types of fluids.

It is important that the vanishing positive dispersion corresponds to the density region where the thermal diffusivity D_T and kinematic viscosity D_L have their smallest values. These two transport quantities define the damping of long-wavelength collective excitations and main hydrodynamic correlation times. Hence the results [35] for soft sphere fluids and previously for the Lennard-Jones ones [5, 34, 59] allow to conclude that the dynamic crossover between the "liquid-like" and "gas-like" states of fluids takes place similarly for fluids with and without the gas-liquid binodal in the region where thermal diffusivity and kinematic viscosity have their minima as functions of density. For supercritical fluids the correspondence between the dynamic crossover in PSD and the Widom line can be obtained by approaching the critical point [34] and using the relation between thermal diffusivity and specific heat at constant pressure.

References

1. http://www.nobelprize.org/nobel_prizes/physics/laureates/1910/waals-lecture.html
2. W. Götze, M.R. Mayr, Phys. Rev. E **61**, 587 (2000)
3. F.A. Gorelli, M. Santoro, T. Scopigno, M. Krisch, G. Ruocco, Phys. Rev. Lett. **97**, 245702 (2006)
4. F.A. Gorelli, M. Santoro, T. Scopigno, M. Krisch, T. Bryk, G. Ruocco, R. Ballerini, Appl. Phys. Lett. **94**, 074102 (2009)
5. G. Simeoni, T. Bryk, F.A. Gorelli, M. Krisch, G. Ruocco, M. Santoro, T. Scopigno, Nature Phys. **6**, 503 (2010)
6. J.R.D. Copley, S.W. Lovesey, Rep. Progr. Phys. **38**, 461 (1975)
7. J.-P. Boon, S. Yip, *Molecular Hydrodynamics* (McGraw-Hill, New York, 1980)
8. R.D. Mountain, Rev. Mod. Phys. **38**, 205 (1966)
9. C. Cohen, J.W.H. Sutherland, J.M. Deutch, Phys. Chem. Liq. **2**, 213 (1971)
10. N.H.W. Lekkerkerker, W.G. Laidlaw, Phys. Rev. A **7**, 1332 (1973)
11. A.B. Bhatia, D.E. Thornton, N.H. March, Phys. Chem. Liq. **4**, 97 (1974)
12. R.D. Mountain, Collective Excitations In Classical One-Component Liquids. *in Dynamics of solids and liquids by neutron scattering*. ed. by. Lovesey S.W. and Springer T. (Springer, Berlin, 1977)
13. L.D. Landau, E.M. Lifshitz, *Fluid Mechanics* (Pergamon Press, Oxford, 1959)
14. J.-P. Hansen, I.R. McDonald, *Theory of Simple Liquids* (Academic, London, 1986)
15. L.D. Landau, G. Placzek, Physik. Z. Sowjetunion. **5**, 172 (1934)
16. R. Zwanzig, *Nonequilibrium Statistical Mechanics* (University Press, Oxford, 2001)
17. T. Bryk, I. Mryglod, Condens. Matter Phys. **11**, 139 (2008)
18. I.M. Mryglod, V. Kuporov, Ukr. J. Phys. **55**, 1172 (2010)
19. I.M. de Schepper, E.G.D. Cohen, C. Bruin, J.C. van Rijs, W. Montfrooij, L.A. de Graaf, Phys. Rev. A **38**, 271 (1988)
20. I.M. Mryglod, I.P. Omelyan, M.V. Tokarchuk, Mol. Phys. **84**, 235 (1995)
21. I.M. Mryglod, I.P. Omelyan, Phys. Lett. A **205**, 401 (1995)
22. T. Bryk, I. Mryglod, Phys. Rev. E **64**, 032202 (2001)
23. T. Keyes, D. Kivelson, J. Chem. Phys. **54**, 1786 (1971)
24. D. Kivelson, T. Keyes, J. Chem. Phys. **57**, 4599 (1972)
25. I.M. de Schepper, P. Verkerk, A.A. van Well, L.A. de Graaf, Phys. Rev. Lett. **50**, 974 (1983)
26. S.W. Lovesey, Phys. Rev. Lett. **53**, 401 (1984)
27. T. Bryk, I. Mryglod, J. Phys. Condens. Matter **13**, 1343 (2001)
28. T. Bryk, Eur. Phys. J. Spec. Top. **196**, 65 (2011)
29. T. Bryk, I. Mryglod, J. Phys. Condens. Matter **12**, 6063 (2000)
30. T. Bryk, I. Mryglod, J. Phys. Condens. Matter **14**, L445 (2002)

31. T. Bryk, I. Mryglod, J. Phys. Condens. Matter **17**, 413 (2005)
32. T. Bryk, I. Mryglod, J. Phys. Condens. Matter **16**, L463 (2004)
33. A.Z. Akcasu, E. Daniel, Phys. Rev. A **2**, 962 (1970)
34. T. Bryk, I. Mryglod, T. Scopigno, G. Ruocco, F. Gorelli, M. Santoro, J. Chem. Phys. **133**, 024502 (2010)
35. T. Bryk, F. Gorelli, G. Ruocco, M. Santoro, T. Scopigno, Phys. Rev. E **90**, 042301 (2014)
36. T. Bryk, I. Mryglod, Phys. Rev. E **62**, 2188 (2000)
37. T. Bryk, I. Mryglod, Condens. Matter Phys. **7**, 471 (2004)
38. P. Verkerk, J. Phys. Condens. Matter **13**, 7775 (2001)
39. V. Giordano, G. Monaco, PNAS **107**, 21985 (2010)
40. Ya. Chushak, T. Bryk, A. Baumketner, G. Kahl, J. Hafner, Phys. Chem. Liq. **32**, 87 (1996)
41. T. Scopigno, U. Balucani, G. Ruocco, F. Sette, J. Phys.: Condens. Matter **12**, 8009 (2000)
42. T. Scopigno, G. Ruocco, F. Sette, Rev. Mod. Phys. **77**, 881 (2005)
43. L. Calderin, D.J. Gonzalez, L.E. Gonzalez, J.M. Lopez, J. Chem. Phys. **129**, 194506 (2008)
44. S. Sengül, D.J. Gonzalez, L.E. Gonzalez, J. Phys. Condens. Matter **21**, 115106 (2009)
45. J. Souto, M.M.G. Alemany, L.E. Gonzalez, D.J. Gonzalez, Phys. Rev. B **81**, 134201 (2010)
46. J.-F. Wax, T. Bryk, J. Phys. Condens. Matter **25**, 325104 (2013)
47. B.G. del Rio, L.E. Gonzalez, J. Phys.: Condens. Matter **26**, 465102 (2014)
48. T. Bryk, G. Ruocco, Mol. Phys. **109**, 2929 (2011)
49. D.E. Woon, Chem. Phys. Lett. **204**, 29 (1993)
50. J.-M. Bomont, J.-L. Bretonnet, T. Pfleiderer, H. Bertagnolli, J. Chem. Phys. **113**, 6815 (2000)
51. E.W. Lemmon, M.O. McLinden, D.G. Friend, Thermophysical Properties of Fluid Systems. in *NIST Chemistry WebBook, NIST Standard Reference Database 69* (National Institute of Standards and Technology, Gaithersburg MD, 2004). URL http://webbook.nist.gov
52. T. Bryk, I. Mryglod, G. Kahl, Phys. Rev. E **56**, 2903 (1997)
53. L. Calderin, L.E. Gonzalez, D.J. Gonzalez, J. Phys. Condens. Matter **25**, 065102 (2013)
54. T. Bryk, G. Ruocco, Mol. Phys. **111**, 3457 (2013)
55. T. Bryk, S. De Panfilis, F.A. Gorelli, E. Gregoryanz, M. Krisch, G. Ruocco, M. Santoro, T. Scopigno, A.P. Seitsonen, Phys. Rev. Lett. **111**, 077801 (2013)
56. K. Nishikawa, K. Kusano, A.A. Arai, T. Morita, J. Chem. Phys. **118**, 1341 (2003)
57. L.V. Woodcock, Fluid Phase Equil. **351**, 25 (2013)
58. J.L. Finney, L.V. Woodcock, J. Phys. Condens. Matt. **26**, 463102 (2013)
59. F.A. Gorelli, T. Bryk, M. Krisch, G. Ruocco, M. Santoro, T. Scopigno, Sci. Rep. **3**, 01203 (2013)
60. D.E. Wetzler, P.F. Aramendia, M.L. Japas, R. Fernandez-Prini, Int. J. Thermophys. **19**, 27 (1998)
61. J.V. Sengers, R.A. Perkins, M.L. Huber, B. Le Neindre, Int. J. Thermophys. **30**, 1453 (2009)
62. P. Jany, J. Straub, Int. J. Thermophys. **8**, 165 (1987)

Chapter 5
Viscous Flow of Glass-Forming Liquids and Glasses

Olexandr Bakai

Abstract Continuous liquid-to-glass transition is characterized by dramatic changes of the liquid structure, thermodynamics and dynamics in a comparatively narrow temperature range. The viscous flow modes of the matter are changing within this temperature range too. The interplay of the structural, thermodynamic, mechanical and dynamic parameters at the viscous flow is still a challenge. Near the glass transition temperature the cooperative diffusion and sliding determine the shear viscosity of the liquid and glass as well as the fragility and the strain rate sensitivity of the flow. In this chapter the theoretical aspects of the physics of viscous flow of glass-forming liquids and glasses are considered within the framework of the heterophase fluctuations model (HPFM), providing a mesoscopic description of the heterophase liquid states. Newtonian and non-Newtonian, Arrhenius and non-Arrhenius flow modes are considered as well as the crossover from the flow to inhomogeneous deformation of glass. The fragility, the strain rate sensitivity, and the fragile-to-strong liquid transformation are described.

Abbreviations

AG	Adam-Gibbs
CD	correlated domains
CRD	cooperatively rearranging domain
DVF	diffusion-viscous flow
HPF	heterophase fluctuations
HPFM	heterophase fluctuations model
SRO	short-range order
SRS	strain rate sensitivity

O. Bakai (✉)
The National Scientific Center Kharkiv Institute of Physics and Technology
(Formerly the Ukrainian Institute of Physics and Technology),
1, Akademichna Str, Kharkiv 61108, Ukraine
e-mail: bakai@kipt.kharkov.ua

L. Bulavin and N. Lebovka (eds.), *Physics of Liquid Matter: Modern Problems*,
Springer Proceedings in Physics 171, DOI 10.1007/978-3-319-20875-6_5

STZ shear transformation zone
VFT Vogel-Fulcher-Tamman
WAXS wide-angle X-ray scattering

5.1 Introduction

The Navier-Stokes equation is in use for the description of the Newtonian liquid flow. The high temperature creep of a solid (which is heterogeneous on small or medium scales) can be treated as a highly viscous homogeneous flow on macroscopic scales. At that the viscosity depends on the topology and kinetic properties of the structural heterogeneities.

These two modes of the viscous flow of the condensed matter are interconvertible at the liquid-glass transition. The difficulties of a unified description of the viscous flow of the glass-forming liquids and glasses arise from the uncertainty of the structural and kinetic properties of the matter in these states. On a deeper level, this uncertainty is connected with non-equilibrium of the super-cooled liquids and glasses. Therefore, the structure, properties and flow of a super-cooled liquid and glass depend on the thermal history and applied external stresses. For this reason, different reasonable models of the structure and kinetics of the glass-forming liquids and glasses are used for describing the flow.

The structure of the glass-forming liquids is essentially heterogeneous on mesoscopic scales (see, e.g. [1–4] and the references quoted). The structural heterogeneities generate the heterogeneity of the relaxational dynamics. The evolution of these heterogeneities at liquid cooling determines the temperature dependence of the viscosity. It occurs that the super-cooled liquid flow becomes non-Newtonian within the glass transition range and the temperature dependence of viscosity does not obey the Arrhenius law. Schematically the flow states map on the (shear rate, temperature)-plane is shown in Fig. 5.1 where the data presented in [5–7] are accounted for.

The region of the non-Arrhenius dynamics lies below the temperature T_A which is, as a rule, above the crystallization temperature, T_m. Transition from the Newtonian to non-Newtonian but macroscopically homogeneous flow takes place in the vicinity of the glass transition temperature, T_g. The crossover from the homogeneous flow to the inhomogeneous deformation occurs with the temperature decrease. The depicted in Fig. 5.1 left boundary separates the regions of the homogeneous flow and inhomogeneous plastic deformation of the glass.

Our goal is to describe the flow modes shown in Fig. 5.1. The structural, thermodynamic and kinetic properties of matter are described within the framework of the heterophase fluctuations model (HPFM) [4, 8–13]. The consideration of the structure and flow of glass is based on results of the theory of polycluster amorphous solids [4].

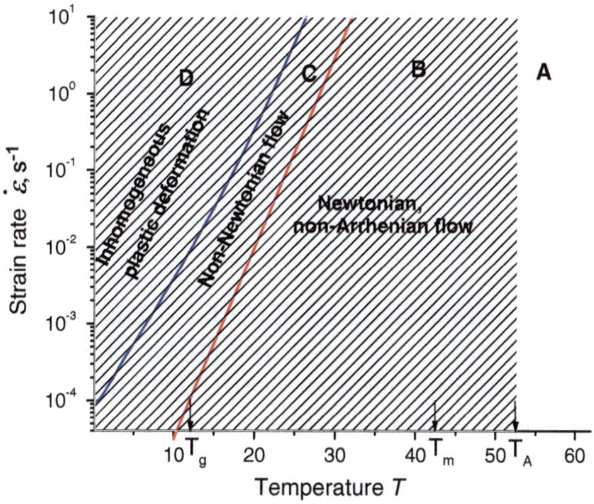

Fig. 5.1 Schematic map of the flow modes of the glass-forming liquids and glasses. Above the crossover temperature T_A (region A) the flow is Newtonian with Arrhenian viscosity. Below T_A (region B) the flow is Newtonian but the viscosity is non-Arrhenian. Within the region C the flow is non-Newtonian. Inhomogeneous plastic deformation of glassy state takes place within the region D. T_m and T_g is the crystallization temperature and glass transition temperature respectively

The chapter is organized as follows. As a background, the viscous flow of the Newtonian liquid and basic relation of viscosity and diffusion are recounted in Sect. 5.2. Characteristic features of the flow of glass-forming liquids are considered in Sect. 5.3. The fragility parameter, as a measure of non-Arrhenian behavior of the diffusion and viscosity, is introduced in Sect. 5.3.1. The strain rate sensitivity is a key quantity characterizing the non-Newtonian flow. It is considered and discussed in Sect. 5.3.2. Basics of the mesoscopic theory of structure, thermodynamics and cooperative relaxational dynamics of liquids below T_A are formulated in Sect. 5.4. Section 5.4.5 is devoted to the diffusion-viscous flow of the polycluster glass and to the crossover from the non-Newtonian viscous flow to the inhomogeneous plastic deformation of glass. The role of the shear banding in establishing the non-Newtonian mixed viscous flow is elucidated. General discussion of the theoretical results and brief comparisons with some experimental data are placed in Sect. 5.5. Conclusions are formulated in Sect. 5.6.

5.2 Flow of Normal Homophase Liquids

Above T_m a liquid is in the thermodynamically stable (normal) state. The methods of the statistical physics and kinetics, hydrodynamic approximation and computer simulations are powerful tools for the research of this state. The discovery of basic laws of flow of normal liquids became possible after the formulation of the principles and equations of motion of the Newtonian mechanics, the Gibbs statistics, and Boltzmann's kinetics.

In his second book of *Principia* (1687) Newton derived the relation between the shear strain rate, $\dot{\varepsilon}$, and the frictional shear stress, σ_{fr}, for liquids, $\sigma_{fr} = \eta\dot{\varepsilon}$, η is the viscosity. At the stationary liquid flow the frictional stress is balanced by the external shear stress, σ. At that the equation of the homogeneous liquid flow is

$$\sigma = \eta\dot{\varepsilon}. \tag{5.1}$$

Navier has formulated the equations of the liquid continuum flow taking into account the viscosity (1822). Later Stokes used the Navier-Stokes equation, solved the problem of the flow past a sphere and derived the expression for the frictional force exerted on a sphere of radius R [14],

$$F_\eta = 6\pi\eta R. \tag{5.2}$$

Calculating the diffusional flux of molecules under the influence of the force (5.2), Einstein in [15, 16] has obtained the formula determining the connection of the viscosity and the diffusion coefficient, D,

$$\eta = \frac{T}{6\pi RD}. \tag{5.3}$$

The formula (5.3) is the Stokes-Einstein relation. The same relation was also deduced by Smoluchowski at almost the same time (for details see his overview paper [17] and Smoluchowski-Einstein correspondence of that time).

For a regular liquid, the radius R in the Stokes-Einstein relation is replaced by the molecular radius, a. As a result, the proportionality of the viscosity to the ratio T/aD is stated,

$$\eta aD/T = \text{Const}. \tag{5.4}$$

Diffusion is directly connected with the relaxation time, τ. For many liquids, in a wide temperature range above T_m, the diffusion coefficient has the form $D \sim a^2/\tau$ with the Arrhenius dependence of τ on temperature,

$$\tau = \tau_0 \exp(-E_{ac}\beta), \quad \beta = 1/T. \tag{5.5}$$

Hereafter the Boltzmann constant, k_B, is equal to 1. E_{ac} is the activation energy; the pre-exponential factor τ_0 is proportional to the period of molecule oscillation.

Thus, if the relation (5.5) is valid, the viscosity can also be presented in the Arrhenian form,

$$\eta = \eta_0 \exp(-E_{ac}\beta). \tag{5.6}$$

The basic relations (5.1)–(5.6) are useful for the rationalization and analysis of the experimental data above T_A, but below T_A they are not valid as it is shown in Fig. 5.1.

5.3 Features of the Glass-Forming Liquids Flow

5.3.1 *Non-Arrhenius Viscous Flow and Fragility*

Rationalization of experimental data using expressions (5.1), (5.6) leads to conclusion that below T_A the activation enthalpy is deviating from a constant and the activation energy depends on T, $E_{ac} = E_{ac}(T)$. The deviation of the E_{ac} from a constant value is considerable for all glass-forming liquids but for many of them it increases dramatically with the temperature decrease below T_m.

Schematically the typical behavior of $\log \eta(T)$ versus $X = T_g/T$ (Angell plots [18, 19]) of liquids is shown in Fig. 5.2. Two types of liquids are distinguished—strong and fragile ones [19]. The viscosity of strong liquids is close but somewhat higher than that given by (5.6) (curve 1 in Fig. 5.2). Fragile liquids are very much non-Arrhenian (curve 2 in Fig. 5.2). The parameter,

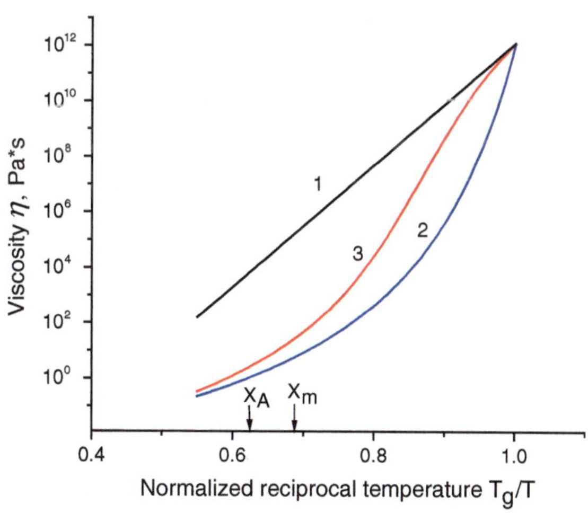

Fig. 5.2 Schematic representation of the T_g-scaled viscosity of three liquids with $\eta(T_g) \approx 10^{12} Pa \cdot s$. Curve 1 presents Arrhenian viscosity. Liquids with small deviation of the viscosity from the Arrhenius one are strong. Liquids with pronounced non-Arrhenian behavior (curve 2) are regarded as fragile. Some liquids (curve 3) at high temperatures look like fragile ones but in the vicinity of T_g they are strong

$$\widehat{m} = \left[\frac{d(\log \tau)}{d\beta}\right]_{T_g} \sim \left[\frac{d(\log \eta)}{d\beta}\right]_{T_g}, \quad (5.7)$$

determines the measure of the kinetic fragility. This parameter is equal to 10–20 for strong liquids. For fragile liquids, $\widehat{m} \sim 50 - 100$.

There are some liquids with $\log \eta(T)$ having a sigmoid shape (like the curve 3 in Fig. 5.2). The viscosity of such liquid at high temperatures is similar to that of typically fragile liquids but the fragility parameter (5.7) (which is a low-temperature distinctive characteristic) is small, like that of strong liquids. In such liquids the fragile-to-strong transformation takes place.

The fragility parameter is an informative quantity. It depends on the structure, thermodynamics and the peculiarities of the relaxation dynamics of the glass-forming liquids. The correlation of this parameter with different physical properties of a material is in use at the rationalization of experimental data [20].

A qualitative explanation of the non-Arrhenius viscosity is based on the fact that below T_A the structural rearrangements of a liquid are cooperative and an elementary rearrangement involves many molecules. The number of molecules within the cooperatively rearranging domain (CRD) is the cooperativity parameter. Assuming that it depends on temperature, one obtains a non-Arrhenius law of the relaxation. The most successful choice of the cooperativity parameter belongs to Adam and Gibbs (AG).

In the AG model [21] the cooperativity is increasing with the temperature decrease $\sim (T - T_0)^{-1}$. Assuming that T_0 is the Kauzmann temperature [22], Adam and Gibbs connected the cooperativity parameter with the configurational entropy and deduced a formula for the cooperative α-relaxation time,

$$\tau_\alpha = \tau_0 \exp\left(-\frac{E^0_{ac}}{T - T_0}\right). \quad (5.8)$$

With E^0_{ac} equal to the activation energy of the normal liquid this formula reproduces the expression (5.5) at $T \gg T_0$. On the other hand, the AG formula reproduces the Vogel-Fulcher-Tamman (VFT) [23–25] empiric law for the relaxation time. Thus, this formula with just two free parameters, E^0_{ac} and T_0, establishes the interconnection of the thermodynamics and the relaxation kinetics at the glass transition. It reproduces the empiric VFT law and has a correct high temperature asymptotic. The intriguing point of this model is that T_0 is a practically unachievable temperature because it is below T_g. Therefore the singularities of the cooperativity parameter and relaxation time, which are interconnected with the Kauzmann "entropy crisis" by Adam and Gibbs, are the hypothetical properties which could presumably be attributed to the "ideal" (equilibrated below T_g) glass.[1]

---

[1]Mesoscopically equilibrated glasses have non-zero residual configurational entropy at $T \to 0$ [12].

As it follows from (5.7), (5.8), in AG model

$$\widehat{m} = T_g^{-1}\left[\frac{d(\log\tau)}{d\beta}\right]_{T_g} = \frac{E_{ac}^0}{(T_g - T_0)^2 \ln 10}. \tag{5.9}$$

It is just one of many phenomenological expressions proposed for the description of the fragility [26]. Investigations [27–29] show that the AG model does not correctly describe the typical features of $\tau_\alpha(T)$ of glass-forming liquids. Nevertheless, it is in use to fit the experimental data which are close to the curves of type 2 shown in Fig. 5.2. Evidently, it has nothing to do with the fragile-to-strong transformations (curve 3 in Fig. 5.2).

A constitutional issue is whether the Stokes-Einstein relation is valid or not below T_A. The experimental data on the viscosity and diffusion (see e.g. [27–29]) show that the relation (5.5) is not satisfied when the temperature is approaching T_g, especially within the region C, at the non-Newtonian flow. It means that within this temperature range the flow mode is changing and that there the diffusion coefficient is not the only the controlling quantity.

5.3.2 *Non-Newtonian Flow. Strain Rate Sensitivity*

The liquid flow is non-Newtonian if the relation (5.1) is not valid. The strain rate sensitivity (SRS) parameter,

$$m_{srs} = \frac{\partial \ln \sigma}{\partial \ln \dot{\varepsilon}} \tag{5.10}$$

is the measure of the strain rate sensitivity. It is less than 1 for the non-Newtonian liquids.

A schematic representation of SRS parameter m_{srs} versus strain rate $\dot{\varepsilon}$ near T_g is shown in Fig. 5.3. With the rate increase the flow becomes non-Newtonian (Fig. 5.1).

The parameter m_{srs} is connected with the activation volume of flow,

$$v_{ac} = T\frac{\partial \ln \dot{\varepsilon}}{\partial \sigma}. \tag{5.11}$$

As it follows from (5.10), (5.11)

$$m_{srs} = \frac{T}{\sigma v_{ac}} = \frac{T}{\eta \dot{\varepsilon} v_{ac}}. \tag{5.12}$$

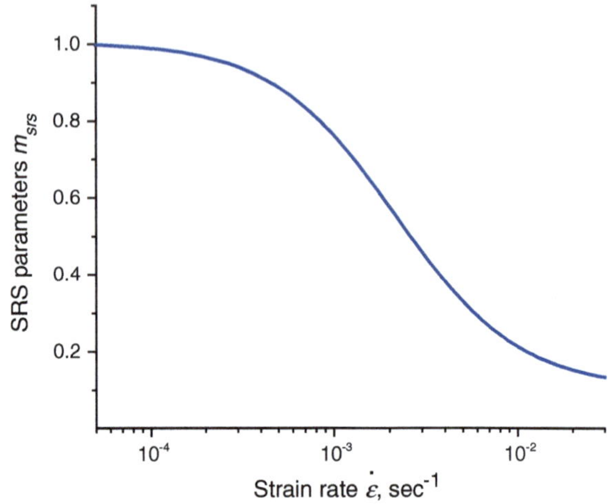

Fig. 5.3 Schematic presentation of the SRS parameter m_{srs} versus the strain rate $\dot{\varepsilon}$ near T_g

The activation volume of a normal liquid is $\sim a^3$. In glass-forming liquids and in a glass near T_g it increases due to the cooperativity of structural rearrangements at the shear strain.[2]

5.3.3 *Maxwell Relation*

In the crossover region C a liquid is highly viscous and behaves as a solid if the shear time, t_{sh}, is shorter than the stress relaxation time, τ_{rel}.

The Maxwell relation [30]

$$\tau_{rel} = \eta/\mu \tag{5.13}$$

connects the shear modulus of glass, μ, with τ_{rel} and the viscosity near T_g. The interpolation formula connecting t_{sh}, τ_{rel}, μ and η, is as follows (see e.g. [31]),

$$\sigma = \frac{\mu\varepsilon}{1 + \frac{t_{sh}}{\tau_{rel}}}. \tag{5.14}$$

At $t_{sh} \gg \tau_{rel}$ and $\varepsilon = \dot{\varepsilon} t_{sh}$ the homogeneous viscous flow (5.1) is settled in while at $t_{sh} \ll \tau_{rel}$ the elastic deformation, $\sigma = \mu\varepsilon$, takes place.

[2]In Argon's model of homogeneous creep of metallic glasses the activation volume is taken $\approx 30a^3$ [28].

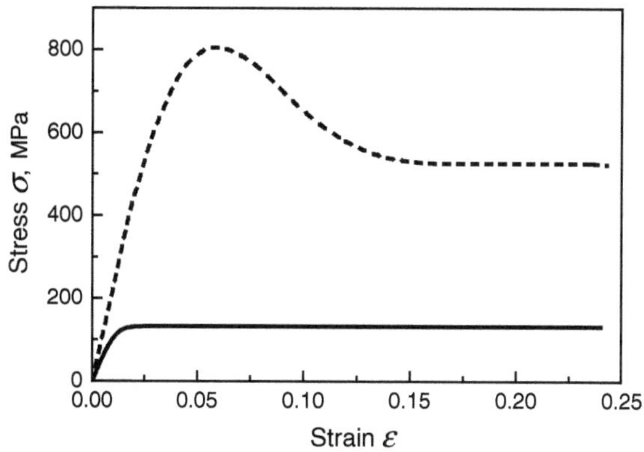

Fig. 5.4 Schematic representation of stress-strain curves observed in metallic glasses at different strain rates ($\dot{\varepsilon} = 10^{-2}$ s^{-1} and $\dot{\varepsilon} = 10^{-3}$ s^{-1} are the *dashed* and *solid lines*, respectively) near T_g [6, 7, 56, 57, 63–65]. With the strain rate increase the shear strain mode is changing from the homogeneous flow (as in the region B) to the inhomogeneous plastic deformation (region D). The solid line represents the flow within the region C

It is evident that the Maxwell relation implies two tacit assumptions. Firstly, the inhomogeneous plastic deformation is absent at $t_{sh} \ll \tau_{rel}$. Secondly, the viscous friction does not depend on the shear strain rate. As the behavior of the stress-strain curves (Fig. 5.4) shows, these assumptions are unjustified in the crossover region C. At low strain rate (solid line in Fig. 5.4, region B in Fig. 5.1) the stress behavior obeys the relation (5.15). At high strain rate (dashed line, region D) the inhomogeneous shear strain and fracture of specimens takes place, as it has to occur with solids. The transient shear strain mode (solid line, region C) is distinguished by the stress overshoot before the homogeneous flow is established.

The next sections are devoted to the consideration of the viscous flow of amorphous matter below T_A.

5.4 HPFM: Thermodynamics, Structure and Dynamics of the Heterophase Liquids

5.4.1 *Order Parameter and Equation of State*

Considering solidification of liquids, Ubbelohde [32] articulated that the observed pre-crystallization anomalies of the heat capacity and viscosity of liquids above T_m appear due to the heterophase fluctuations (HPF). HPF are transient solid-like embryos in fluid. The temperature T_A, at which the anomalies are noticeable, is some tens of degrees above T_m. Ubbelohde attributed the glass transition to the increase of the solid-like fraction of HPF in a liquid. In effect, this scenario of the glass transition is described within the framework of HPFM [4, 8–13].

The description of the thermodynamics and structure of heterophase liquids is based on the equilibrium statistics of mesoscopic species—transient solid-like and fluid-like clusters (s- and f-fluctuons), which are characterized by the short-range order (SRO). The fluctuon size, r_0, is equal to the SRO correlation length, $r_0 \cong \xi_{SRO} \sim a$ which is comparable with the molecular interaction range and with the molecule size a. Thus a glass-forming liquid is considered as an ensemble of mutually transforming s- and f-fluctuons.

Let us denote by N the total number of molecules and by $N_f, N_1, \ldots, N_m$ the numbers of molecules belonging to f-fluctuons and to m types of s-fluctuons respectively,

$$N_f + N_1 + \cdots + N_m = N. \tag{5.15}$$

The (m + 1)-component order parameter of the heterophase liquid is

$$\{c\} = (c_f, c_1, \ldots, c_m); \quad c_i = \frac{N_i}{N} \geq 0; \; i = f, 1, \ldots, m. \tag{5.16}$$

Evidently,

$$c_f + c_1 + \cdots + c_m \equiv c_f + c_s = 1, \tag{5.17}$$

c_s is the total amount of a solid-like fraction.

The HPFM is a model based on the bounded statistics of the f- and s-fluctuons. In the bounded statistics the crystalline states are excluded. As a result, just a non-crystalline (glassy) solid state can be formed while the liquid is cooling down. There are kinetic conditions which determine the time range of this method applicability,

$$\tau_{LRO} \gg \tau_{obs} \gg \tau_{SRO} \sim \tau_\alpha. \tag{5.18}$$

These conditions connect the observation time, τ_{obs}, the crystallization time, τ_{LRO}, and the time of equilibration of SRO of the s- and f-fluctuons, τ_{SRO}. The last one is equal to τ_α.

In practice, the equilibrium glassy state can not be formed due to the dramatic slowing down of the structure relaxation. The kinetic glass transition temperature can be properly estimated as the temperature at which $\tau_\alpha(T) \approx 10^3$s.

The phenomenological equation of state with the multi-component order parameter (5.16) in the simplest form is as follows [12, 13],

$$\mu_i = \mu_i^0 + \sum_k c_k g_{ik} + T \ln c_i, \tag{5.19}$$

$$\mu_f(P, T) = \mu_1(P, T) = \cdots = \mu_m(P, T), \tag{5.20}$$

μ_i is the free energy of the ith fluctuon, μ_i^0 is its part irrespective of the order parameter; coefficients g_{ik} describe the fluctuonic pair interactions.

5.4.2 Two-State Approximation

In the two-state approximation (the solid-like and the fluid-like states are implied)

$$(1 - 2c_s)\tilde{g}_{sf} + T\ln\frac{c_s}{1 - c_s} = h_{sf}. \tag{5.21}$$

Here

$$\begin{aligned} \tilde{g}_{sf} &= g_{sf} - g_{ss}/2,\ h_{sf} = \mu_f^0 - \mu_s^0 - g_{ss}/2, \\ \mu_s^0 &= \sum_k c_k^* g_k^0 + T\sum_k c_k^* \ln c_k^*, \\ g_{ss} &= \sum g_{ik} c_i^* c_k^*,\ c_i^* = c_i/c_s. \end{aligned} \tag{5.22}$$

In this approximation the coefficients g_{sf} and g_{ss} are considered to be constants and the differences between the thermodynamic properties of the s-fluctuons of different types are not taken into account. Both g_{sf} and g_{ss} are positive. The former is the interaction free energy of s- and f-fluctuons, the latter one describes the interaction of s-fluctuons. It is the structural frustration parameter of the solid-like fraction [12, 13].

The “external field” $h_{sf}(P, T)$ can be presented in the form of series expansion near the temperature T_e at which $h_{sf}(P, T) = 0$. In the linear approximation

$$h_{sf}(P, T) = (s_f - s_s)(T_e - T) \equiv \Delta s_{s,f}(T_e - T), \tag{5.23}$$

s_f, s_s is the entropy of the f- and s-fluctuon respectively.

The parameters of (5.21)–(5.23) depend on the chemical composition of a liquid and vary in a wide range of values. Since these coefficients determine the equilibrium liquid and solid states, the phase diagram can be depicted in terms of the coefficients and T (parametric phase diagram) [13].

Besides the parameters g_{sf}, g_{ss} and $\Delta s_{s,f}$, there are the specific temperatures T_e, T_e^0, T_e^1, determined by the following equations,

$$\begin{aligned} g_f^0(P, T_e^0) &= g_s^0(P, T_e^0), \\ g_f^0(P, T_e) &= g_s^0(P, T_e) + g_{ss}/2, \\ g_f^0(P, T_e^1) &= g_s^0(P, T_e^1) + g_{ss}. \end{aligned} \tag{5.24}$$

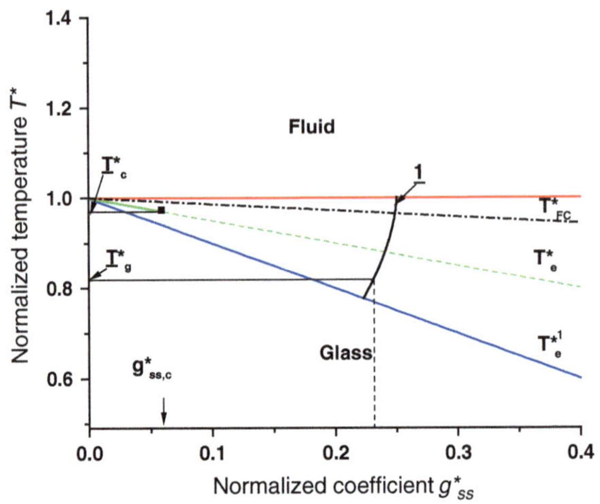

Fig. 5.5 Parametric phase diagram on the plane (T^*, g^*_{ss}), $T^* = T/T_e^0$, $g^*_{ss} = g_{ss}\Delta s_{s,f} T_e^0$. The phase coexistence lines $T = T_e^0$; $T = T_e$; $T = T_e^1$ are shown. The critical end point is on the line $T = T_e$ at $g^*_{ss,c}$. Heavy line 1 schematically represents a continuous evolutional phase curve of the equilibrium liquid. In reality it is terminated at T_g^* where the bounded statistics becomes invalid

The physical meaning of these temperatures is as follows. T_e is the coexistence temperature of two heterophase liquid states with $c_s = c_f = 1/2$. T_e^0 is the temperature at which the fluid and the heterophase liquid coexist. T_e^1 is the coexistence temperature of equilibrated liquid and (theoretically) equilibrated glass. The equilibration kinetics is regarded as terminated at the kinetic glass transition temperature. Therefore, $T_e^1 < T_g$.

The parametric phase diagram (the two-state approximation) is shown in Fig. 5.5. In addition to the specific temperatures (5.26), the temperature T_F^*, which bounds from above the Fischer cluster temperature range [11, 13, 33], is depicted. The Fischer cluster is the fractal aggregation of s-fluctuons. Typically it has the correlation length $\sim 10^2$–10^3 nm. It has no impact on the α-relaxation and viscosity.

It is worth noting that at $g^*_{ss} < g^*_{ss,c} = 2g_{sf}/\left(\Delta s_{s,f} T_e^0\right) - 4T_e^*(P)$ the 1st order liquid-liquid phase transition takes place on the line T_e^*.

It is convenient to introduce the temperature scaled by T_e^0 and the frustration parameter scaled by $\Delta s_{f,s} T_e^0$:

$$T^* = T/T_e^0,\ g^*_{ss} = g_{ss}/\Delta s_{f,s} T_e^0, \tag{5.25}$$

At that

$$T_e^{0*} = 1, \quad T_e^* = 1 - g_{ss}^*/2, \quad T_e^{1*} \approx 1 - g_{ss}^*. \tag{5.26}$$

Continuous ($g_{ss}^* > g_{ss,c}^*$) and discontinuous ($g_{ss}^* < g_{ss,c}^*$) solutions $c_s(T)$ of (5.21) are shown in Fig. 5.6a, b. The loss of the equilibrium of liquid (because of the violation of the condition (5.18)) restricts the range of validity of the solutions $c_s(T)$ in the vicinity of T_g.

The experimentally determined temperature dependence of the order parameter of salol (Fig. 5.6c) confirms this statement.

The schematic picture of the equilibrated liquid and glass structure at different temperatures is presented in Fig. 5.7a–c. In Fig. 5.8 the heterophase structures similar to those in Fig. 5.7, but with several types of s-fluctuons, are depicted. In Fig. 5.8b a fragment of the intercluster boundary, possessing a defective SRO, is shown. A grainy (polycluster) structure with mosaic amorphous grains composed by the frozen mesoscopic s-fluctuons is formed due to the loss of equilibrium and a huge increase of the structure relaxation time at the liquid vitrification. The polycluster structure of metallic glasses is visualized by means of the ion field microscopy [34–37]. The typical size of the grains is $\sim 10^1$ nm.

In [4, 8, 13, 38] different aspects of the physics of the polycluster matter are considered. The ideas of a grainy mosaic structure of glass-forming liquids near T_g are also used in [39–42] in the investigations of the relaxational dynamics and viscous flow.

5.4.3 *Phase Transitions Within the Solid-Like Fraction*

In fact, in the two-state approximation the multiplicity of the SRO types of s-fluctuons is ignored. Meanwhile, phase transformations within the solid-like fraction can induce considerable changes of the order parameter $c_s(T)$. In this case the two-state approximation fails to describe the features of the heterophase liquid and more complicated equations of state (5.20) have to be considered.

The simplest phase transition within the solid-like fraction is induced by the continuous or discontinuous change of the SRO of s-fluctuons. A minimal model of this phenomenon is developed in [12, 13]. To describe the transition, at least two types of s-fluctuons have to be included in consideration. At that, along with (5.21), the equations for the fractions of s-fluctuons of types 1 and 2 have to be considered. As follows from (5.20), in this case

$$\begin{gathered}(1 - 2c_1^*)c_s\tilde{g}_{12} + T\ln\frac{c_1^*}{1 - c_1^*} = h_{12}, \quad c_1^* + c_2^* = 1, \\ \tilde{g}_{12} = g_{12} - c_s\frac{g_{22} - g_{11}}{2}, \quad h_{12} = \mu_2^0 - \mu_1^0 + c_s\frac{g_{22} - g_{11}}{2}.\end{gathered} \tag{5.27}$$

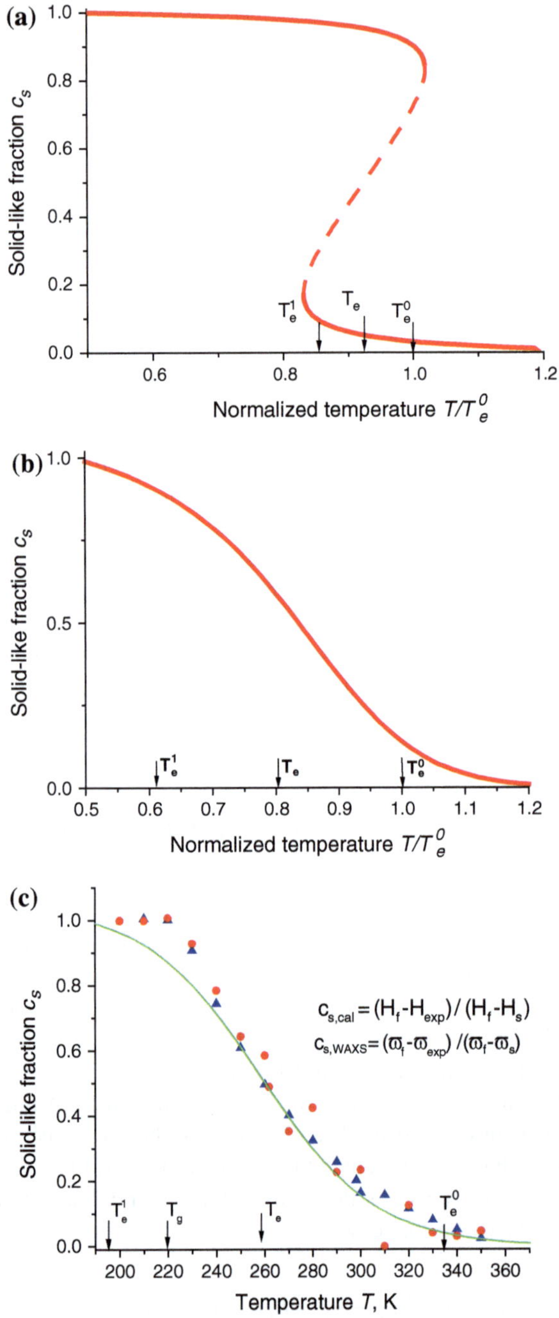

Fig. 5.6 Graphic representation of the discontinuous (**a**) and continuous (**b**) temperature evolution of the total amount of the solid-like fraction $c_s(T)$; **c** $c_s(T)$ for salol recovered from calorimetric (cal) and structural (wide-angle X-ray scattering, WAXS) experimental data, the *solid line* represents the equilibrium curve, similar to that shown in (**b**)

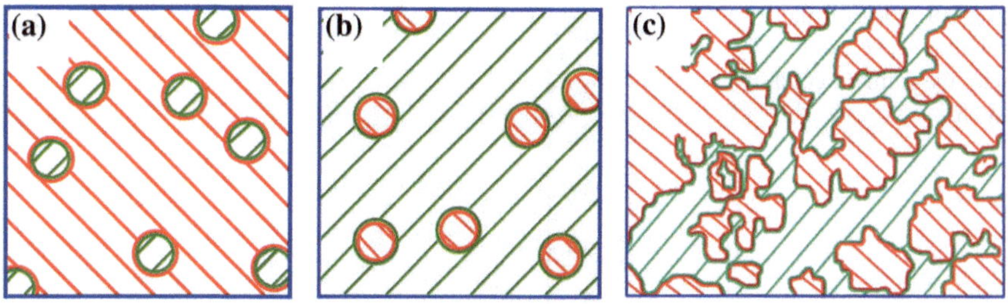

Fig. 5.7 Schematic representation of the heterophase liquid structure **a** near T_e^0, with rare s-fluctuons; **b** above T_e^1, with f-fluctuons in a glass; **c** near T_e

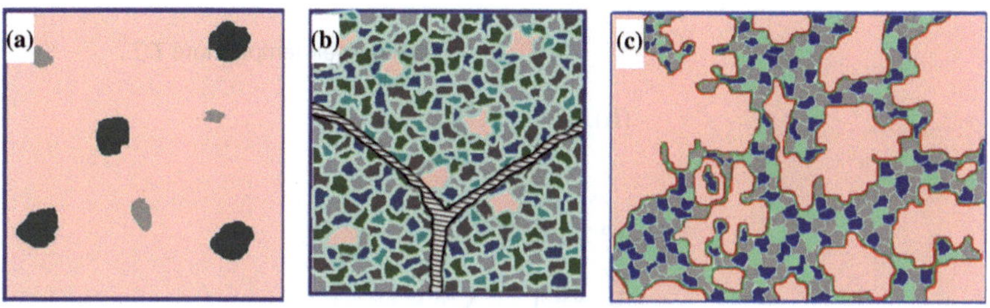

Fig. 5.8 The same structure as in Fig. 5.7, but here several types of non-spherical s-fluctuons and triple joint of the cluster boundaries in a glass (in **b**) are shown

This equation is similar to (5.21) but its coefficients depend on c_s. Therefore, joint solutions of (5.21), (5.27) have to be found.

For certainty let us consider the case when the solid-solid coexistence temperature, T_{12}, is below T_e, $c_s(T_{12})$ is less than 0.5.

At

$$c_s(T_{12})\tilde{g}_{12} > 2T_{12}, \quad (\mu_2^0 - \mu_1^0|_{T_{12}} = 0), \tag{5.28}$$

$c_s(T)$ is discontinuous at T_{12} while at

$$c_s(T_{12})\tilde{g}_{12} < 2T_{12}, \tag{5.29}$$

it is a continuous function.

A graphic representation of the solutions of (5.20) and (5.27) with the condition (5.28) is shown in Fig. 5.9. The step of $c_s(T)$ at $T = T_{12}$ is

$$\Delta c_s(T_{12}) \approx (s_1 - s_2)c_s(T_{12}), \tag{5.30}$$

s_1, s_2 is the entropy of s-fluctuons 1 and 2 respectively.

Fig. 5.9 View of the solutions $c_1^*(T)$ and $c_s(T)$ with the fulfilled condition (5.28)

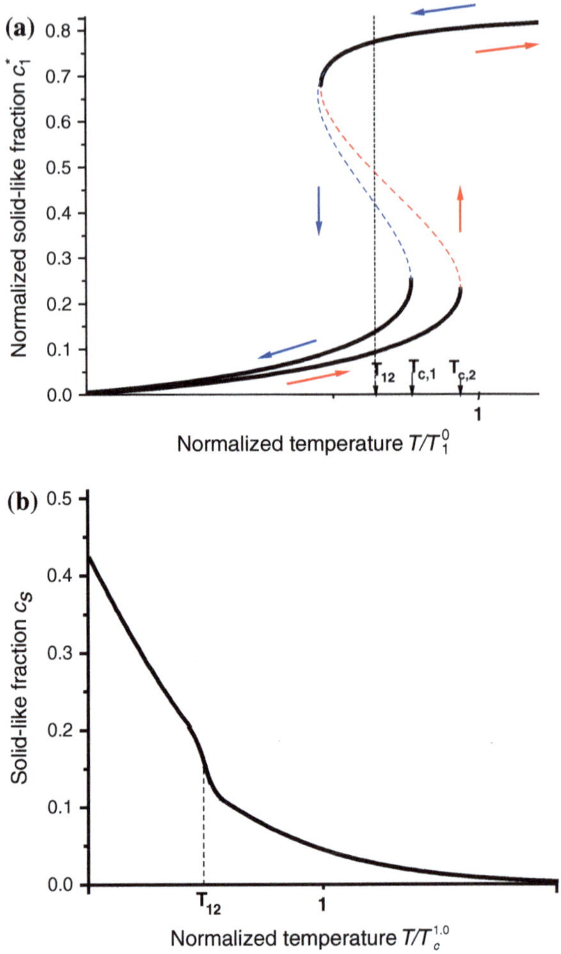

The heat of the phase transition is equal to

$$\Delta H = \Delta c_s(T_{12}) H_{fs}(T_{12}) + c_s(T_{12}) H_{12}. \tag{5.31}$$

Here $H_{fs}(T_{12})$ is the heat of the fluid-solid transformation at $T = T_{12}$ and $H_{12} = (s_1 - s_2) k_0^{-1} T_{12}$ is the heat of the $1 \leftrightarrow 2$ solid-solid phase transformation. As it follows from (5.30), (5.31)

$$\Delta H = c_s(T_{12}) \frac{s_1 - s_2}{k_0} \left(H_{fs} + T_{12} \right). \tag{5.32}$$

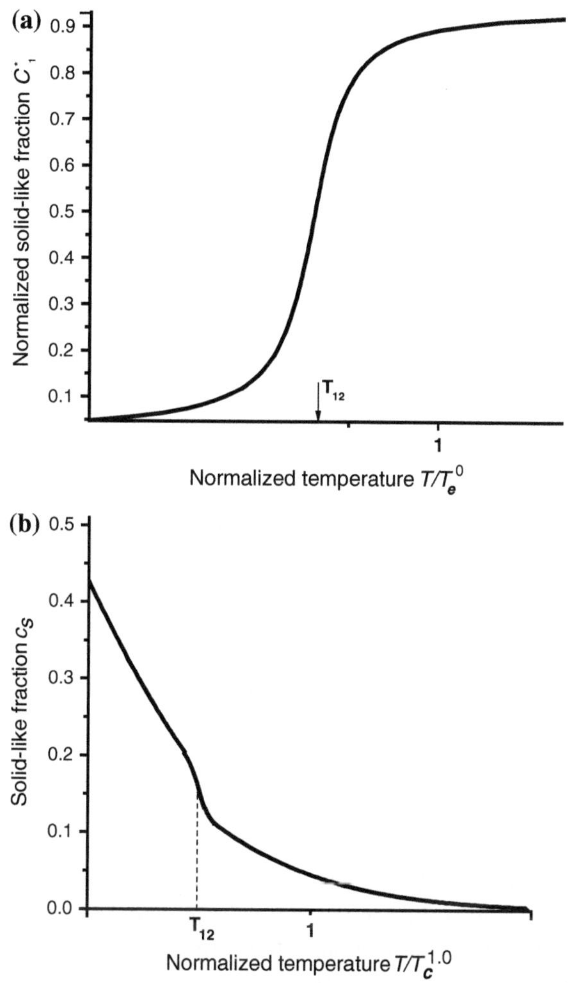

Fig. 5.10 Stepwise evolution of $c_1^*(T)$ and $c_s(T)$ at the fulfilled condition (5.29)

The multiplier $(s_1 - s_2)k_0^{-1}$ in the right-hand side of (5.32) is the entropy difference per molecule.

A continuous phase transformation takes place if the condition (5.29) is satisfied. Instead of a sharp step, a continuous sigmoidal jog appears on $c_s(T)$ with the inflection point at $T = T_{12}$ (Fig. 5.10). The jog height can be estimated using (5.30). The phase transformation heat is determined by the (5.32).

Noting that the entropy jump at the crystallization is usually ~ 1 and that at the solid-solid phase transformations this quantity is $\sim 10^{-1}$, from (5.32) we have that at $c_s(T_{12}) = 0.2 - 0.4$ the phase transformation heat is $\sim 10^{-1} H_{fs}$.

5.4.4 Cooperative Dynamics of the Heterophase Liquid, Fragility and Fragile-to-Strong Transition

Fluctuons are correlated due to their interactions. The direct correlation function of fluctuons has a typical for the Ising-type models correlation length, $\xi_{fl} \approx 2\xi_{SRO}$. The length ξ_{fl} determines the size of correlated domains (CD) and the size of the CRD [10, 11, 13]. The number molecules per CD (cooperativity parameter) is $z_{CD} \sim (\xi_{fl}/a)^3$.

Cooperative rearrangement is thermally activated. Its activation energy depends on the order parameter and can be presented in the form of an expansion in its powers [11],

$$E_{ac} = E^0_{ac} + E^1_{ac}c_s + E^2_{ac}c^2_s + \cdots \tag{5.33}$$

Fischer and Bakai [11] have developed a phenomenological model based on the assumption that CD can be rearranged when all its molecules are in fluid-like state. This assumption leads to a conclusion that $T_A \approx T^0_e$ and that E^0_{ac} is equal to the activation energy of the homophase fluid state above T_A. The resulting expressions for the α-relaxation time and activation energy are as follows,

$$\tau_\alpha = v_0^{-1} \exp\left[-\frac{E^0_{ac}}{T - T_K} - z_{CD}c_sH_{fs}\beta\right], \tag{5.34}$$

$$E_{ac} = \frac{E^0_{ac}}{(1 - T_K/T)^2} + z_{CD}c_sH_{fs} + O(c^2_s). \tag{5.35}$$

The first term in the right-hand side of (5.35) is the activation energy of the homophase fluid in the free volume model [43, 44], T_K is the Kauzmann temperature. The contribution of this term to E_{ac} below T_e, where the flow is essentially non-Arrhenian, is ~10 % [45]. Therefore, the non-Arrhenius behavior of the diffusion and viscosity is determined mainly by the second term.

The diffusion coefficient below T_A is $D_\alpha \sim a^2/\tau_\alpha$, but the Stokes-Einstein relation (5.4) holds valid just at low values of c_s. At $T/T_g < 1.1 - 1.25$, i.e. at $1 - c_s \ll 1$, it is violated (see e.g. [41, 46–48]) due to the impact of the mesoscopic structural heterogeneities. In this temperature range the mode of liquid fluidity is changing.

The validity of the relations (5.34), (5.35) is confirmed by numerous experimental data [49]. As an example, the experimentally measured activation energy of salol versus the reciprocal temperature and a theoretical curve (5.35) are shown in Fig. 5.11 [10].

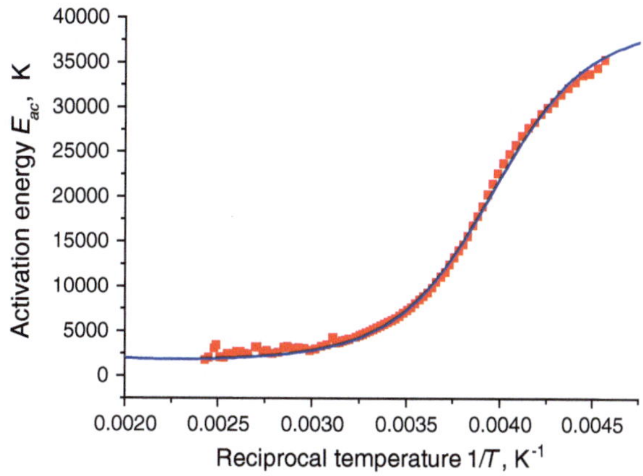

Fig. 5.11 The activation energy E_{ac} of salol versus the reciprocal temperature $1/T$. Closed squares depict the experimental data [10]. Line represents (5.35)

The kinetic fragility parameter (5.7) is proportional to $E_{ac}(T_g)$. Combining (5.7) and (5.35), we have

$$\widehat{m} = \frac{1}{T_g}\left[\frac{E_{ac}^0}{\left(1 - T_K/T_g\right)^2} + z_{CD}H_{fs}\right],\ H_{fs} = \frac{\Delta s_{f,s}T_e}{k_0}. \tag{5.36}$$

In particular, for salol $\widehat{m} \approx 67$ [10]. This value is in accord with that obtained by others, e.g. in [50].

If the two-state approximation is applicable and c_s is changing continuously, the curves of type 2 shown in Fig. 5.2 can be obtained using (5.35). If this approximation fails (e.g. due to the phase transformation within the solid-like fraction (Sect. 5.4.2)), the relation (5.35) can be used to estimate the jump of the fragility parameter at the fragile-to-strong transformation. In this case the two-state approximation (but with different thermodynamic parameters) is applicable for the approximation of $c_s(T)$ in the vicinity of T_A (state 1) and near T_g (state 2). The parameters E_0, T_k, z_{cd} for both states are nearly the same but the values of H_{fs} and T_g are different. The Angell-plot in this case looks like the curve 3 in Fig. 5.2.

A general view of the asymptotic solutions of (5.22) for states 1 and 2 along with a solution of (5.22), (5.27) (curves 1, 2 and 3 respectively) is shown in Fig. 5.12. It is noteworthy that $T_{g1} < T_{g2}$ and $\widehat{m}_1 > \widehat{m}_2$. As follows from (5.36),

$$\widehat{m}_1 - \widehat{m}_2 \approx \widehat{m}_1 \frac{T_{g,2} - T_{g,1}}{T_{g,1}}\left(1 + \frac{2E_0}{T_{g,1}\left(1 - T_K/T_{g,1}\right)}\right). \tag{5.37}$$

It is worth pointing out that any phase transformation within the solid-like fraction induces a fragile-to-strong transition with a larger or smaller jump of the

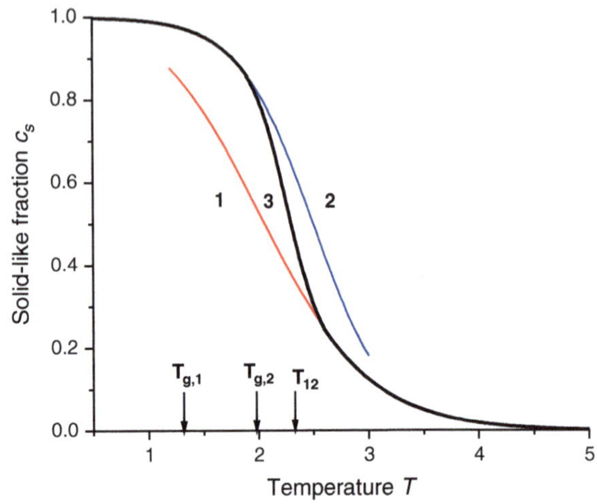

Fig. 5.12 Behavior of the solid-like fraction $c_s(T)$ at the fragile-to-strong transition (curve 3). Curves 1 and 2 represent the asymptotic solutions $c_s(T)$ above T_A and below T_2 respectively

fragility parameter. To ensure that, let us consider the impact of such phase transformation on the phase trajectory on the parametric phase diagram (Fig. 5.5). The thermodynamically driven transformations lead to the decrease of the free energy of the solid-like fraction. At that the entropy, s_s, and/or the frustration parameter, g_{ss}, is decreasing [13]. As a result, g_{ss}^*, T_e^1 and T_g increase gradually if continuous phase transformations, equilibration and ordering of the solid-like fraction take place. In this case the phase trajectory looks like the curve 1 in Fig. 5.5. A stepwise or discontinuous evolution of $c_s(T)$, like that in Figs. 5.10 and 5.12, leads to a congruent behavior of the phase trajectory.

5.4.5 *Diffusion-Viscous Flow of a Glass*

At the diffusion-viscous flow (DVF) the polycrystalline solids behave as highly viscous liquids [51–54]. This mode is inherent in the homogeneous plastic deformation of glasses possessing the polycluster structure [4, 8].

Glass inherits heterogeneities of the heterophase liquid. The features of the polycluster mosaic structure, like those shown in Fig. 5.8b, depend on the composition and cooling rate. Some possible forms of the polycluster structure are shown in Fig. 5.13.

As it was noted, in metallic glasses the locally ordered amorphous grains ~ 10 nm are separated by the boundary layers with a width ~ 1–3a [34, 35]. The topologic order within the boundaries is poor, like that within the large-angle boundaries of polycrystalline alloys. Due to the large density of the boundaries, the Coble creep of a glass is the dominating DVF mode near T_g at low strain rates [4, 8].

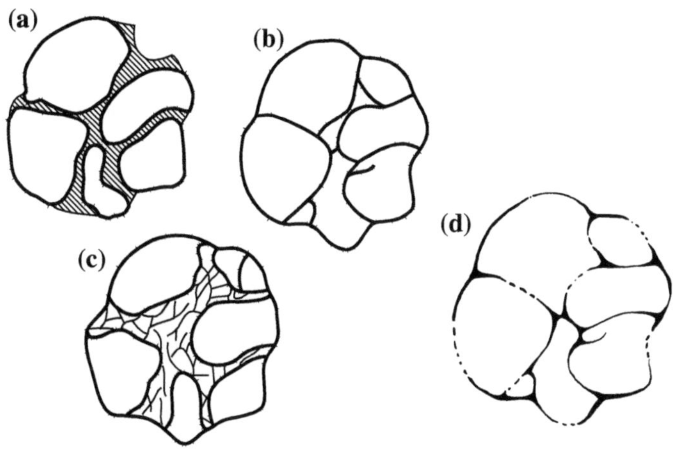

Fig. 5.13 Schematic picture of the polycluster fragments. Locally ordered solid clusters are separated by boundary layers of different structure and width. In **c** the density of free volume in the boundaries of structure **b** is schematically indicated by the intensity of the boundary lines

5.4.5.1 Conservative Diffusion-Viscous Flow of Glass

At the isostructural or conservative DVF the number of grains and the volume of each of them do not change. DVF takes place at a low stress, $\sigma \to 0$, when the viscosity, which is the linear response of flow on the stress, is independent of σ.

It is shown in the Lifshits theory of the conservative DVF [54] that the diffusion creep is accompanied by the grain boundary sliding. Diffusion provides normal to grain surface deformation and the self-consistent adjustment of the normal stress while the tangent stress induces the grain boundary sliding. Therefore, the viscosity coefficient includes two summands relating to the diffusion viscosity and the sliding friction respectively,

$$\eta = \eta_d + \eta_{sl}, \tag{5.38}$$

η_d is the diffusion viscosity, η_{sl} is the sliding resistance viscosity,

$$\eta_d \sim \frac{T}{aD_s}\left(\frac{L}{h_s}\right)^3, \quad \eta_{sl} \sim \gamma_{sl}\left(\frac{L}{h_s}\right). \tag{5.39}$$

Here, L is the mean size of grains, h_s is the boundary width, D_s is the coefficient of the grain boundary diffusion, γ_{sl} is the sliding friction coefficient. The geometric factors, depending on the grains geometry, are omitted in (5.39).

Applying the Lifshits theory of conservative DVF to polyclusters we have to take that L in (5.39) is the mean size of blocks of the locally ordered clusters surrounded by connective boundary layers of the mean width h_s [4]. The diffusion within the boundary layer is much faster than that in the blocks. L is comparable with or larger than the mean size of the cluster, l_{cl}.

The coefficients D_s and γ_{sl} have to be specified to determine η_d and η_{sl}.

At low temperatures $(T < 0.8T_g)$ the boundary diffusion in a glass is a thermally activated hopping of single atoms in vacant neighboring holes with the volume $\approx a^3$. At higher temperatures, typically at $T > T_{c1} \sim 0.8 - 0.9T_g$, a cooperative mode of the boundary diffusion is dominating [4, 8]. This mode is connected with the cooperative fluctuations in the form of disc-like shear loops of size $R_l \sim 4 - 5$a.[3] Due to the elastic loop-loop interactions, the loop activation energy, U_{coop}, essentially decreases with the increase of their number and at $T > T_{c1}$,

$$U_{coop} \sim \frac{\mu_s a^2 R_l}{1 - \nu_P}, \tag{5.40}$$

μ_s is the shear modulus within the boundary, ν_P is the Poisson coefficient. At that the diffusion coefficient is

$$D_s \sim D_{s,coop} \sim 10^{-2} a^2 \nu_c \exp(-U_{coop}\beta). \tag{5.41}$$

The pre-exponential coefficient $\sim 10^{-2}$ appears due to the configurational entropy of the loop.

The sliding friction viscosity controlled by the cooperative diffusion is [4]

$$\eta_{sl} \sim \frac{T}{aD_s}\frac{L}{h_s}\left(\frac{a}{R_l}\right)^2. \tag{5.42}$$

As it follows from (5.11), (5.38), at conservative DVF the SRS parameter is

$$m_{srs} = \frac{\alpha_d \eta_d + \alpha_{sl}\eta_{sl} m_{srs}^{sl}}{\alpha_d \eta_d + \alpha_{sl}\eta_{sl}}; \quad m_{srs}^{sl} = \frac{\partial \ln \sigma}{\partial \ln \dot{\varepsilon}_{sl}}, \; \dot{\varepsilon}_{sl} = \frac{L}{h_s}\dot{\varepsilon}. \tag{5.43}$$

Comparing (5.39) and (5.43), one can see that $\eta_{sl}/\eta_d \sim (R_l L/ah_s)^2 \ll 1$, i.e. the sliding viscosity is negligibly small. At that, $m_{srs} \approx 1$. Thus the conservative DVF near T_g is Newtonian (region B).

5.4.5.2 Non-Newtonian Diffusion-Viscous Flow

The conservative DVF is not only the mode of the glass flow. Sliding within the intercluster boundary is blocked in the triple joint. At the increasing stress the slip

[3] Cooperative structural fluctuations of such type were introduced by Argon [28]. Later they were termed "shear transformation zones" STZs [53].

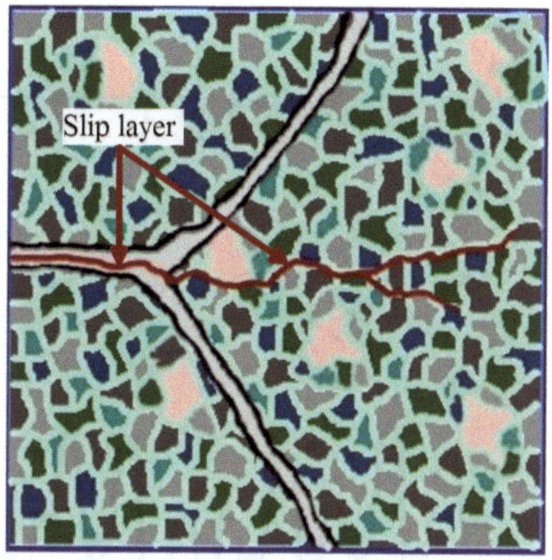

Fig. 5.14 The slip layer formed within the boundary penetrates into the blocking cluster body if the stress in the triple joint exceeds the cluster strength

layer can propagate into the blocking cluster body as it is shown in Fig. 5.14. The fragmentation of clusters by the propagating slip layers can lead to establishing the non-conservative (mixed) DVF or to the inhomogeneous deformation and fracture of a glass. Since the slip rate is much larger than the diffusional creep, the boundary diffusion can not considerably impact the slipping kinetics. It means that the diffusion and slipping kinetics are decoupled.

Considering the crossover from the conservative DVF to the mixed flow, let us denote by σ^* the critical value of the external stress at which the propagation of slip layers and shear banding occurs. The necessary condition of the crossover to the non-Newtonian mixed flow is $\sigma > \sigma^*$[4, 8], i.e.

$$\eta\dot{\varepsilon} > \sigma^*. \tag{5.44}$$

To determine σ^*, the slip kinetics and shear banding have to be considered. The scenario of the glass shear banding is as follows. The disordered boundaries are the regions of easy slipping which is blocked in the triple joints until the concentrated local shear stress and thermal activation provide the propagation of the slip layer into the blocking cluster. The propagating slip layer is branching and initiates the shear band formation due to the translational and rotational movements of the solid fragments [55].[4]

The shear band width, h_{sb}, is much larger than that of the intercluster boundary. It is found that $h_{sb} \sim 10$–100 nm [57], i.e. h_{sb} is comparable with or larger than l_{cl}. The viscosity within the shear band is essentially lower that that in the surrounding matrix due to the formation of multiple slip layers and enhanced diffusion.

[4]Comprehensive overviews of the inhomogeneous deformation and shear banding of metallic glasses are presented in [6, 56]. The scenario with two consecutive stages of the shear band formation is considered in [56] and the references quoted.

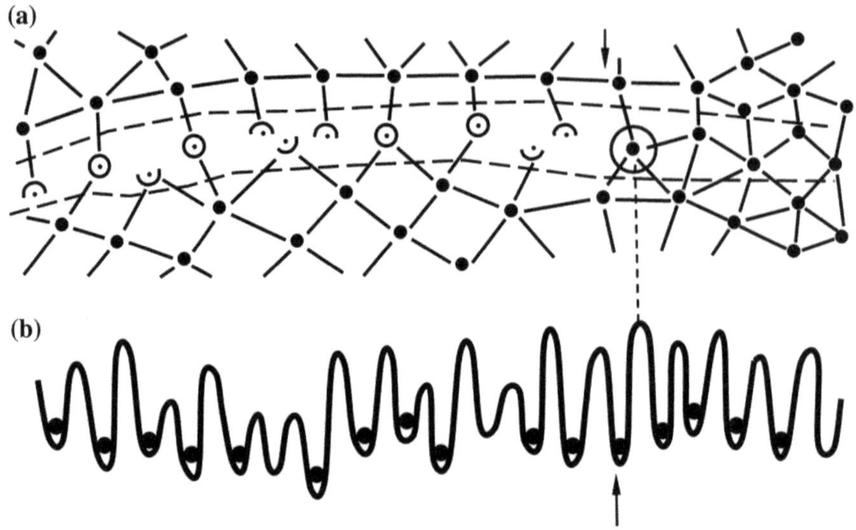

Fig. 5.15 **a** Fragment of the slip layer blocked in the cluster body at the point marked by an *arrow*; **b** the potential relief of the slip layer. Atoms are depicted by full *circles*

Therefore the shear banding facilitates the flow of a glass. It signifies the crossover to the stable non-Newtonian mixed flow, i.e. the crossover from the region D to the region C at $\eta\dot{\varepsilon} \approx \sigma^*$ due to the increase of the strain rate (Fig. 5.1).

The slip layer is specified by the distribution function, $g(\sigma_{cr})$, of the local critical shear stress, σ_{cr}, which is randomly changing from site to site. The theory of the slip within a layer of atomic width with the arbitrary homogeneous distribution $g(\sigma_{cr})$ is developed in [4, 55]. The ultimate strength and the thermally activated strain rate at fixed shear stress are the functionals of $g(\sigma_{cr})$.

The equations of the homogeneous boundary slip are deduced and solved for specific distributions $g(\sigma_{cr})$ in [4]. The exact solution for an arbitrary distribution $g(\sigma_{cr})$ is published in [55].

In Fig. 5.15 the potential relief of a layer (it includes a boundary strip and surrounding cluster body) is shown. The local critical shear stress σ_{cr} is proportional to the difference of the potential energy values in the neighboring maxima and minima.

Since the boundary contains regular and defective sites, as it shown in Fig. 5.15, the distribution $f(\sigma_{cr})$ within the boundary layer is bimodal while within the cluster body, containing the locally ordered sites, it can be taken in a unimodal form (Fig. 5.16).

The solutions of the equations of a homogeneous boundary slip with the distributions $f(\sigma_{cr})$ shown in Fig. 5.16 are presented graphically below.

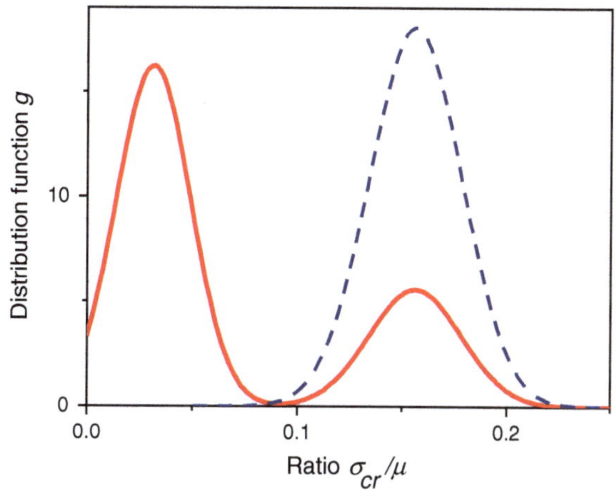

Fig. 5.16 Examples of the unimodal (*dashed line*) and bimodal (*solid line*) distribution function g of the ratio of the critical shear stress and shear modulus σ_{cr}/μ

The stress-strain rate curves at different temperatures near T_g are shown in Fig. 5.17a. T_g is estimated as $T_g = 0.012\,\mu_0 a^3$ [58]. Solid and dashed lines correspond to bimodal and unimodal distributions respectively. The stress is given in the shear modulus units.

The strain rate decreases with the increase of the fraction of sites with large critical stress. At that the activation volume decreases (Fig. 5.17b). Noteworthy is that at all temperatures the stress-strain rate curves are merging at the yield stress of the slip layer, $\sigma^*_{sl} \sim 10^{-2}\,\mu$. At stresses higher than σ^*_{sl} the slip is athermic.

The yield stress of the cluster body, σ^*_{cl}, calculated using the unimodal distribution, is $\sim 10^{-1}\,\mu$. It is the counterpart of the theoretical strength of crystals [59].

The growth of the activation volume at the increase of the fraction of the sites with low local critical stress indicates the increase of the cooperativity due to the stress concentration on the strong sites after dropping the weak ones.

The short-dashed curve in Fig. 5.17b corresponds to a narrower unimodal distribution. It demonstrates that the cooperativity effect decreases with diminishing of the width of distribution $f(\sigma_{cr})$.

The slipping resistance coefficient, $\gamma_{sl} = \sigma/\dot{\varepsilon}_{sl}$, in units $\mu \cdot s$ is shown in Fig. 5.18a. At large strain rates ($\sim 10^0$–10^1 s^{-1}) the γ_{sl} becomes rather small and almost independent of temperature. It is much larger at the strain rates less than 10^{-2} s^{-1}.

The strain rate sensitivity parameter m^{sl}_{srs} is represented in Fig. 5.18b. Qualitatively the dependence of SRS on the strain rate is compatible with that shown in Fig. 5.3. It is worth noting that SRS increases with the temperature growth and at the strain rate decrease. At $T = 1.2T_g$ and $\dot{\varepsilon}_{sl} < 10^{-3}\,s^{-1}$ it approaches 1. This result indicates that with the temperature increase the homogeneous slip gradually ceases to play any role.

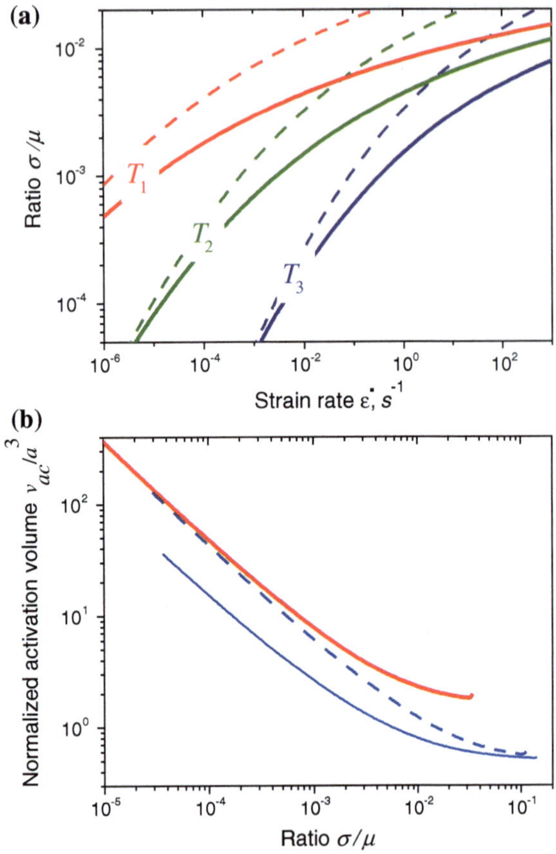

Fig. 5.17 **a** Stress-strain rate curves at temperatures $T_1 \approx 0.8 \cdot T_g$, $T_2 \approx T_g$ and $T_3 \approx 1.2 \cdot T_g$. *Solid* and *dashed lines* correspond to bimodal and unimodal distributions respectively. **b** Normalized activation volume v_a/a^3(**a** is a molecular radius) versus the ratio of the shear stress and shear modulus σ/μ. The thin dot line corresponds to a unimodal distribution $g(\sigma_{cr})$ with the smaller width [55]

Sliding in the boundary layer of length L leads to the concentration of the stress on the layer edge stress $\sim (L/2a)^{1/2}\sigma$. Therefore, at

$$\sigma > \sigma^* = \left(\frac{2a}{L}\right)^{1/2} \sigma^*_{cl} \tag{5.45}$$

the slip layer propagates in the blocking cluster. The equation

$$\eta\dot{\varepsilon} = \sigma^* = \left(\frac{2a}{L}\right)^{1/2} \sigma^*_{cl} \tag{5.46}$$

determines the boundary between regions B and C shown in Fig. (5.1).

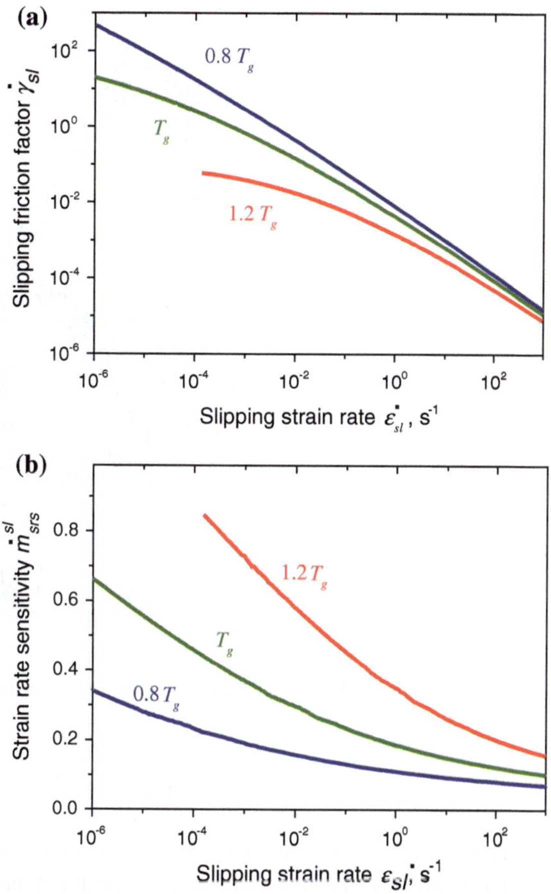

Fig. 5.18 **a** The slipping friction factor γ_{sl} in units $\mu \cdot s$ and **b** the SRS parameter of sliding m^{sl}_{srs} versus the slipping strain rate $\dot{\varepsilon}_{sl}$

The development of the shear banding within the region C leads to the formation of the glass block structure like that shown in Fig. 5.19. The polycluster blocks are separated by shear bands. The viscosity of the shear banded glass, η_{sb}, is given by (5.38) with similar to (5.39) relations

$$\eta_d = \eta_{d,sb} = \frac{T}{aD_{sb}}\left(\frac{L_{sb}}{h_{sb}}\right)^3, \tag{5.47}$$

$$\eta_{sl} = \eta_{sl,sb} = \gamma_{sl}\left(\frac{L_{sb}}{h_{sb}}\right)\rho_{sl}h_{sb}. \tag{5.48}$$

The index "sb" denotes the quantities concerning the shear bands; ρ_{sl} is the density of slip layers within the shear band.

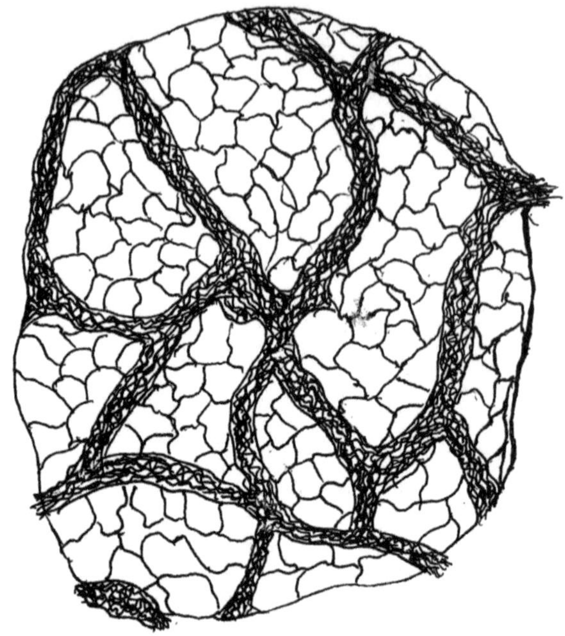

Fig. 5.19 Shear banded piece of a polycluster. *Dark strips* represent the shear bands

The condition of the mixed flow stability is as follows,

$$\eta_{sb}\dot{\varepsilon} = \frac{1}{aD_{sb}\beta}\left(\frac{L_{sb}}{h_{sb}}\right)^3 \dot{\varepsilon} < \sigma^*. \tag{5.49}$$

The multiplication of the shear bands is impossible if this requirement is fulfilled. The viscosity at the mixed flow is less than that at the conservative DVF, $\eta_{sb} < \eta_{DVF}$. Therefore, the conservative DVF is unstable if the condition (5.45) is satisfied.

As it follows from this inequality and (5.38), (5.49), the shear band spacing at the mixed flow is

$$L_{sb} \leq L^*_{sb} = \frac{h_{sb}l_{cl}}{a}(\rho_{sl}h_{sb})^{1/3} \sim \frac{h_{sb}l_{cl}}{a}. \tag{5.50}$$

One can see that with $h_{sb} \sim$ 10–100 nm and $l_{cl} \sim$ 10 nm the shear band spacing is of micron and submicron scale.

The decrease of the viscosity due to the shear banding leads to the decrease of the stress at a constant strain rate. As a result, a peak of the stress (stress overshoot) appears on the stress-strain curve (see Fig. 5.4). The stress overshoot, $\Delta\sigma_{over}$, is equal to

$$\Delta\sigma_{over} = \sigma^* - \sigma_{flow} = \sigma^* - \eta_{sb}\dot{\varepsilon}. \tag{5.51}$$

To determine the location of the boundary between regions C and D we have to formulate the condition at which the formation of shear bands, provided by the inequality (5.49), leads to the inhomogeneous glass deformation and fracture. The fracture occurs at large strain rates when there is no time for the initiation of the DVF and the consecutive mixed flow. The stress-strain curve is changing in the process (see Fig. 5.4). The ultimate strain, ε_{ult}, is relatively small, typically it does not exceed 10^{-1}.

The establishment of DVF requires the formation of the self-consistent secondary stress field which provides adapting of the diffusion-governed deformation and boundary sliding [54, 60]. The stress field relaxation depends on the kinetics features [54, 60]. In polyclusters near T_g, as it is noted in Sect. 5.4.3, the diffusional relaxation of the normal stress on the boundaries is the slowest relaxation mode. Its characteristic time is

$$\tau_{rel}^{d} \sim L^2/D_s. \tag{5.52}$$

If the ultimate strain is attained at a time much shorter than τ_{rel}^{d}, the relaxation of the concentrated stress does not occur during the initiation of the boundary slip layers and the formation of the catastrophic shear bands. Thus, the inequality

$$\dot{\varepsilon}\tau_{rel}^{d} > \varepsilon_{ult} \sim 10^{-1} \tag{5.53}$$

determines the range of strain rates at which the crossover from region C to region D takes place.

In essence, the inequality (5.53) reproduces the Maxwell condition of the crossover from the viscous flow to the mechanical deformation in accord with the interpolation formula (5.14).

The relations (5.42), (5.47), (5.52) and (5.53) determine the temperature range of the region C, ΔT_C.

Assuming that $U_{coop} \sim 1\,eV \sim 10^4\,K$ [61, 62], $T_g/a^3 \sim 10^{-2}\,\mu$ [58], $L/a \sim 10^2$, and $\sigma^*/\mu \sim 10^{-2}$ [7], we have

$$\Delta T_C \cong 0.04 T_g. \tag{5.54}$$

This estimation is in harmony with the experimental data for metallic glasses [7, 56, 63–65].

The strain rate sensitivity parameter (5.44) becomes small ($\sim 10^{-1}$) at the crossover from region C to region D.

5.4.5.3 Diffusion Crossover Near T_g

The α-relaxation in the region B is connected with the "spherical" cooperative rearrangements of CDs. Incidentally the diffusion coefficient is $D_\alpha \sim a^2/\tau_\alpha$.

Considering the diffusion in a glass near T_g, we have taken into account that the elementary cooperative fluctuation is the disc-like shear loop within the intercluster boundary. At that the diffusion coefficient is determined by (5.41). Therefore the liquid-glass transformation is accompanied by the diffusion mode changing within the crossover range of regions B and C. To get the diffusion crossover temperature, T^*_{diff}, it is necessary to equate the diffusion coefficients in the glassy and liquid states,

$$\frac{10^{-2}a}{L}\exp\left(-U_{coop}\beta\right) = \exp\left(-z_{CD}H_{fs}\beta\right)\Big|_{T^*_{diff}}. \tag{5.55}$$

The solution of this equation is

$$T^*_{diff} = \frac{z_{CD}H_{fs} - U_{coop}}{2\ln 10 + \ln(L/a)} \cong 10^{-1}\left(z_{CD}H_{fs} - U_{coop}\right). \tag{5.56}$$

Evidently, this solution makes sense if $z_{CD}H_{fs} > U_{coop}$, i.e. if the activation enthalpy of the boundary diffusion in a glass is less than that of the bulk diffusion in a liquid. The experimental data on the diffusion in metallic glasses [61] show that the mean frequency of the cooperative rearrangements in a glass at T_g has a small pre-exponential factor, in (5.41) it is $\sim 10^{-2}(a/L)$, but the activation enthalpy is considerably less than that in a liquid. This feature is in accord with the adduced above expressions for D_s and D_α and estimations of $z_{CD}H_{fs}$ and U_{coop}.

Equation (5.56) can be taken as an alternative definition of the kinetic glass transition temperature, $T^*_{diff} \approx T_g$.

Recovering of the ergodicity above T_g is connected with the rearrangements of the solid-like clusters. At that ξ_{fl} and $\xi_{SRO} \sim r_0$ become the only scales of the structural heterogeneity of the liquid. To get an interpolative expression for the viscosity in the crossover range, L in (5.38) has to be considered as a quantity decreasing with temperature. With $L \to \xi_{fl}$ and $h_s \to r_0$ the geometric factor $(L/h_s)^3$ becomes not very large as compared with (L/h_s), and the ratio η_{sl}/η_d can not be considered as a small quantity.

5.5 Discussion

The changes of the viscous flow mode shown in Fig. 5.1 are caused by the liquid structure evolution of the relaxation kinetics. There are thermodynamically and mechanically driven structure rearrangements. The equilibrated fluid state above T_A (region A) is characterized by just one structural characteristic—SRO-correlation length, $\xi(A) = \xi_{SRO} \sim a$. In this region the viscous flow is isostructural because the structure relaxation on the scale $\xi(A)$ is much faster than the shear strain. Therefore, the viscosity is independent of the strain rate. The diffusion, characterized by the dominating type of the local rearrangements of SRO, obeys the Arrhenius law. At that the Newtonian law and the Stokes-Einstein relation are valid.

In the region B the thermodynamically driven equilibration of the local order and mesoscopic structure is still independent of the shear strain and the flow is isostructural. Along with that, the heterophase state is specified by the order parameter $\{c\}$ (5.16) and by the mesoscopic correlation length, $\xi(B) = \xi_{fl} \sim 8a$, characterizing the fluctuonic SRO. The state of the heterophase liquid is determined by the interaction coefficients of the fluctuations (Fig. 5.5). Its flow is Newtonian but the relaxation time depends on $c_s(T)$ (5.34) and for this reason it is non-Arrhenian.

The mechanically driven cooperative structural rearrangements dominate in a glass (region D) while in the heterophase liquid they become considerable just in the vicinity of the glass transition temperature in the region C at a large strain rate initiating the mixed flow. The interplay of the thermodynamically and mechanically driven structural rearrangements leads to the shear banding of the liquid and to the appearance of the new correlation length determining the shear band spacing, $\xi(C) = L^* \sim h_{sb}l_{cl}/a_{cl}$. On the scales $l \gg L^*$ the liquid flow is isostructural. The macroscopic viscosity of the shear banded liquid (5.48), which is much less than that within the block, essentially depends on the strain rate.

The decoupling of the viscosity and diffusion (the violation of the Stokes-Einstein relation), the appearance of the strain rate sensitivity of viscosity, the non-Arrhenian relaxations and non-Newtonian flow of glass-forming liquids are the principal features of the viscous flow of the glass-forming liquids and glasses. They are caused by three kinds of cooperative structural rearrangements of different spatial scales:

(i) Cooperative α-relaxation in the heterophase liquid controlled by the rearrangements of CDs of size ξ_{fl};
(ii) Cooperative disc-like shear fluctuations of size 4–5a within the inter-cluster boundaries;
(iii) Initiation and coherent cooperative propagation of slip layers which induces the formation and propagation of the shear bands. The shear band spacing, $L^*_{sb} \sim h_{sb}l_{cl}/a$, is much larger than the regular cluster size l_{cl}.

The cooperativity (i) is responsible for the non-Arrhenian character of the viscosity and for the increase of fragility due to the changes of the heterophase structure at the temperature evolution. In particular, the fragile-to-strong transformation is induced by the phase transformation of the solid-like fraction. The depression of this kind of cooperativity (as it is in the covalently bonded liquids) decreases the fragility.

The theory of the fragile-to-strong transformation of glass-forming liquids (Sect. 5.4.2) is in accord with the experimental data obtained in [66–70] at investigation of the glass transition in multicomponent metallic glasses.

The equation of state in the two-state approximation (5.21) provides the abbreviated description of the heterophase liquid. It can be used to get the approximate solution $c_s(T)$ above T_g when the effects such as fragile-to-strong transformation are not important. On the other hand, it is useful in piecewise approximations at derivation of asymptotic solutions like those shown in Fig. 5.12.

Let us notice that the piecewise approximations in [66–68] are based on the AG relation while in [70, 71] another heuristic model is used.

It has to be noted that (5.21) belongs to the class of equations which is used in different two-state models of the continuous and discontinuous phase transformations including the gas-to-fluid and fluid-to-glass transition (e.g. in [71–85]). The equations of HPFM, describing the mesoscopic fluid-solid heterophase states with multiple SRO types, enable to consider the heterophase systems beyond the range of the two-state approximation.

The cooperative fluctuations within the boundary layers (cooperativity (ii)) in the absence of the external shear field can be interpreted as the β-relaxation weakly impacting the diffusion until the dislocation melting of the boundaries occurs (above the temperature T_{c1}). Then these fluctuations determine the boundary diffusion and the sliding velocity under stress. The β-relaxation described in [62] and its interplay with the α-relaxation and diffusion in metallic glasses reveals itself below T_g within the temperature range $T_g - T_{c1} \cong 0.3 - 0.2T_g$. The diffusion coefficient of a metallic glass is continuously changing with temperature near T_g but a kink of the activation enthalpy and a pre-exponential factor [61] indicates the crossover of the diffusion mode near T_g (Sect. 5.4.5.3).

The formation and propagation of slip layers (cooperativity (iii)) and consecutive formation of the shear bands in the heterophase liquid and in a glass displays itself as the softening of matter and settling of the non-Newtonian flow with stress overshoot on the stress-strain curve. It is worth noting that the strain rate sensitivity is directly connected with slipping (5.43). Hence the decrease of m_{srs} is a manifestation of the increasing contribution of the slip friction factor in the viscosity. In particular, the strain rate sensitivity parameter of the inhomogeneously deforming metallic glasses (region D) is usually much less than 1 [56, 63–65, 86].

The interconnection of the shear banding of a liquid and the non-Newtonian flow is substantiated by direct observations of the shear band formation at the shear flow of non- Newtonian liquids [87].

To check the relevance of the obtained theoretical results to the flow properties in the region C, one can compare them with the experimental data (see overviews [7, 57] and the references quoted). In [63, 64] the comprehensive researches of the plasticity and viscous flow of metallic glasses near T_g (region C) are performed. The comparisons confirm the relevance of the theoretical predictions (Sect. 5.4.5). In particular, taking the strain rate sensitivity and the stress overshoot as attributes of the mixed flow, one can empirically determine the boundaries of the region C using the data of [63]. The results are in harmony with the estimations (5.46) and (5.53).

The validity of the stress overshoot formula (5.51) is confirmed by the comparison with data represented in Fig. 5.12 of [63]. The experimentally observed sigmoidal shape in $\log \sigma_{flow}$ versus $\log \dot{\varepsilon}$ is explained by the relation (5.45) taking into account that the contribution of the sliding friction into the viscosity increases at shear banding and the m^{sl}_{srs} decreases as it is shown in Fig. 5.17b. The relation

$$\frac{\partial \eta_{sb}}{\partial \ln \dot{\varepsilon}} = m_{srs} - 1, \tag{5.57}$$

which follows from the definition of σ_{flow} (5.51), is in accord with the data represented in Figs. 5.8 and 5.11 in [63]. These data also allow estimating the contribution of the sliding friction into viscosity using the following formula,

$$\frac{\eta_{sl}}{\eta_{sb}} = \frac{1 - m_{srs}}{1 - m_{srs}^{sl}}. \tag{5.58}$$

This relation follows from (5.44) and (5.57). In particular, as it is shown in Fig. 5.17, near T_g at $\dot{\varepsilon}_{sl} \sim 10^{-1} - 10^{0} s^{-1}$ the m_{srs}^{sl} is considerably less than 1 and

$$\eta_{sl} \approx \eta_{sb}(1 - m_{srs}). \tag{5.59}$$

Taking into account that $m_{srs} \approx 0.5$ at these values of the $\dot{\varepsilon}_{sl}$, one can conclude that the diffusion viscosity and the sliding resistance of the shear-banded liquid are comparable.

5.6 Conclusions

1. The validity of classic laws of the Newtonian liquid flow with Arrhenian structure relaxation and the Stokes-Einstein relation fails due to the appearance of the HPF at $T < T_A$.
2. The transformation of the mesoscopic structure of a heterophase liquid from fluid to glassy state takes place within the $[T_A, T_g]$ temperature range (region B). At that the liquid flow is Newtonian but non-Arrhenius due to the changes of the phase structure. The fragility is scaled by the cooperativity parameter, $z_{CD} \sim (\xi_{fl}/a)^3$, and by the enthalpy of the solid-fluid transformation.
3. The continuous and discontinuous phase transformation in the solid-like fraction (seen as a week phase transformation of a liquid) induces the continuous sigmoidal or stepwise change of the viscosity and the fragile-to-strong transformation of the glass-forming liquid.
4. The domination of the solid-like fraction near T_g induces the increase of the contribution of the solid-solid sliding resistance in the viscosity at a growing strain rate.
5. The crossover from the Newtonian DVF to the non-Newtonian mixed flow near T_g (region C) is accompanied by the shear banding of the solid-like fraction. As a result, the stress overshoot on the stress-strain curve appears.
6. The sliding resistance, occurring in the slip layers, is a strain rate sensitive quantity. Therefore, the strain rate sensitivity parameter becomes less than 1 at the mixed flow.

7. The fact that the derivative of viscosity with respect to $\ln \dot{\varepsilon}$ is equal to the deviation of the strain sensitivity parameter from 1, allows determining the contribution of the sliding friction in viscosity.

Acknowledgments I express my thanks to Dr. N.P. Lazarev for valuable comments.

References

1. H. Sillescu, J. Non-Cryst. Solids **243**, 81 (1999)
2. R. Richert, J. Phys. Chem. B **101**, 6233 (1997)
3. R. Richert, J. Non-Cryst. Solids **235–237**, 41 (1998)
4. A.S. Bakai, Polycluster amorphous solids (Energoatomizdat, Moscow, 1987). (In Russian)
5. E.W. Fischer, Phys. A **201**, 183 (1993)
6. J. Lu, G. Ravichandran, W.L. Johnson, Acta Mat. **51**, 3429 (2003)
7. C.A. Schuh, T.C. Hufnagel, U. Ramamurty, Acta Mater. **55**, 4067 (2007)
8. A.S. Bakai, in *Glassy Metals III, Topics in Applied Physics*, vol. 72. Ed. by H. Beck, H.-J. Guentherodt (Springer, Heidelberg, 1994), p. 209
9. A.S. Bakai, Low Temp. Phys. **24**, 3 (1998)
10. E.W. Fischer, A.S. Bakai, slow dynamics in complex systems, in *AIP Conference Proceedings*, vol. 469. ed. by M. Tokuyama, I. Oppenheim (1999), p. 325
11. A.S. Bakai, E.W. Fischer, J. Chem. Phys. **120**, 5235–5252 (2004)
12. A.S. Bakai, J. Chem. Phys. **125**, 064503 (2006)
13. A.S. Bakai, Cond. Mat. Phys. **17**(43701), 1–24 (2014)
14. G.G. Stokes, Trans. Cambridge Philos. Soc. **8** (1845)
15. A. Einstein, Inaugural Dissertation (Zuerich Universitaet, 1905)
16. A. Einstein, Ann. Phys. **19**, 289 (1906)
17. M.S. Smoluchowski, Proc. Math. 5-th Int. Congress Math. **2**, 192, (1912)
18. R. Bohmer, C.A. Angell, Phys. Rev. B **45**, 10091 (1992)
19. C.A. Angell, Science **267**, 1924 (1995)
20. A.P. Sokolov, Endeavour **21**, 109 (1997)
21. G. Adam, J.H. Gibbs, J. Chem. Phys. **43**, 139 (1965)
22. W. Kauzmann, Chem. Rev. **43**, 219 (1948)
23. H. Vogel, Phys. Z. **22**, 645 (1921)
24. G.S. Fulcher, J. Am. Ceram. Soc. **8**, 329 (1925)
25. G. Tamman, W.Z. Hesse, Anorg. Allgem. Chem. **156**, 245 (1926)
26. A.L. Greer, K.F. Kelton, S. Sastry (eds.), *Fragility of Glass-Forming Liquids, TRiPS series* (Hindustan Book Agency, 2014)
27. S.F. Stickel, E.W. Fischer, R. Richert, J. Chem. Phys. **102**, 6251 (1995)
28. F. Stickel, E.W. Fischer, R. Richert, J. Chem. Phys. **104**, 2043 (1996)
29. F. Stickel, in *PhD thesis D77 Mainz University* (Verlag Shaker Aachen, 1995)
30. J.C. Maxwell, Phi l. Trans. Royal Soc. **157**, 49 (1867)
31. L.D. Landau, E.M. Lifshits, *Statistical Physics* (Elsevier, 1980)
32. R. Ubbelohde, *Melting and Crystal Structures* (Clarendon Press, 1965)
33. A.S. Bakai, Materialovedenie **6**, 2 (2009)
34. A.S. Bakai, V.V. Kul'ko, I.M. Mikhailovskij, V.B. Rabukhin, O.A.J. Velikodnaya, Non-Crystal. Solids **182**, 315 (1995)
35. A.S. Bakai, I.M. Mikhailovskij, T.I. Mazilova, N. Wanderka, Low Temp. Phys. **28** (2002)
36. F. Vurpillot et al., Surf. Interface Anal. **39**, 273 (2007)
37. A.S. Bakai, E.V. Sadanov, V.A. Ksenofontov, S.A. Bakai, J.A. Gordienko, I.M. Mikhailovskij, Metals **2**, 441 (2012)

38. A.S. Bakai, Polycluster amorphous structures and their properties, in *Prep. KIPT 84–33, TsNIIatominform* (1984) (in Russian)
39. F.H. Stillinger, J. Chem. Phys. **89**, 6461 (1988)
40. M. Oguni, J. Non-Cryst. Solids **210**, 171 (1997)
41. I. Chang, H. Sillescu, J. Phys. Chem. B **101**, 8794 (1997)
42. P.G. de Gennes, Compt. Rend. Phys. **3**, 1263 (2002)
43. R. Hall, P.G. Wolynes, J. Chem. Phys. **86**, 2943 (1987)
44. G. Wolynes, Phys. Rev. A **40**, 1045 (1989)
45. E.W. Fischer, A. Bakai, A. Patkowski, W. Steffen, L. Reinhardt, J. Non-Cryst. Solids **307–310**, 584 (2002)
46. E. Roessler, Phys. Rev. Lett. **65**, 1595 (1990)
47. J.A. Hodgdon, F.H. Stillinger, Phys. Rev. E **48**, 207 (1994)
48. G. Tarjus, D. Kivelson, J. Chem. Phys. **103**, 3071 (1995)
49. E.W. Fischer, *Private Communication* (1999)
50. A.P. Sokolov, A. Kisliuk, D. Quitmann, A. Kudlik, E. Roessler, J. Non-Cryst. Solids **172–174**, 138 (1994)
51. F. Nabarro, Report of the conference on the strength of solids. Phys. Soc. Lond. (1948)
52. C. Herring, J. Appl. Phys. **21**, 437 (1950)
53. R. Coble, J. Appl. Phys. **34**, 1679 (1963)
54. I.M. Lifshitz, Sov. Phys.-JETP **17**, 909 (1963)
55. N.P. Lazarev, A.S. Bakai, J. Mech. Behav. Mater. **22**(3–4), 119 (2013)
56. S.X. Song, T.G. Nieh, Intermetalics **17**, 762 (2009)
57. A.L. Greer, Y.Q. Cheng, E. Ma, Mat. Sci. Eng. **74**, 71 (2013)
58. A.V. Granato, U. Huesen, H. Kopf, Appl. Phys. Lett. **97**, 171911-3 (2010)
59. J. Frenkel, Zeitschrift für Physik **37**, 572 (1926)
60. I.M. Lifshits, V.B. Shikin, Sov. Phys. Solid State **6**, 2211 (1965)
61. F. Faupel, W. Frank, M.-P. Macht, V. Naundorf, K. Raetzke, H.R. Schober, S.K. Sharma, Rev. Mod. Phys. **75**, 237 (2003)
62. H.B. Yu, W.H. Wang, K. Samwer, Mater. Today **16**, 183 (2013)
63. Y. Kawamura, T. Nakamura, A. Inoue, T. Masumoto, Mater. Trans. JIM **40**, 794 (1999)
64. M. Heilmaier, J. Eckert, Adv. Eng. Mater. **7**, 833 (2005)
65. F.H. Dalla Torre, D. Klaumuenzer, R. Maaß, J.F. Loeffler, Acta Mater. **58**, 3742 (2010)
66. Chris Way, P. Wadhwa, R. Busch, Acta Mater. **55**, 1977 (2007)
67. Z. Evenson, T. Schmitt, M. Nicola, I. Gallino, R. Busch, Acta Mater. **60**, 4712 (2012)
68. S. Wei, F. Yang, J. Bednarcik, I. Kaban, A. Meyer, R. Busch, Nat. Commun. **4**, 2083 (2013)
69. C.Z. Zhang, L.N. Hu, Y.Z. Yue, J.C. Mauro, J. Chem. Phys. **113**, 014508 (2010)
70. L. Hu, C. Zhou, C. Zhang, Y. Yue, J. Chem. Phys. **138**, 174598 (2013)
71. J.D. Van der Waals, *in Ph. D. Thesis, University of Leiden* (1873)
72. V.I. Yukalov, Phys. Rep. **208**, 395 (1991)
73. P.B. Macedo, W. Capps, T.A. Litovitz, J. Chem. Phys. **44**, 3357 (1966)
74. E. Rapoport, J. Chem. Phys. **46**, 2891 (1967)
75. C.A. Angell, K.I. Rao, J. Chem. Phys. **57**, 470 (1972)
76. M.H. Cohen, G.S. Grest, Phus. Rev. **B20**, 1077 (1979)
77. M.H. Cohen, G.S. Grest, ibid **26**, 6313 (1982)
78. V.I. Yukalov, Phys. Rev. B **32**, 436 (1985)
79. E.G. Ponyatovsky, O.I. Barkalov, Mat. Sci. Rep. **8**, 147 (1992)
80. J.T. Bendler, M.F. Shlessinger, J. Stat. Phys. **53**, 531 (1988)
81. J.T. Bendler, M.F. Shlessinger, J. Chem. Phys. **96**, 3970 (1992)
82. D. Kivelson, S.A. Kivelson, X. Zhao, Z. Nussinov, G. Tarjus, Phys. A **219**, 27 (1995)
83. H. Tanaka, J. Chem. Phys **111**, 3163–3175 (1999)
84. H. Tanaka, Phys. Rev. E **62**, 6968 (2000)
85. V.I. Yukalov, Int. J. Mod. Phys. B **17**, 2333 (2003)
86. J. Gilman, J. Appl. Phys. **46**, 1625 (1975)
87. P.D. Olmsted, Rheol. Acta **47**, 283 (2008)

Chapter 6
The Generalized Similarity Laws and Isocontours in the Thermodynamics of Simple Liquids

Evgeny Apfelbaum and Vladimir Vorob'ev

Abstract Several new similarity relations regarding the universal lines on the density-temperature plane are described. The first of them—the line of the unit compressibility factor or the Zeno-line—gives rise to the general equation for the liquid binodal branch and the universal correlation between the critical and the Zeno-line parameters. The latter relations have allowed us to estimate the critical points of metals, which can not be measured up to now. Besides, there is projective transformation between the linear elements of the lattice gas phase diagram and that of the continuous systems. The relation for the saturation pressure has been obtained on the basis of this correspondence. The other regularities concern to the lines of the ideal enthalpy, the enthalpy minima and the isothermal compressibility maxima. Although initially they were obtained for the van der Waals equation, they have appeared to be valid for the real substances and models described by completely different equations of state.

Abbreviations

vdW	Equation or system—van der Waals equation or system
LJ	Lennard–Jones system
CP	The Critical Point
TP	The Triple Point
Z-line	Zeno Line
LG	Lattice Gas
H-line	Line of ideal enthalpy (*H*)

E. Apfelbaum (✉) · V. Vorob'ev
Joint Institute for High Temperatures of Russian Academy of Sciences,
Izhorskaya 13 bldg. 2, Moscow 125412, Russia
e-mail: apfel_e@mail.ru

V. Vorob'ev
e-mail: vrbv@mail.ru

L. Bulavin and N. Lebovka (eds.), *Physics of Liquid Matter: Modern Problems*,
Springer Proceedings in Physics 171, DOI 10.1007/978-3-319-20875-6_6

6.1 Introduction

The similarity laws are the scalable volumetric rules which observed for a wide class of substances and models. The development of the similarity laws has started almost simultaneously with the appearance of the equations of states—namely, with the famous van der Waals equation (vdW). The principle of corresponding states is the most known example of this kind of regularities. It was initially derived from vdW equation [1, 2]. Then the law of rectilinear diameter was formulated by Cailletet, Mathias on the experimental basis for the real substances [3]. Later both these laws have been generalized for many other models and real materials, although there are exclusions, like metals [4]. After that many other similarity relations were obtained like, for example, Pitzer factor [5]. To find the theoretical grounds for various similarity laws the powerful techniques of statistical physics (see, for instance the monograph by I.Z. Fisher [6]) can be used [7]. In general the similarity conception is appeared to be very fruitful. In particular the scaling concepts, developed by many scientists (see the review by M. Fisher [8] for details), are close in ideology to similarity laws. But in present study we would like to pay attention to some "geometrical" peculiarities of the similarity laws. Let us note that the law of rectilinear diameter points on some general line at the density-temperature plane. So a natural question has arisen—do the other analogous universal lines exist?

The positive answer was given more than 100 years ago in the research by Batchinskii [9]. Namely, if one considers the curve of unit compressibility factor for the vdW system, then this curve is appeared to be the straight line over the all density-temperature plane. The compressibility factor Z is defined as the ratio of the pressure to that of an ideal gas. That is $Z = P/(nT)$, where P is the pressure, n is the particle density, T is the temperature (measured in energy units). This is one of the most important dimensionless parameter of thermodynamics. When $Z = 1$ then the pressure coincides with the pressure of an ideal gas. In [9] some measurements had been done as well, to confirm the linear dependence of $Z = 1$ contour for the real substances. Nevertheless, during next 50 years this fact was considered as some curiosity. But then it was experimentally confirmed for many real substances. Among them there are the noble gases, the hydrocarbons [10–13], the water (partially), and other nonmetallic materials [14]. So the idea that this property is of general nature was supposed in study by Holleran [12]. In our studies we have also confirmed that the phase diagrams of two metals (mercury and cesium) possess this property at least for liquid domain [15–19]. As a result, the linear contour $Z = 1$ has appeared to have a wider area of applicability than the corresponding states principle or the law of rectilinear diameter. And now the line $Z = 1$ is referred to as the Zeno-line [14] (below also the Z-line) or the Batchinskii line. It extends from the Boyle point of the dilute supercritical gas (where $n \rightarrow 0$ and the second virial coefficient vanishes) down to the triple point. So, if the Z-line remains straight along all the range, it means that the pressure of corresponding substance is described by the ideal gas equation even in the domain of dense liquid (of course, along this line only). This property can be very important when the equations of state are considered [12].

The subsequent investigations of the Z-line have resulted in some additional properties of this contour. It was shown in study by Filippov [20] and later in our researches [16–19] that the Z-line is the tangent to the liquid branch of the liquid-gas binodal when $T \to 0$. More exactly, it is the tangent to the extension of binodal beyond the triple point. This property is kept even in the case when Z-line is not the straight over all the plane density-temperature! So it is possible to derive the equation for the liquid branch of the binodal, which will be discussed below. Besides, to check the considered findings and to establish the limits of their applicability one can use the model systems with known interaction potentials. Corresponding analysis has been done in [21]. Its results will be also discussed below. But here we should note that there are at least two cases when Z-line is not the straight. This occurs when the attractive part of the interaction potential is too short-ranged or too long-ranged.

Another direction of investigations was found by Kulinskii [22, 23]. He noted that in the famous Ising model for the lattice gas (LG) the Zeno-line and the binodal diameter are the straight lines exactly. Farther, using this fact, he established the isomorphism between the linear elements of the phase diagrams of the lattice gas and the continuous systems (including the real substances too). This approach has some limitations, because there is no unambiguous correspondence between any arbitrary point or line of real and lattice systems [24]. But it allows one to find the interrelation between different regularities, like, for instance, the Zeno-line and the Pitzer acentric factor [25]. It also provides the natural explanation for the linearity of the Zeno-line and the binodal diameter, their correlation, and the asymptotic congruence with the liquid binodal branch [22, 23]. Moreover, as we will see below, we can apply the above isomorphism to investigate various thermodynamic properties.

Besides the Zeno line, the vdW equation generates many other "ideal curves" in the form of the straight lines [26]. "Ideal curves" are the contours, where some property of the system coincides with that of an ideal gas. So, in the vdW system there are the straight lines of the ideal enthalpy, the ideal internal energy and others [27]. In real substances the linear dependence for these "ideal curves" are less frequent than the Zeno line. Nevertheless all they are used in the analysis and construction of the model and reference equations of state [12, 25–27].

Along any of "ideal curves" corresponding thermodynamic value is described by the same dependence as in an ideal gas (even in the high-density region). Then, as it is stated in [27], "plots of certain "ideal curves" are useful in assessing the behavior of an equation of state in regions away from the available data as well as in revealing inconsistencies in the available data sets". Besides they can be used to check consistence of various theoretical models for fluids. For example, in recent study [25] it was shown that the Peng-Robinson equation of state has significant curvature for its own $Z = 1$ contour. This model is widely used in chemical engineering. But the substances (like argon, carbon dioxide etc.) where it is used have accurate linear Z-line. This evident discrepancy points to an inexactness of the Peng-Robinson model and necessity to correct it [25]. Thus, the investigated lines are very important not only in the theory but in practical applications too.

One more important remark should be noted. In supercritical region there are many other remarkable curves that initially were obtained for the vdW system [28–30]. Some of them start at the critical point (CP), so they are connected with the divergence of various physical values at CP. The volume expansion coefficient, the isothermal compressibility and others are the examples of these values. Another one, as the Joule–Thompson inversion curve and others, do not cross the critical point and do not diverge. Nevertheless, there is the link between the "ideal curves" discussed above and some of the supercritical curve. The interrelation will be also considered below.

The subsequent text will be organized as the following. At first, we will consider the properties of the Zeno line and show how they correlate with the binodal position and critical point coordinates. The limits of applicability of these properties will be discussed as well. The CP estimates for a number of metals will be presented either. Then we will show how the Kulinskii isomorphism can be applied to the real substances. Finally we will consider the other "ideal curves" and their property. The conclusions will complete present communication.

6.2 Zeno Line and Its Properties

The linear dependence in the Zeno line contour was shown for the first time for the vdW equation [9]. Let us rewrite the latter as

$$P = \frac{nT}{1 - bn} - an^2 = nT + n^2 \frac{bT + abn - a}{1 - bn}. \tag{6.1}$$

Here *a*, *b* are the usual material constants of the vdW equation [1, 2]. Evidently, $Z = P/(nT) = 1$, when

$$bT/a + bn = 1. \tag{6.2}$$

One can see that (6.2) describes the linear contour at *T*–*n* plane, which cut on the axes *T* and *n* the segments [0, *b/a*] and [0, 1/*b*] respectively. Of course, the vdW equation does not describe the real substances with acceptable accuracy. But for the real substances at low densities the virial expansion can be applied. So when $n \to 0$ it is possible to show exactly [16, 20] that the Zeno line is also the straight. It is described by the following equation:

$$T/T_B + n/n_B = 1. \tag{6.3}$$

The parameters T_B, n_B are defined by means of the second (B_2) and the third (*C*) virial coefficients. Namely, T_B is appeared to be the well-known Boyle temperature. That is why the subscript "B" is used for the *Z*-line parameters. We will also call the parameter n_B as the Boyle density. Their definitions are [21]:

$$T_B : B_2(T_B) = 0, \quad n_B = \left(\frac{dB_2}{dT}\right)_{T_B} \frac{T_B}{C(T_B)}, \tag{6.4}$$

So, for given interparticle potential the Boyle parameters can be calculated directly. It allows us to find the Z-line for model systems and to check whether it is the straight line at all *T*–*n* plane. But, at first, we would like to draw your attention to additional regularities related to the Z-line. The analysis has shown that the liquid branch of the binodal (or its extension beyond the triple point) and the Z-line should coincide at the limit $T \rightarrow 0$. Moreover, this "tangential" behavior is kept independently of the Z-line form [16, 20]. This fact has allowed us to connect the binodal liquid branch and the Z-line [19]. We have used the famous 3-term equation for the binodal *n*(*T*):

$$n(T) = n_c + A_1\tau + A_2\tau^{\beta}, \ \tau = 1 - T/T_c. \tag{6.5}$$

In (6.5) the subscript "c" relates to the critical point, while the index β defines the class of criticality of the system [2]. This equation was introduced by van der Waals himself with index $\beta = 0.5$. Later it has been expanded for real substances with various values of β [31, 32]. The constants A_1, A_2 can be defined by different ways. Generally the liquid and the gas binodal branches should have each own values, i.e. $(A_1^{liq}, A_2^{liq}), (A_1^{Gas}, A_2^{Gas})$ But if the binodal is symmetrical relative to its diameter, then $A_1^{liq} = A_1^{Gas}, A_2^{liq} = -A_2^{Gas}$. And, moreover, when $n^{Gas}(T \rightarrow 0) \rightarrow 0$, then $A_2^{Gas} + A_1^{Gas} = -n_c$ [20]. Besides, one can use simple fitting to experimental data. As an example, in [33] the binodals for 108 substances were fitted this way. However, the fitting can be used only when the coexistence curve is known. So we decided to use the above asymptotical properties of the Zeno line and binodal. Namely, we have required for the liquid binodal branch that at $T \rightarrow 0$ (6.5) should be transformed into (6.3) for the Z-line [17–19, 21]. It unambiguously defines the coefficients as:

$$A_1^{Liq} = \frac{(T_c/T_B - \beta)n_B + \beta n_c}{1 - \beta}, \ A_2^{Liq} = \frac{(1 - T_c/T_B)n_B - n_c}{1 - \beta}. \tag{6.6}$$

One can see that the coefficients (6.6) and the binodal (6.5) depend upon both the critical and the Boyle parameters. This fact points on possible link between these two sets of parameters. Corresponding relation was considered initially in [16–19, 21]. The experimental data analysis for real materials has shown that it is possibly also linear, namely:

$$x\frac{T_c}{T_B} + y\frac{n_c}{n_B} = 1. \tag{6.7}$$

In (6.7) "*x*" and "*y*" are the numerical factors. At first in [16] we have set $x = 1$, $y = 0.5$, so the corresponding line was the median in the triangle composed by the

n and T axes and the Zeno line. But more detailed analysis, applied to the model systems, has shown [19] that it should be sooner the line parallel to the Z-line:

$$\frac{T_c}{T_B} + \frac{n_c}{n_B} = S_1. \tag{6.8}$$

The application of (6.8) to the Lennard-Jones (LJ) system as well as to some real substances [17, 19] have resulted in $S_1 \approx 0.67$. But for the vdW system $S_1 = 17/27 \approx 0.63$. Note, that LJ and real substances belong to the so-called "Ising" class of criticality, while vdW belongs to the "mean-field" models [2]. Then we have investigated various models with given potentials [21], belonging to different criticality classes. In result the values of S_1 have varied in dependence of the criticality class. So we supposed that that parameter S_1 should depend on it. Farther, for symmetrical binodal $A_1^{Liq} - A_2^{Liq} = n_c$. The substitution (6.6) into the latter equation gives

$$\frac{T_c}{T_B} + \frac{n_c}{n_B} = \frac{1+\beta}{2}. \tag{6.9}$$

For Ising class $\beta \approx 0.326$. It gives $S_1 \approx 0.663$, which is very close to 0.67. But for vdW ($\beta = 0.5$) (6.9) gives 0.75, which is far from 0.63. That means that dependence of S_1 on kind of substance is more complicated than given by (6.9). But we have decided to kept the general dependence $S_1 = S_1(\beta)$. Schematically the phase diagram density-temperature of a simple liquid is presented in Fig. 6.1 in the units reduced to the Boyle parameters.

To find out whether Fig. 6.1 corresponds to the real situation we have investigated the phase diagram for many substances [16–19] and models with given potentials [21]. Corresponding pictures have been published earlier in our works

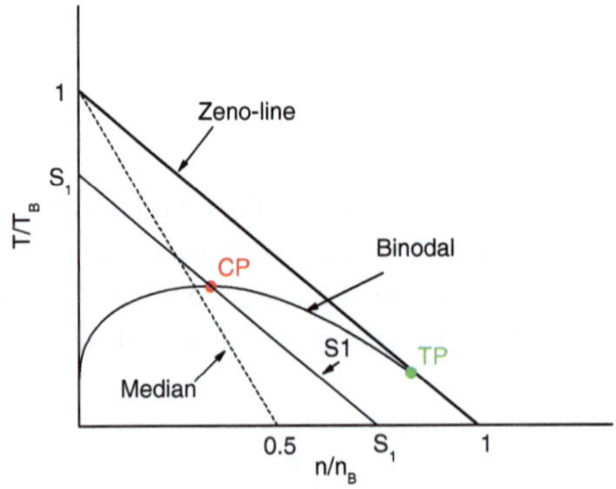

Fig. 6.1 The phase diagram density-temperature (n-T) of a simple liquid in the units reduced to the Boyle parameters T_B and n_B defined by (6.4). CP is the critical point, TP is the triple point. The Zeno-Line is defined by (6.3). The critical points line, marked S_1, and the value S_1 are defined by (6.8). "Median" is defined by (6.7)

[16–19, 21]. It was found that the phase diagrams of the investigated substances are qualitatively corresponds to that one presented in Fig. 6.1, although there are exclusions (see below). To find the quantitative agreement we have also compared (6.3)–(6.8) with the available data of measurements and calculations. For real gases and liquids we have used the NIST database [34]. For models there is variety of the numerical simulations data (see [21] and references therein). Unfortunately, among the metal the binodals were measures only for Hg and Cs, because they have the critical point located at relatively low temperatures [35, 36]. In result, we have come to the following conclusions:

1. The contour defined by the Zeno-line is appeared to be the straight for many real substances. But there are exclusions like, for instance, water and helium. The former does not obey to any similarity relations, while the latter discovers quantum properties. More detailed discussion is presented in [19].
2. The model systems with pair-wise central potentials have discovered two cases when the Zeno-line is not linear. (We do not consider the third—evident—case of purely repulsive potential, when always $Z > 1$). The first one is when the attractive part of the potential is too short-ranged. In this case the liquid-gas transition can be metastable with respect to the solid–liquid transition (see [21] and references therein). The opposite situation—too long-ranged (attraction) potential—is the second case. Here the triple point has too high temperature approaching to the critical point.
3. Equations (6.5) and (6.6) describes the liquid binodal branches of the real substances (with $\beta = 0.326$) within few percents accuracy. Exclusion is Hg. We will discuss it in the section devoted to the critical points of metals. The same accuracy is found for the model systems with different β as well (if the Z-line is the straight).
4. Equation (6.8) was initially tested in [16–19, 21] and good agreement was obtained. In [37] corresponding pictures for the model and real systems were constructed. In present study we have included additional substances and models into consideration. The results are given in Fig. 6.2a, b. The description of the models presented in Fig. 6.2b one can find in [21]. They are defined by corresponding potentials i.e. “Square Well” means the system of particles interacting via the Square Well potential, “Morse”—via the Morse potential etc. (see [21]). The value $S_1 = 0.67$ corresponds to the LJ system (with $\beta = 0.326$), while $S_1 = 0.63$ corresponds to the vdW equation (with $\beta = 0.5$). The other S_1 values are for the intermediate β values. One can see that (6.8) is confirmed for all substances, excluding for the water and hydrogen. For model systems the dispersion is greater, because of possible inaccuracy in the critical point coordinates and because the Z-line is not always straight in case of model systems [21]. This fact can also points on possible limitations for (6.8).

All above conclusions have been obtained under assumption that the considered system has the same chemical composition when the density changes from gaseous to liquid values. That is the interparticle potential does not change in all gas-liquid and fluid domains. Otherwise there is the problem with definition of what the

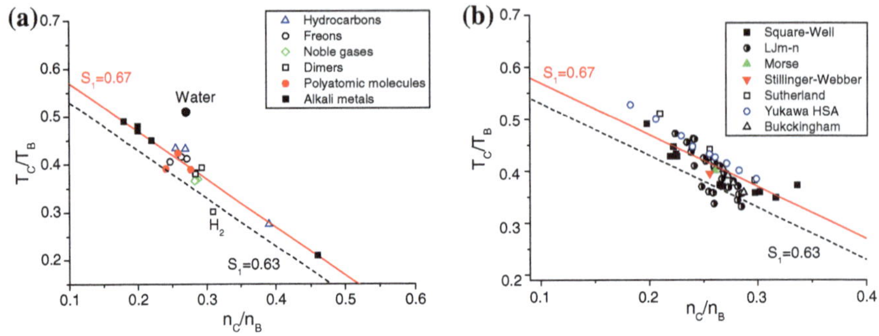

Fig. 6.2 The ratio of temperatures T_C/T_B versus the ratio of densities n_C/n_B. The comparison of prediction of (6.8) with the data for the real substances (**a**) and with the data for the models (**b**) (the models are described in the text). The subscripts "C" and "B" correspond to the critical and the Boyle parameters respectively. The *lines* S_1 = 0.67 and S_1 = 0.63 are described by (6.8) with the values of parameter S_1 corresponding to LJ and vdW models respectively (see text)

density is and how the compressibility factor should be defined. We will return to this problem when the critical points of metals will be discussed.

6.2.1 The Critical Pressure and the Zeno-Line

The Zeno-line equation is an explicit *T*–*n* relation. It does not include the pressure directly. That is why it is interesting to obtain the relation for the critical pressure analogous to (6.8). It would link the latter with the Boyle parameters. Initially such a relation was offered in [17, 19]. It was denoted as S_2 and had the form: $S_2 = (n_c T_c - P_c)/(n_B T_B)$. For LJ $S_2 = 0.076$, while for the vdW system $S_2 = 5/81 \approx 0.062$. For the real substances we have set $S_2 = 0.076$ too. We have also supposed the dependence of the criticality class: $S_2 = S_2(\beta)$ (like in case of S_1). But the subsequent analyses [37] have shown that S_2 relation is less accurate in comparison with S_1. So we have addressed to the Timmermans relation [38]. Although initially it was obtained as purely empirical basis there are some theoretical grounds for it (see [20, 54] for details). It connects the critical point compressibility factor with the Boyle density, namely: $Z_c \equiv P_c/(n_c T_c) = n_c/n_B$. In [39] we have corrected it and have presented in the form

$$\frac{P_c}{n_c T_c} = \left(\frac{n_c}{n_B}\right)^{1+p}. \tag{6.10}$$

Here p is the small fitting factor. It was introduced in [39] only as the empirical parameter on the basis of analysis of experimental data. The least squares fitting for different substances gives us p = 0.019, which also corresponds LJ system. If we take into account S_1 relation (6.8) then

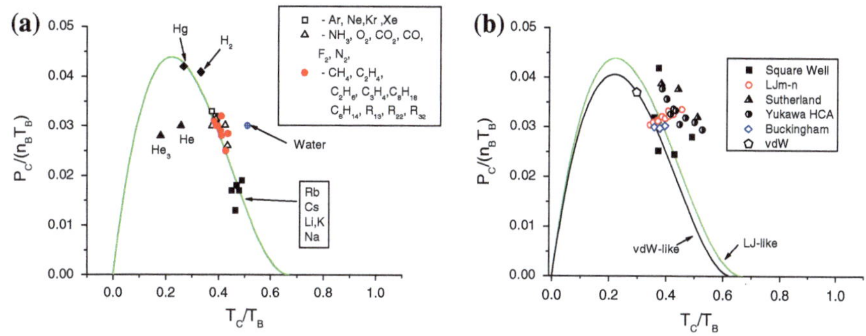

Fig. 6.3 The ratio of the reduced critical pressure versus the reduced critical temperature. The comparison of prediction of (6.11) with the data for the real substances (**a**) and with the data for the models (the models described in the text) (**b**). The *solid lines* correspond to (6.11) with various values of S_1 (see text)

$$\frac{P_c}{n_B T_B} = \left(S_1 - \frac{n_c}{n_B}\right)\left(\frac{n_c}{n_B}\right)^{2+p} = \left(S_1 - \frac{T_c}{T_B}\right)^{2+p}\left(\frac{T_c}{T_B}\right). \tag{6.11}$$

To check whether (6.11) is applicable to real substances in [39] we have constructed corresponding pictures. In present report we have extended one of these pictures by inclusion of additional substances (Fig. 6.3a). (Remind, $S_1 = 0.67$ for real substances like previously). The quantum liquids (He, He^3 and partially H_2) and water fall out from the dependence under study. We should also note that for alkali metals (besides Cs) we have presented only the estimates, as far as there are no final data for their critical points. But other substances demonstrate very good agreement with (6.11).

We have also constructed analogous picture for the model systems (Fig. 6.3b). At first, we have marked, that for the vdW equation the correction factor is negative, namely $p = -0.107$. So, we have considered two cases: first one is the LJ-like systems with $S_1 = 0.67$ and $p = 0.019$, the other one is the vdW-like systems with $S_1 = 0.63$ and $p = -0.107$. One can see in Fig. 6.3b, that the agreement is worse in comparison with the case of real substances. The situation is analogous to the results presented in Fig. 6.2a, b. So, the inaccuracy in the critical points coordinates can be important. But the example of the vdW system has shown that the problem can be also in correction factor p which has been introduced only by means of fitting. Possibly this parameter should be variable too. This question requires further studies.

Besides the relation for the critical point, it is interesting to connect the Boyle parameters with the whole saturation curve $P_{sat}(T)$, analogously to (6.5), (6.6). Corresponding attempt was made in [40]. We have succeeded in substantiation of the Riedel equation [41] in partial case. But the accuracy of the obtained expression was too low. So this problem is still waiting for its solution. Below we will return to it.

6.2.2 The Critical Point Coordinates Estimates for Some Metals

The similarity relations considered in the previous sections have allowed us to make the estimates of the critical points coordinates for a number of metals. These metals are Al, Cu, W, U, Zr, Fe, Be. They are not amiable to the measurements at the temperatures higher than approximately 5000 K. As a result, there are only low-temperature liquid data along the isobars or other curves [42]. Nevertheless, it is well-known fact that the liquids at low temperatures have a very low compressibility. The latter circumstance, in turn, results in the fact that the isobars almost coincide with the liquid binodal branch (see, for instance, Fig. 6.2 in [36]). In this case the liquid binodal can be reconstructed by various approaches [20, 43, 44]. Consequently, the critical point coordinates can be estimated as well. The similarity relations presented above have been also used for this purpose. Corresponding procedure are described in details in our researches [17–19, 45]. So here we present only the results. But, before proceeding with it, we should make the following important remark.

Usually in the measurements the mass density ρ is the input data. For an one-component system the particle density n is defined simply as $n = \rho/m_a$, where m_a is the atom (or molecule) mass of a single particle. Then, the compressibility factor is $Z = P\, m_a/(\rho T)$. Thus, for the most of non-metallic substances (which are one-component in the region under study) the relation between n and ρ is unambiguous. But the metals are generically the two-component systems (electrons and positive ions). At low temperatures and gaseous densities the electrons recombine with ions [46]. Under these conditions we can consider a metal as an atomic one-component system with the above relation between n and ρ. But when someone starts to compress this metallic gas along the isotherm, then the ionization starts too. When the state of liquid or dense fluid is achieved, the metal is really the two component system with $\rho = m_e n_e + m_i n_i$, where the subscript "e" or "i" denotes the electrons or ions. In this case there is no single particle density n, so the definition of Z becomes ambiguous. To overcome this problem it is possible to consider a metal as the one-component system consisting of "effective" atoms with the particle density, defined just likes in the case of gaseous state. These effective atoms interact via some "effective" potential. The problem is that the "effective" potential are usually completely different from the potential between the real atoms in gaseous phase. Moreover, the most exact "effective" potentials in liquid phase are not pair-additive [47]. Thus, the definition of the Zeno-line becomes also unambiguous. In particular, we have shown [15] that for liquid Hg the Zeno-line is the straight if we suppose that Hg consists of above "effective" atoms. But our study was limited from the down by the densities higher than the critical one. The extrapolation to the lower densities gives rise to $T_B = 6350$ K for Hg. Meanwhile, the spectroscopic data analysis at low densities has shown that Z-line has sharp turn when the density increases, which give rises to $T_B = 2874$ K [48]. The close value was also obtained in the analysis of the potential for the mercury dimer [49]. The analogous situation is possible for other metals too. In our studies of metals (besides Cs and Hg) we can

Table 6.1 The estimates of the critical point coordinates for the metals. Likalter [50], Fortov [43], Hess [51], Filippov [20], Beutl [52], Lomonosov [53]

Metal	T_c (K)	ρ_c (g/cm^3)	P_c (atm)	Method
Al	6380	0.45	2174	Present work
	8860	0.28	4680	Likalter
	8000	0.64	4470	Fortov
Cu	7090	1.95	4262	Present work
	7620	1.40	5770	Likalter
	8390	2.40	7460	Fortov
W	12,400	4.92	6978	Present work
	12,500	4.52	11,000	Hess
	21,010	5.87	15,830	Fortov
U	7000	3.30	1972	present work
	9000	2.80	5000	Likalter
	11,600	5.30	6100	Fortov
Zr	15,200	1.00	3432	Present work
Fe	6500	1.36	4620	Present work
	7650	1.63	1534	Filippov
	9250	1.21	8856	Beutl
Be	5400	0.26	1917	Present work
	9200	0.35	12,200	Likalter
	8080	0.55	11,700	Fortov
	8877	0.398	2870	Lomonosov

use the data only for the liquid state. So the Boyle temperature is obtained by extrapolation to the zero density, which can be different from "gaseous" T_B.

The results for the critical points of the metals mentioned above presented in the Table 6.1. For P_c we have used (6.11) like in [30] (not S_2).

Our estimates for the critical pressure are lower in comparison with the data of other investigators. As far as ρ_c is of the same magnitude of value with these data, the difference in pressure can be ascribed to inaccuracy in our definition of ρ_B for metals (see [45] for details). But the critical pressure of other researchers presented in Table 6.1 has big inaccuracy too. The critical pressure is usually calculated by means of a subtraction of two big quantities responsible for thermal and potential part. This effect is especially evident in metals. The subtraction has been used in the estimates of other researchers but not in our approach. So it is difficult to say now which of the estimates is the most reliable.

6.3 The Isomorphic Transformation Between the Lattice Gas and the Real Substances

For the substances and model systems considered previously there is no strict prove that the Z-line is the straight. But there is the model where it is truly straight. It is the famous Ising model for the lattice gas (LG) [2]. Moreover it has also strictly linear binodal diameter. The binodal of the model is also absolutely symmetrical relative to the diameter. These facts are valid both for 2D and 3D cases, but below we will consider only 3D system. On the basis of the symmetry of Ising 3D model, Kulinskii [22, 23] has offered the isomorphic transformation between the linear elements of LG and corresponding contours of a continuous system. The linear elements of LG are the critical isotherm, the Zeno-line and the diameter. The latter should be transformed into the median of the triangle at phase diagram of continuous model (see Fig. 6.1). The form of Kulinskii transformation is:

$$n = n_B \frac{x}{1 + at}, \quad T = T_B \frac{at}{1 + at}, \quad a = \frac{T_c}{T_B - T_c}. \tag{6.12}$$

Here (n, T) are the particle density and the temperature of a continuous system; (x, t) are the analogous values for LG. The LG critical point is $(x_c, t_c) = (0.5, 1)$. The Zeno line for LG is described by equation $x = 1$, while the equation $x = 0.5$ corresponds to the binodal diameter. Consequently, the Boyle temperature is $t_B = \infty$ for LG and the binodal diameter coincides with the median. Unfortunately, (6.12) is not valid for any points at phase diagram. For instance, it does not transform the critical point of the lattice gas to that one of a continuous system. Namely, the substitution of $(x_c, t_c) = (0.5, 1)$ into (6.12) gives rise to the relation: $n = 0.5\, n_B (1 - T_c/T_B)$ and $T = T_c$. So there is the correspondence only between the critical temperatures. To remove this inconsistency we have changed (6.12) as [24]

$$\begin{aligned} n &= n_B \frac{x^\gamma}{1 + at}, \quad T = T_B \frac{at}{1 + at}, \\ a &= \frac{T_c}{T_B - T_c}, \gamma = -\log_2[n_c(1 + a)/n_B]. \end{aligned} \tag{6.13}$$

The exponent γ has been fitted to provide the transformation between the critical points. Note that (6.13) is no more isomorphism in comparison with (6.12). Both transformations are also limited by $T < T_B$.

To apply (6.12) or (6.13) to any other lines (besides the initial linear elements) it is necessary to know the equation of states (EOS) for LG. That is if $\pi = \pi\,(x,t)$ is the equation for the LG pressure π, then it is possible to consider the LG isotherm. Then one can check whether the pressure at this isotherm (in reduced units) would correspond to the pressure of a continuous system at the isotherm, obtained from the above transformations. The same relates to the isochors, isobars, binodals and other lines. Unfortunately, there is no analytical EOS for 3D Ising models. But there are several approximated EOS. In our study [24] we have used the well-known Bragg-Williams and Bethe-Guggenheim EOS. We have checked the isotherms and binodals for the models (vdW, LJ) and substances (Hg, NH_3 and others). As a result we have found that only the critical isotherm can be reproduced within sufficient accuracy. The more the difference $|T - T_c|$ was, the greater the discrepancy in pressure along the isotherms was observed. The same concerns the isobars. To illustrate this situation we present below in the Fig. 6.4 the binodals for SF_6

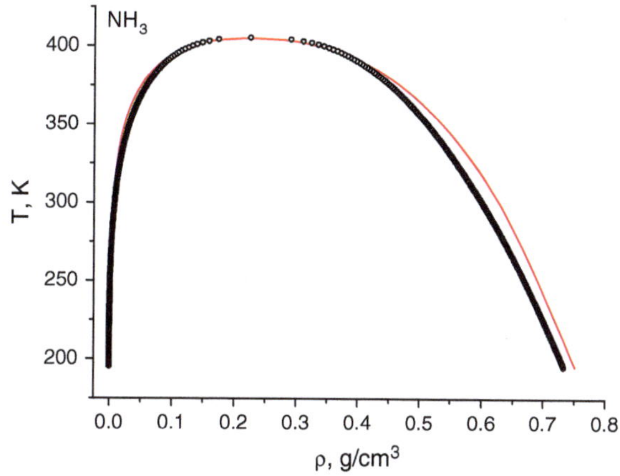

Fig. 6.4 The liquid-gas coexistence curve for SF_6. *Symbols* are the experimental data [34], *solid line* is constructed by (6.13)

constructed on the basis of the experimental data [34] and by means of the transformation (6.13). The details of the latter are presented in [24].

So we can see that the transformation between LG and continuous systems can not reproduce exactly any line at the phase diagram with acceptable accuracy. But it can be applied for the substantiation of the various similarity laws, which previously have only semi-empirical grounds. For instance, the critical points line, like as (6.7), (6.8), can be deduced from the Kulinskii transformation (6.12) [22]. Recently, it was used to estimate the critical points coordinates [54]. It also was used in [55] to investigate the volume expansion coefficient (VEC). This value is denoted as ξ and is defined as

$$\xi = -\frac{1}{n}\left(\frac{\partial n}{\partial T}\right)_P. \tag{6.14}$$

We have discussed above that at low temperatures the isobars almost coincide with the liquid binodal branch. So we can take the derivative along this branch, using 3-term (6.5), (6.6). It gives rise to the expression

$$\xi = \frac{A_1 + A_2\beta\tau^{\beta-1}}{T_c(n_c + A_1\tau + A_2\tau^\beta)}, \quad \tau = 1 - T/T_c. \tag{6.15}$$

The coefficients A_1, A_2 are defined by (6.6), while β is the critical index (as previously). So VEC can be also expressed by means of the Boyle parameters. Let us formally consider the limit $T \to 0$. We have already mentioned above, that the application of all Zeno-line relations in this limit is unphysical, because below the triple point a liquid becomes a crystal. However, it gives the lower boundary for the VEC under conditions $T \to 0$, $P \to 0$. Thus, (6.15) gives

$$\xi_0 = \lim_{T\to 0, P\to 0} \xi(T) = 1/T_B. \tag{6.16}$$

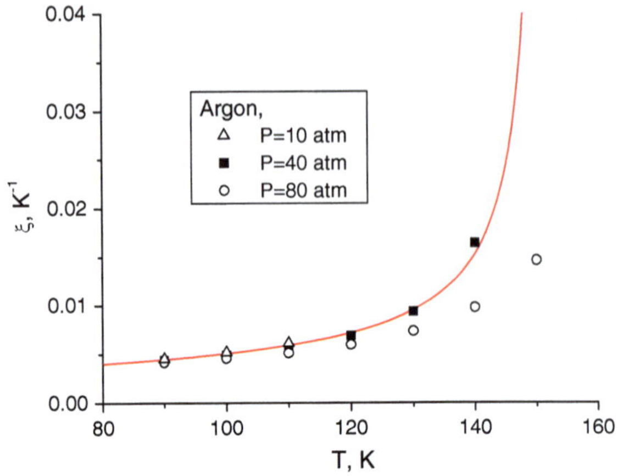

Fig. 6.5 Volume expansion coefficient ξ for Ar at different pressures P. The *solid line* is obtained by (6.15), the *symbols* are the experimental data [56]

The latter relation has no rigorous theoretical grounds. But the isomorphism (6.12) allows us to solve this problem. The limit $T \to 0$ corresponds to $t \to 0$ as well. In this case, the density derivative over the temperature can be expressed by means of (6.12) as

$$\frac{dn}{dT} = \frac{n'_x x'_t + n'_t}{T'_x x'_t + T'_t} \Rightarrow -\frac{1}{n}\frac{dn}{dT} = \frac{1+at}{T_B} + \frac{(1+at)^2}{aT_B}\left(-\frac{1}{x}\frac{dx}{dt}\right)$$
$$\underset{t\to 0}{\Rightarrow} -\frac{1}{n}\frac{dn}{dT} = \frac{1}{T_B} + \frac{1}{aT_B}\left(-\frac{1}{x}\frac{dx}{dt}\right). \qquad (6.17)$$

The term $\left(\frac{dx}{dt}\right) \to 0$ at $T \to 0$ according to the third law of thermodynamics [55]. Below in Fig. 6.5 we present the comparison of experimental data for VEC of Ar at various pressures and the calculated values according to (6.15).

When the isobars really coincide with the binodal, (6.15) is in good agreement with the experimental data at lower temperatures. The inconsistence starts at higher temperatures, as expected.

6.3.1 LG Transformations and the Saturation Pressure

We have already said above (end of the Sect. 6.2.1) that we have not succeeded yet to construct the equation for the saturation pressure $P_{sat}(T)$. But we may check what the considered transformations can offer for this task.

Let us start from the symmetric expression for binodal of LG as suggested in [24]. Here we somewhat modify it by introducing an additional parameter α_1 (in [24] the condition $\alpha_1 = \beta$ has been used). We present the lattice binodal as

$$x_{l,g}(t,\alpha_1,\beta) = \frac{1}{2}\left(1 \pm \left(1 - e^{-\frac{(1-t^{1/\beta})}{\alpha_1 t}}\right)^{\beta}\right). \tag{6.18}$$

Here α_1 и β are determined by the shape of LG binodal. The subscripts "l", "g" as well as the sign "±" relate to the liquid or gas binodal branch.

The expansion of (6.18) near the critical point gives $|x - 0.5| \sim (1 - t)^{\beta}$. Consequently, the parameter β is the corresponding critical exponent for LG. On the other hand, the gas density along the binodal at $t \to 0$ changes according to (6.18) as $x_g(t) = 0.5\ \beta\ \exp(-1/(\alpha_1 t))$. So, it decreases exponentially with the temperature. This asymptotic form corresponds to the general dependence on the temperature for the gas density along the binodal at $t \to 0$ for an arbitrary system. The parameter α_1 determines the rate of this decrease.

Our task is to describe the dependence of the saturation density and pressure on the temperature in a wide range of variation of the latter (not only in the CP vicinity). Therefore, now the parameters β and α_1 will play the role of fitting parameters. Their values will be found from the condition of a mean deviation minimization of the calculated pressure from the experimental (or tabulated) one. The ability to use (6.18) for different lattice models have been established earlier [24]. Comparison of the binodal curves, calculated by means of the corresponding choice of a parameter $\beta = \alpha_1$, has shown good agreement with the analytical results [22] and simulations [57, 58]. We should also note, that the projective transformation offered in [22] can map with some accuracy LG binodal $x_{\rm Bin}(t)$ into the real fluid binodal $n_{\rm Bin}(T)$. But it gives no correspondence for the critical densities. So, it is more correct to make basis on our transformation (6.13). Now we need some relation for the pressure. Let us boldly supposed that the Timmermans relation is valid not only for the compressibility factor at the critical point Z_c but along the all binodal, namely

$$Z_{l,g}(n,T) \equiv \frac{P_{sat}(T)}{n_{l,g}(T)T} = \frac{n_{l,g}(T)}{n_B}. \tag{6.19}$$

Here $n_{l,g}(T)$ denotes the binodal density (at both branches), while $P_{sat}(T)$ is the desired expression of the saturation pressure. The direct application of (6.19) to the real substances and models does not give enough accuracy. Moreover, near the critical point (6.19) gives rise to incorrect asymptotic behavior [40]. But what will happen if we express $n_{l,g}(T)$ through $x_{l,g}(t)$ by means of (6.13), (6.18)? Corresponding consideration was given in [59]. After some transformation (see details in [59]) the following equation for the saturation pressure can be obtained

$$P(T,\alpha_1,\beta) = Tn_B(1 - T/T_B)[x_g(t,\beta,\alpha_1)]^{\gamma}[x_l(T/T_c,\beta,\alpha_1)]^{\frac{\ln Z_c}{\ln(1/2)}}. \tag{6.20}$$

This expression depends on the critical and Z-line parameters as well as on two fitting parameters α_1 and β. The values of these parameters are found from the

Table 6.2 The values of the calculated parameters for (6.20)

Model					Substances		
	L-J	SW			Ar	Hg	Cs
		λ = 1.5	1.75	2			
T_c	1.31	1.217	1.809	2.68	150.86	1751	1938
n_c	0.31	0.308	0.265	0.251	0.536	5.8	0.39
Z_c	0.308	0.248	0.263	0.294	0.29	0.39	0.2
T_B	3.418	2.846	4.842	7.484	393	6650	4120
n_B	1.14	1.375	1.0	0.844	1.97	14.4	1.96
α_1	0.5	0.67	0.57	0.6	0.485	0.49	0.756
β	0.545	0.435	0.48	0.405	0.55	0.435	0.555
ε (%)	3.4	1.2	2.8	3	2.7	2.4	2.8

condition of a mean deviation minimization of the calculated pressure from the experimental (or tabulated) one. The mean deviation is found as

$$\varepsilon(\beta,\alpha),\ \% = \frac{\sum_{i=1}^{n}\left|\frac{P(\beta,\alpha)}{P_{\exp}} - 1\right|}{n}\cdot 100, \tag{6.21}$$

where $P_{\exp}$ is the experimental or tabulated saturation pressure. The sum in (6.21) is taken over entire massive of the experimental (table) data along the liquid-gas coexisting line.

The detailed investigation of (6.20) is given in [59]. So here we give only as example the parameters, entering (6.20), for two models and three real substances. We have chosen LJ and Square-Well (SW) systems as the models. The latter is characterized by the well-width λ. In Table 6.2 there are 3 values of λ. The temperatures and the densities for the model systems are given in dimensionless LJ units (see [21]). For substances they are presented in Kelvins and g/cm^3. The critical point data for the models one can find in [21, 60].

The data given in Table 6.2 show that the method suggested above facilitates a unified way to calculate the saturation pressure in a wide temperature range for the different model systems and substances. In additional, it covers a wide temperature range, which could not be described with similar accuracy by the approaches known earlier.

6.4 The Other "Ideal" Curves

Now we proceed with the other remarkable curves. Some of them can be also deduced from the vdW equation. In recent study [26] there is the description of several "ideal" curves besides the Zeno-line. Among them let us mark the ideal enthalpy line (below the H-line) and the ideal internal energy line (the U-line). The definition of this contours are analogous to the Zeno-line. Namely, corresponding properties (H or U) at these contours are the same as that one of an ideal gas. The enthalpy H and internal energy U for vdW and an ideal gas ("id") are:

$$\begin{aligned} H^{vdW} &= \frac{3}{2}NT - 2aNn + \frac{NT}{1-bn}, H^{id} = c_P^{id}NT, \\ U^{vdW} &= \frac{3}{2}NT - aNn, U^{id} = c_V^{id}NT. \end{aligned} \tag{6.22}$$

Here the temperature is measured in energy units as previously, N is the particle number ($n = N/V$, where V is the volume), c_P^{id}, c_V^{id} are the heat capacities of an ideal gas per particle at a constant pressure and at a constant volume respectively. The vdW model is supposed to be a mono-atomic system ($c_P^{id} = 2.5$, $c_V^{id} = 1.5$). Corresponding U-line in this case is expressed simply, as $n = 0$. It is of course the trivial result, because any system should be ideal at zero density. Moreover, the analysis of the measurements data from [34] has not discovered any other U-line, besides this trivial one. The situation with the H-line is much more interesting. Equation (6.22) gives the following relation for it:

$$bn + bT/(2a) = 1. \tag{6.23}$$

Thus, the H-line in the vdW system is also the straight as like as the Zeno-line. Its parameters below are marked by subscript "H". For vdW $n_H = n_B = 1/b$, while $T_H = 2T_B = 2a/b$ (see (6.2)). Farther, we have analyzed the virial EOS for the enthalpy [30] (just like in case of the Zeno line). If we suppose that the H-line is described by the linear dependence:

$$n/n_H + T/T_H = 1, \tag{6.24}$$

then the virial expansion gives rise to the following expressions for T_H and n_H:

$$T_H : \left[B_2(T) - T\frac{dC(T)}{dT}\right]_{T=T_H} = 0, \quad n_H = \left[\frac{-T^2\frac{d^2B_2(T)}{dT^2}}{C(T) - 0.5T\frac{dC(T)}{dT}}\right]_{T=T_H}, \tag{6.25}$$

Here B_2 and C are the virial coefficients, like previously. Now we can test (6.24) for the real substances and model systems. But before, we would like to pay attention to several other curves in supercritical region. These curves are not the straight, but there are indications, that they can also have a universal character. Initially these curves have also been obtained for the vdW system [28–30]. It is the line of the minima for the enthalpy isotherms (below $H_{\min}$ line). This curve coincides as well with the inversion curve for the Joule-Thompson effect, as far as the latter are defined by the enthalpy minimum [61]. Its equation for the vdW system directly follows from (6.22), (6.25):

$$T_{H_{\min}} = 2T_B(1 - n/n_B)^2, \tag{6.26}$$

Another line is the line of maxima of isothermal compressibility. The latter is defined as $\varsigma = T(\partial n/\partial P)_T$. It is also the same as the value of the structure factor at the zero wave vector $S(0)$ [2]. That is why we denote this line as S_0. The maximum of ς corresponds to the minimum of $1/\varsigma$. So it is defined by condition $(\partial^2 P/\partial^2 n)_T = 0$. For the vdW equation it gives

$$T_{S_0\min} = T_B(1 - n/n_B)^3, \tag{6.27}$$

One more line is the line of maxima for VEC. VEC is defined by (6.14). Line of its maxima along the isotherms for the vdW system is expressed as

$$T_{VEC_{\max}} = T_H(1 - 2n/n_B)(1 - n/n_B)^2, \tag{6.28}$$

All these lines are described in details in [28, 30]. The resulting phase diagram of the vdW system with all these lines are presented in Fig. 6.6. Note: the lines (6.26)–(6.28) are not the straight contours! Nevertheless their form can be the same for some other system as we will see below.

Now let us consider how the above lines are located in the models and real substances. Corresponding picture for LJ system was constructed in [30]. For the LJ system the H-line is also straight with T_H = 6.431 and n_H = 1.24. So qualitatively the LJ phase diagram has the same features as the vdW phase diagram. Here we present also the phase diagram for another model system—Exp-6 (or Buckingham) potential. This potential has already been studied by us, when Z-line was investigated [21]. It has the form

$$U(r) = \begin{cases} \dfrac{\varepsilon}{1 - 6/\alpha_1}\left[\dfrac{6}{\alpha_1}\exp\left(\alpha_1\left[1 - \dfrac{r_m}{r}\right]\right) - \left(\dfrac{r_m}{r}\right)^6\right], & r \geq r_{\max}, \\ 0, & r < r_{\max}. \end{cases} \tag{6.29}$$

Here ε, r_m, and α_1 are the model parameters. Parameter r_m is the distance where the potential reaches its minimum. It is usually used as the length unit, while ε is used as the energy unit. The cutting distance $r_{\max}$ is necessary because the potential has unphysical maximum near $r = 0$. The hard wall is introduced for the segment $[0, r_{\max})$ to remove this incorrect behavior. The parameter α_1 generates the family of potentials, which can be applied to describe the noble gases and various organic matters [62]. In particular, it was shown in [62] that the methane (CH_4) binodal can be reproduced if $\alpha_1 = 15$. So we have also chosen this value. The phase diagram of Exp-6 system with $\alpha_1 = 15$ is presented in Fig. 6.7. The binodal and the critical point (CP) were calculated in [62], while the Zeno-line and H-line were obtained by means of MC NVT simulation in present research. The simulations details are the same as in our study for LJ [30]. The results presented in Fig. 6.7, confirm that both lines are the straight ones.

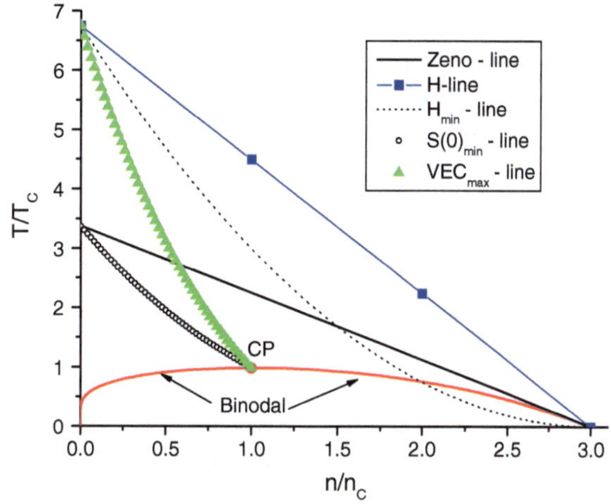

Fig. 6.6 The phase diagram density-temperature of vdW system in the units reduced to the critical temperature T_C and density n_C. CP is the critical point. H-Line is defined by (6.24). H_{min}-line is defined by (6.26). $S(0)_{min}$-line is defined by (6.27). VEC_{max}-line is defined by (6.28)

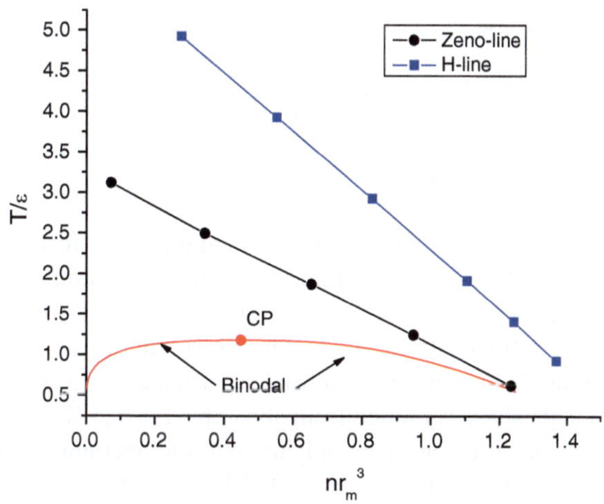

Fig. 6.7 The phase diagram of Exp-6 system in reduced variables. CP is the critical point

Note that all above *H*-line studies are valid for mono-atomic systems only. For dimers the specific heat is $c_P^{id} = 3.5$, for more complicated molecules (with more than 2 atoms) this value is $c_P^{id} = 4$. It is necessary to take into account this circumstance. It is well known fact that the LJ and Exp-6 potentials describe well enough the mono-atomic systems (like argon) with $c_P^{id} = 2.5$. The comparison of *Z* and *H*-lines for the LJ model and argon in [30] have shown a good agreement. Besides, we have said above that the Exp-6 potential reproduces the CH_4 binodal with good accuracy. But the *H*-line for methane does not correspond to that one, presented in Fig. 6.7, because the methane consists of polyatomic molecules. The same picture arises under application of central potential models to the many-atomic materials. Namely, the contours, like the binodals or *Z*-lines, can be reproduced in such models, while the contours, like *H*-line, can not. To illustrate this situation in

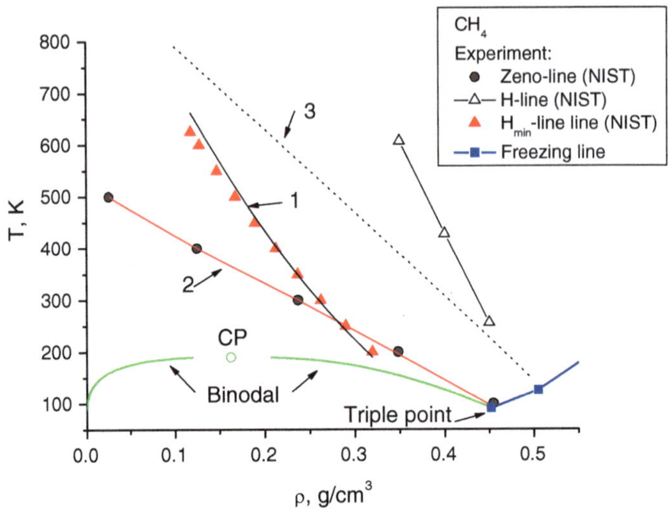

Fig. 6.8 The phase diagram of Methane. Experimental data: CP (the critical point, *empty circle*), Zeno-line (*filled circles*), H-line (*solid line with empty triangles*), H_{min}-line (*filled triangles*), Freezing line (*solid line with filled squares*). NIST [34], Freezing Line [63]. Our calculations: Line 1 is H_{min}-line obtained by (6.26) (*solid line*). Line 2 is the Zeno-line (*solid line*). Line 3 is the H-Line (*dotted line*). Lines 2, 3 are the results of present MC NVT calculations with Exp-6 potential (see text)

Fig. 6.8 we present the phase diagram of the real methane and calculations of the Z and H lines with the Exp-6 potentials. According to [62], the following values for its parameters were chosen: ε/k_B = 160.3 K, r_m = 4.188 Angstrem, α_1 = 15, $r_{max} = 0.168245 r_m$. "k_B" is the Boltzmann constant. The experimental binodal, the Z and H-lines were taken from NIST database [34]. Unfortunately the temperature in the database is limited from above by T = 625 K. So we can reconstruct H-line from the measurements data only up to this temperature. The least squares fitting gives for methane $T_H = 1834$ K, $\rho_H = 0.523$ g/cm^3. But the database has allowed us to find the inversion curve of Joule- Thompson effect. It is also the curve H_{min} (see (6.21) and its description.). In Fig. 6.8 we have presented the experimental line H_{min} and that one given by (6.26) (line 1). The maximum error was 5.8 % at higher temperatures. Besides we have drawn the freezing line according to data of [63].

One can see that the Exp-6 potential can reproduce the Z-line but fails with H-line as expected.

6.5 Conclusions

In this study we have presented several new similarity laws related to the existence of the universal lines on the density-temperature plane. The lines can be divided in two groups. The first one includes the "ideal" curves (Z-line, H-line and the others).

The latter are defined as the contours, where corresponding property (pressure, enthalpy etc.) are the same as in an ideal gas. These lines can be the straight ones over all plane density-temperature. Another group is the "supercritical" curves, which initially have been obtained for the vdW system (the line of the enthalpy minima along the isotherms, the line of maxima of isothermal compressibility and others).

The Z-line (the line of unit compressibility factor) is appeared to be the most fruitful among others. Its geometrical property (the straightness) has allowed us to obtain the general equation for the liquid binodal branch and to establish the relation between the critical and Zeno-line parameters. After that the estimates of the CP coordinates (temperature and density) have been made for a number of metals, where it is impossible to carry out measurements. Together with the Timmermans relation the offered Zeno-line relations allow us to estimate the critical pressure as well. Both these relations have also resulted in the expression for the saturation pressure close to the Riedel formula. But it is appeared to be not very accurate. The inaccuracy witnesses about the possible limitations of the considered similarities. We have found as well the other limitations of our relations. They concern the range of the interaction potentials for models and the quantum properties for substances.

Besides, the study of the Z-line for the lattice gas has given rise to the global isomorphism (introduced by Kulinskii [22, 23]) between two types of systems (continuous and LG). Although the transformation does not establish complete correspondence between any point or line of lattice and continuous systems it allows one to investigate various thermodynamic properties (like the volume expansion coefficient). The modification of initial transformation has resulted in the more accurate relation for the saturation pressure. But it contains two fitting parameters.

We have also marked that besides the Z-line there are other universal curves. The relations, describing them, initially were derived for the vdW system. But, if these relations are expressed in the Boyle parameters (6.24)–(6.28), then they are valid for other systems too. Thereupon we would like to emphasize the especial role of the Z-line parameters (the Boyle temperature and density). They can be used as the reduced units. In this case the variation of the saturation pressure in the chosen reduced units is very small—for all substances the value of the critical pressure changes in a rather narrow range from about 0.02 to 0.04. The same relates to the critical density and temperature.

Finalizing, we would like to point upon the task which is still waiting for the solution. It is the problem of construction of the relation for the saturation pressure with acceptable accuracy analogously to (6.5), (6.6) for the liquid binodal branch without fitting parameters.

Acknowledgments The work is supported by RFBR Grants No 14-08-00536, 14-08-00612, 13-08-12-248.

References

1. J.D. van der Waals, Verhan. Konink. Akad. van Wetens **20**, 1 (1880)
2. R. Balescu, *Equilibrium and Non-equilibrium Statistical Mechanics* (Wiley-Interscience Publication, New York, USA, 1975)
3. L.P. Cailletet, E.C. Mathias, J. Phys. Théor. Appl. **5**, 549 (1886)
4. F. Hensel, Phil. Trans. R. Soc. London A **356**, 97 (1998)
5. K. Pitzer, J. Chem. Phys. **7**, 583 (1939)
6. I.Z. Fisher, *Statistical Theory of Liquids* (University of Chicago Press, USA, 1964)
7. B.M. Smirnov, Physics -Uspekhi **44**, 1229 (2001)
8. M.E. Fisher, Rev. Mod. Phys. **70**, 653 (1998)
9. A. Batchinskii, Ann. Phys. **324**, 307 (1906)
10. V.A. Rabinovich, A.A. Vasserman, V.I. Nedostup, *Thermophysical Properties of Neon, Argon, Krypton, and Xenon* (Hemispere, Berlin, Germany, 1988)
11. V.I. Nedostup, E.P. Galkevich, *Calculation of Thermodynamic Properties of Gases and Liquids by the Ideal Curves Method* (Naukova Dumka: Kiev, Ukraine, 1986)
12. E. Holleran, J. Chem. Phys. **47**, 5318 (1967)
13. J.G. Powels, J. Phys. C: Solid State Phys. **16**, 503 (1983)
14. M.C. Kutney, M.T. Reagan, K.A. Smith, J.W. Tester, D.R. Herschbach, J. Phys. Chem. B **104**, 9513 (2000)
15. E.M. Apfelbaum, V.S. Vorob'ev, G.A. Martynov, Chem. Phys. Lett. **413**, 342 (2005)
16. E.M. Apfelbaum, V.S. Vorob'ev, G.A. Martynov, J. Phys. Chem. B **110**, 8474 (2006)
17. E.M. Apfelbaum, V.S. Vorob'ev, J. Phys. Chem. B **112**, 13064 (2008)
18. E.M. Apfelbaum, V.S. Vorob'ev, Chem. Phys. Lett. **467**, 318 (2009)
19. E.M. Apfelbaum, V.S. Vorob'ev, J. Phys. Chem B **113**, 3521 (2009)
20. L.P. Filippov, *Methods for Calculating and Predicting the Properties of Substances* (Edition of Moscow State University, Moscow, Russia, 1988). (in Russian)
21. E.M. Apfelbaum, V.S. Vorob'ev, J. Chem. Phys. **130**, 214111 (2009)
22. V.L. Kulinskii, J. Phys. Chem B **114**, 2852 (2010)
23. V.L. Kulinskii, J. Chem. Phys. **133**, 034121 (2010)
24. E.M. Apfelbaum, V.S. Vorob'ev, J. Phys. Chem. B **114**, 9820 (2010)
25. Q. Wei, D.R. Herschbach, J. Phys. Chem. C **117**, 22438 (2013)
26. V.I. Nedostup, High Temp. **51**, 72 (2013)
27. R. Span, E.W. Lemmon, R.T. Jacobsen, W. Wagner, A. Yokozeki, J. Phys. Chem. Ref. Data **29**, 1361 (2000)
28. V.V. Brazhkin, V.N. Ryzhov, J. Chem. Phys. **135**, 084503 (2011)
29. V.V. Brazhkin, A.G. Lyapin, V.N. Ryzhov, K. Tkachenko, YuD Fomin, E.N. Tsiok, Phys. Usp. **55**, 1061 (2013)
30. E.M. Apfelbaum, V.S. Vorob'ev, J. Phys. Chem. B **117**, 7750 (2013)
31. E.A. Guggenheim, J. Chem. Phys. **13**, 253 (1945)
32. R.C. Raid, J.M. Prausnitz, T.K. Sherwood, *The Properties of Gases and Liquids* (McGraw-Hill, New York, USA, 1977)
33. A. Mulero, M.I. Parra, Phys. Chem. Liq. **46**, 263 (2008)
34. http://webbook.nist.gov/chemistry/fluid/
35. W. Goltzlaff, G. Schonherr, F. Henzel, Z. Phys. Chem., Neue Folge **156**, 219 (1988)
36. V.F. Kozhevnikov, JETP **97**, 541 (1990)
37. E.M. Apfelbaum, V.S. Vorob'ev, Ukr. J Phys. **56**, 838 (2011)
38. J. Timmermans, Bull. Soc. Chim. Belges **26**, 205 (1912)
39. E.M. Apfelbaum, V.S. Vorob'ev, J. Chem. Phys. **139**, 046101 (2013)
40. E.M. Apfelbaum, V.S. Vorob'ev, Chem. Phys. Lett. **591**, 212 (2014)
41. L. Riedel, Chem. Ing. Tech. **26**, 83 (1954)
42. R. Gathers, Rep. Progr. Phys. **49**, 341 (1986)
43. V.E. Fortov, A.N. Dremin, A.A. Leont'ev, High Temp. **23**, 984 (1975)

44. I. Iosilevskiy, V. Gryaznov, J. Nucl. Mater. **344**, 30 (2005)
45. E.M. Apfelbaum, J. Phys. Chem. B **116**, 14660 (2012)
46. E.M. Apfelbaum, Phys. Rev. E **84**, 066403 (2011)
47. G.E. Norman, S.V. Starikov, V.V. Stegailov, JETP **114**, 792 (2012)
48. L.V. Fokin, V.N. Popov, High Temp. **51**, 465 (2013)
49. E. Pahl, D. Figgen, C. Thierfelder, K.A. Peterson, F. Calvo, P. Schwerdtfeger, J. Chem. Phys. **132**, 114201 (2010)
50. A.A. Likalter, Phys. Rev. B **53**, 4386 (1996)
51. H. Hess, A. Kloss, A. Rakhel, H. Schneidenbach, Int. J. Thermophys. **20**, 1279 (1999)
52. M. Beutl, G. Pottlacher, H. Jaiger, Int. J. Thermophys. **15**, 6 (1994)
53. I.V. Lomonosov, *Phase Diagrams and Thermodynamic Properties of Metals at High Pressures and Temperatures*. (Doctoral Thesis, Chernogolovka 1999.) (in Russian)
54. V.L. Kulinskii, J. Chem. Phys. **139**, 184119 (2013)
55. E.M. Apfelbaum, V.S. Vorob'ev, J. Phys. Chem B **115**, 10049 (2011)
56. W.B. Streett, Physica **76**, 59 (1974)
57. Q. Yan, H. Liu, Y. Hu, Fluid Phase Equilib. **218**, 157 (2004)
58. Q.L. Yan, H.L. Liu, Y. Hu, East China Univers. Sci. Techn. **22**, 188 (1996), 22 188. (in Chinese)
59. V.S. Vorob'ev, J. Phys. Chem. B **116**, 4248 (2012)
60. F. Lofti, J. Vrabec, J. Fisher, Mol. Phys. **76**, 1319 (1992)
61. C. Kittel, H. Kroemer, *Thermal Physics* (W. H. Freeman and Company, New-York, USA, 1980)
62. J.R. Errington, A.Z. Panagiotopoulos, J. Chem. Phys. **109**, 1093 (1998)
63. E.H. Abramson, High Pressure Res. Int. J. **31**, 549 (2011)

Part III
Colloids

Chapter 7
Structure of Polyglycols Doped by Nanoparticles with Anisotropic Shape

Eduard Lysenkov, Iryna Melnyk, Leonid Bulavin, Valeriy Klepko and Nikolai Lebovka

Abstract The structure of polyglycols (polyethylene glycol and polypropylene glycol) doped by nanoparticles with anisotropic shape (carbon nanotubes and inorganic nanoplatelets) is discussed. Various experimental methods such as X-ray scattering, optical microscopy, impedance spectroscopy, differential scanning calorimetry and electrical conductivity measurements have been used. Introduction of such nanoparticles results in noticeable changes in the polymer structure even at rather small concentration of fillers (0.3–0.5 wt%). The percolation behaviour is typical for fluid and semicrystalline polyglycols doped with carbon nanotubes. In the vicinity of percolation threshold the entangled network of nanotubes is formed. This network has impact on the kinetics of polymer crystallization, degree of crystallinity, electrical and thermal properties of the composites. Different types of electrical conductivity in these systems were identified. The structure of polyglycols doped with nanoplatelets depends upon the type of the filler. The data evidence the presence of partial intercalation of macromolecules inside the interlayer space of nanoplatelets of organo-modified montmorillonite and the complete exfoliation of

E. Lysenkov
V.O. Sukhomlyns'kyi Mykolayiv National University, Nikolska Street 24, Mykolayiv 54030, Ukraine
e-mail: ealisenkov@mail.ru

I. Melnyk · L. Bulavin
Faculty of Physics, Taras Shevchenko National University, Volodymirska ave. 64, Kyiv 01601, Ukraine
e-mail: irinamelnik1989@gmail.com

L. Bulavin
e-mail: bulavin221@gmail.com

V. Klepko (✉)
Institute of Macromolecular Chemistry NAS of Ukraine 48, Kharkivske Road, Kyiv 02160, Ukraine
e-mail: klepko_vv@ukr.net

N. Lebovka
Department of Physical Chemistry of Disperse Minerals, Biocolloidal Chemistry Institute named after F. D; Ovcharenko, 42, Vernadsky av, Kyiv 03142, Ukraine
e-mail: lebovka@gmail.com

L. Bulavin and N. Lebovka (eds.), *Physics of Liquid Matter: Modern Problems*, Springer Proceedings in Physics 171, DOI 10.1007/978-3-319-20875-6_7

organo-modified laponite platelets inside the polymer matrix. The doping by nanoplatelets also improved the dispersion of nanotubes inside polyglycols. The effects of nanofillers on the behaviour of polymer electrolytes on the base of polyglycols are also discussed. The doping allows decreasing the degree of ionic association, resulting in higher concentration of the free ions and increased electrical conductivity.

List of Symbols and Abbreviations

Symbols

a	Aspect ratio
C	Concentration
C_n	Concentration of nanotubes
C_n^c	Percolation concentration
D_f	Fractal dimension
D_f^e	Fractal dimensions of the nanocomposite-electrode interface
D_f^c	Fractal dimensions of the polymer-MWCNT interface
d_i	Interlayer distance in MWCNT
E_a	Activation energy
ΔH_m	Melting enthalpy
ΔH_m^0	Melting enthalpy of 100 % crystalline polymers
h	Thickness of a sample
I	Scattering intensity
K_I	Association constant of individual ions
K_T	Association constant of triplets
k	Thermal conductivity exponent
L_c	Size of a correlation zone
l	Length of nanotubes
l_p	Persistence length
M_w	Molecular weight
n	Avrami exponent
q_i	Charge of the ions
T	Temperature
T_m	Melting temperature
t	Time
t_c	Critical conductivity exponent
V_c	Cooling rate
V_h	Heating rate
Z'	Real part of the impedance
Z''	Imaginary part of the impedance
α_I	The fractions of individual ions
α_P	The fractions of ion pairs
α_T	The fractions of ion triplets
β	Width of peak at half-height
χ	Fractional crystallinity of polymer composites

ϕ_n^c	Volume percolation concentration of nanotubes
λ	Specific thermal conductivity
σ	Electrical conductivity
σ_{dc}	DC-electrical conductivity
σ_0	Electrical conductivity at infinite temperature
θ	Diffraction angle
θ_m	Half of the diffraction angle
φ	Angle in Cole-Cole plot
ω	Cyclic frequency

Abbreviations

CNTs	Carbon nanotubes
CPE	Constant phase element
CTAB	Cetyl-trimethyl ammonium bromide
DC	Direct current
DSC	Differential scanning calorimetry
Lap	Laponite
LapO	Organo-modified laponite
MMT	Montmorillonite
MMTO	Organo-modified montmorillonite
MWCNTs	Multi-walled carbon nanotubes
PE	Polymer electrolyte
PEG/PEO	Polyethylene glycol/oxide
PPG/PPO	Polypropylene glycol/oxide
SAXS	Small-angle X-ray scattering
WAXS	Wide-angle X-ray scattering

7.1 Introduction

Composites based on polyglycols such as polyethylene glycol (PEG) and polypropylene glycol (PPG) doped by nanoparticles of anisotropic shape (e.g., carbon nanotubes (CNTs) and nanoplatelets (silicate layers)), are attractive as materials with improved ionic and thermal conductivity, electrical, optical and mechanical properties [1].

Nanotube-based composites on the base of these polymers are expected to show good performance. Good wetting of multi-walled carbon nanotubes (MWCNTs) was observed using both polymers [2]. Dispersions of surface-modified CNTs in water-soluble (PEG) and water-insoluble (PPG) polymers have been studied [3]. The general guidelines to produce uniform dispersions using a dispersive agent and/or surface treatment in these polymers were discussed. The combination of

PEG with rod-like CNTs allows improvement of dispersion stability in common solvents [4].

Functionalization of CNTs by the so-called PEGylation [5] increases their solubility in water and aqueous solutions [6]. PEG-grafted MWCNTs exhibited excellent hydrophilicity, good bioelectrocatalytic activity and biocompatibility [7]. The high selectivity chemical vapour and gas sensors [8, 9], taste and bio-nanosensors [10–12] based on PEG-functionalized CNTs have been already developed, showing fast response, good reproducibility and high selectivity.

Recently, PEO + MWCNT composites had attracted attention as potential materials for protection of optical sensors from high-intensity beams [13]. Polypropylene grafted CNTs were also synthesized [14, 15], with CNTs surrounded by the polymer chains with an average thickness of $\approx$2 nm. The electro-induced shape memory polymer composites on the base of the poly(butylene succinate)-PEO multiblock copolymer have been synthesized [16]. The fine dispersion of CNTs in the matrix enabled it to form a percolation network at $\approx$0.5 wt% of the filler. Moreover, CNTs served as a nucleating agent for both PBS and PEG segments. It resulted in increase in crystallization temperature and crystallinity of both segments. MWCNTs + polyurethane foams with shape memory properties have been synthesized on the base of polypropylene glycol [17].

PEG and PPG in combination with nanoplatelets (montmorillonite and laponite) have also been intensively studied. It was demonstrated that the combination of these polymers with montmorillonite (MMT) or organo-modified montmorillonite (MMTO) (so-called, PEG + MMT hybrids) improves barrier [18], mechanical, viscoelastic and thermal [19–21] properties of polymeric films.

MMTO + PEG hybrids have been used as effective modifiers of rheological behaviour and mechanical properties of ultrahigh molecular weight polyethylene [22]. Intercalation in MMT was observed only with oxyethylene units containing polyglycols [23]. Oxyethylene sequences of 5–6 units proved to be sufficient for intercalation. The MMT preferentially absorbed the higher mass fractions of PEG [24]. PEG chains can penetrate inside the interlayer of MMT and MMTO galleries [24–26]. Moreover, PEG can be used as a modifier for preparation of organoclays [27]. The PEG + MMT hybrids were shown to be useful for improving the rheological and mechanical properties of polypropylene [27–29]. The effects of PEG and PPG on aqueous laponite suspensions was studied with small-angle neutron scattering (SANS) [30]. It was demonstrated that for comparable molecular weights, M_w, PPO speeds up the gelation more efficiently than PEO.

The high interest in polyglycols doped with Li salts is due to their technological applications as solid polymer electrolytes in electrochemical devices such as lithium rechargeable batteries and fuel cell, display devices and sensors [31, 32]. Their attractive properties are stipulated by the relatively high melting point, good structural integrity, low glass transition temperature, which permits ion transport at ambient temperatures, low toxicity and biocompatibility. However, the application of such polymer electrolytes is restricted by their low ionic conductivity at ambient temperatures limits. The incorporation of nanofillers to the PEO–$LiCF_3SO_3$ allowed improving the ionic conductivity [33]. The nanocomposite polymer electrolytes are rather attractive in applications for lithium batteries [34]. The introduction of MMTO in polymeric nanocomposite electrolytes based on poly (ethylene oxide)–$LiClO_4$ resulted in noticeable increases in ionic conductivity [35]. This effect was attributed to a reduction in the degree of crystallinity. Nanocomposite polymeric electrolyte based on PEO–$LiClO_4$-doped with MWCNTs has also been studied [36]. The addition of MWCNT to PEO–$LiClO_4$ resulted in noticeable enhancement of ionic conductivity. The effect was attributed to enhanced amorphization by local modification of PEO chains from crystalline to disordered arrangements.

This work discusses phase behaviour, microstructure and percolation effects in PEG and PPG doped by MWCNTs, MMT, laponite or their organo-modified species studied by different experimental methods (X-ray scattering, optical microscopy, impedance spectroscopy, differential scanning calorimetry and measurements of electrical conductivity).

7.2 Materials and Methods

7.2.1 Materials

7.2.1.1 Polyglycols

Polyethylene glycols (PEG-300, PEG-400, PEG-1000, PEG-10000) and polypropylene glycol (PPG-400) were the Aldrich products. Polyethylene glycols and polypropylene glycols are the typical polyglycols with many different applications. The term glycol is reserved for low molecular weight polymer while the term oxide is used for high molecular weight polymer. The structures of PEG and PPG are presented in Fig. 7.1.

At room temperature PEGs with a molecular weight, M_w, up to 600 are fluids, PEGs with a molecular weight between 800 and 2000 are pasty, and with molecular weights above 3000 the PEGs are solids. Degree of PEG crystallinity, χ, increases with increasing molecular weight. PEG is a hydrophilic nonionic polymer and it has the remarkable solubility in water [37].

It can be explained by the hydrogen bonding between the ether oxygen atoms of PEG and water molecules. At high temperatures the hydrogen bonds become broken and this results in clouding of PEG aqueous solutions. PEG is also soluble in methanol, ethanol and benzene and is insoluble in hexane.

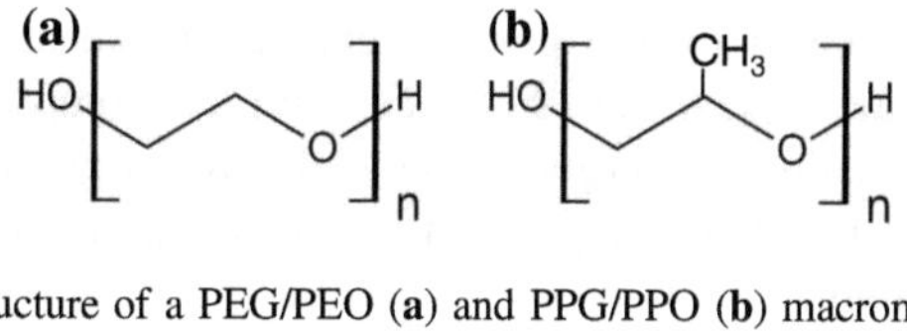

Fig. 7.1 Chemical structure of a PEG/PEO (**a**) and PPG/PPO (**b**) macromolecules

At room temperature PPGs with a molecular weight between 250 and 4000 are fluids and have melting temperatures below ≈250 K. The solubility of PPG in water is lower than that of PEG. PPG of low molecular weights are completely soluble in water. With high molecular weights, they are partially soluble, and their solubility decreases with increasing of temperature or molecular weight of PPG, M_w.

7.2.1.2 Carbon Nanotubes

In this work, multi-walled carbon nanotubes were used. They were prepared from ethylene using the chemical vapour deposition (CVD) method (TM Spetsmash Ltd., Kiev, Ukraine) with FeAlMo as a catalyst [38]. MWCNTs were further treated by alkaline and acidic solutions and washed by distilled water until reaching the distilled water pH values in the filtrate. The typical outer diameter, d, of MWCNTs was ≈30–50 nm [39], their length, l, was ≈5–10 μm and mean aspect ratio was $a = l/d \approx 100–300$.

The specific surface area of the powders determined by N_2 adsorption was $S = 130 \pm 5\ m^2/g$. The electrical conductivity, σ, of the powder of MWCNTs compressed at 15 TPa was about 10 S/cm along the axis of compression. The density of the MWCNTs was assumed to be the same as the density of pure graphite, 2045 kg/m^3.

7.2.1.3 Nanoplatelets (Montmorillonite, Laponite)

The pristine montmorillonite was the MMT of Pygevskii deposit (Ukraine). The crude mineral was preliminarily refined to remove impurities. For facilitation of MMT transfer into the organic form, it was initially transferred into the sodium form through its multiple (5 times) treatment by 0.1 M solution of NaCl during 24 h at 340 K. The solid/liquid ratio was kept as 1/100. Then MMT was centrifuged and washed several times by deionised water to remove surplus NaCl salt. The prepared sample of homoionic clay contained highly pure MMT, which was confirmed by X-ray and chemical analysis, and had the exchange capacity of 1.05 μmol-equ/g and the specific surface (determined by methylene blue adsorption) 640 m^2/g.

The nanoplatelets of laponite RD (Rockwood Additives Ltd., UK) are composed of rigid disk-shaped charged sheets with 0.92 nm thickness and the average diameter about 25–30 nm. It is a synthetic swelling layered clay of 2:1 structural type with molecular formula $Na_{0.7}[(Si_8Mg_{5.5}Li_{0.4})O_{20}(OH)_4]$. Its specific surface area is ≈370 m^2/g, the surface charge density of individual discs is 0.4 e/unit cell, and the specific particle density is 2650 g/cm^3. The sheets are bearing negative surface charge related to isomorphic substitution of Mg by Li atoms. Their negative charge is counterbalanced by the positive charge of exchangeable sodium ions present in the interlayer. Each sheet had a central layer composed of Mg_2^+ cations in octahedral coordination to oxygen atoms or hydroxyl groups. In Laponite RD, the concentration of negative charges, defined as a cation exchange capacity was equal to 0.75 mequiv/g.

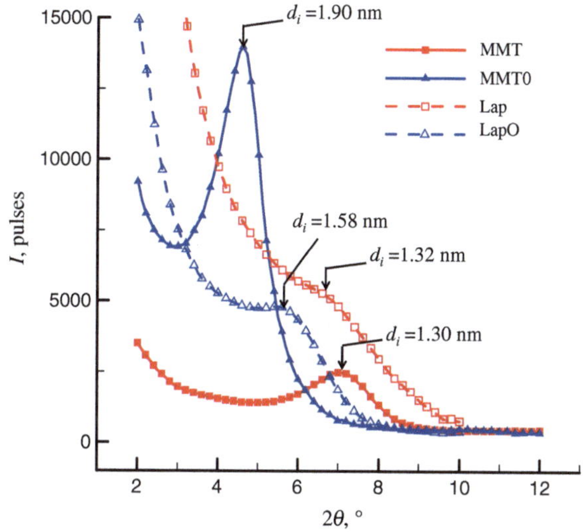

Fig. 7.2 WAXS patterns for MMT, MMTO, Lap and LapO samples in powder air-dry state. T = 298 K

The organo-modified MMT (MMTO) and Laponite (LapO) were prepared by ion-exchange reactions. The long chain cetyl-trimethyl ammonium bromide ions ($C_{16}H_{33}N(CH_3)_3^+ \cdot Br^-$, CTAB, Merck) were chosen as a swelling agent for intercalation into MMT platelets. The aqueous solution of CTAB (1 wt%) was slowly added to the aqueous platelet dispersion (1 wt%) and stirred vigorously 24 h at 360 K. The quantity of solutions corresponded to the required 1:1 stoichiometric ratio of the exchange capacity of clay and CTAB. After incubation, the dispersion was filtered using disc filter funnel and centrifuged. Finally, the obtained samples were freeze-dried for preserving its high dispersion ability in organic media.

The interlayer distance, d_i, between nanoplatelets was measured using the method of wide-angle X-ray scattering (WAXS). Figure 7.2 presents WAXS patterns for MMT, MMTO, Lap and LapO samples in powder air-dry state [40]. The derived interlayer distances were about 1.30, 1.90, 1.32 and 1.58 nm for MMT, MMTO, Lap and LapO samples, respectively. The increase in interlayer distance, Δd_i, in organoclays MMTO ($\Delta d_i \approx 0.6$ nm) and LapO ($\Delta d_i \approx 0.26$ nm) evidenced that CTAB successfully penetrate into the clay interlayer. In the organo-modified samples the alkylammonium cations are strongly bound to the negatively charged platelets.

7.2.1.4 Preparation of the Samples

The composite samples were prepared by adding the appropriate weights of filler (MWCNTs, MMTO, Lap, LapO) to the polymers (PEG or PPG) in the liquid state with subsequent 20 min sonication of the mixture using a UZDN-2T ultrasonic disperser. After sonication, the composites were cooled down to the solid state and kept at room temperature until further experiments.

7.2.2 Methods

7.2.2.1 Differential Scanning Calorimetry (DSC)

The kinetics of isothermal crystallisation was studied using the crystallisation isotherms. These isotherms were obtained using the differential scanning calorimeter DSC-2 Instrument (Perkin Elmer, Waltham, MA, USA) upgraded and supplied with signal processing software by IFA GmbH, Ulm. Before each measurement the sample was heated up to 375 K, incubated for a 3 min and then cooled down to 263 K at the cooling rate, V_c, (1–20 K/min) [41].

The melting temperature, T_m, and melting enthalpy, ΔH_m, were measured with a TA Q2000 Instrument (TA Instruments Ltd. West Sussex, England) in the regime of heating from 273 to 333 K at the heating rate, V_h, of 2 K/min. The samples (20–30 mg) were packed in the aluminium pans.

The fractional crystallinity of polymer composites, χ_{DSC}, was calculated as

$$\chi_{DSC} = \Delta H_m / \Delta H_m^0, \tag{7.1}$$

where ΔH_m is the melting enthalpy of the composite under investigation and ΔH_m^0 is the melting enthalpy of 100 % crystalline polymers (=165.5 J/g for PEG [42, 43]).

Though application of the enthalpy method for estimation of the crystallinity degree is controversial [44], it is widely used for determination of the fractional crystallinity of PEGs [45] and it is useful in applications with small filler content.

7.2.2.2 Thermal Conductivity

The thermal conductivity was measured by the method of dynamic calorimetry with an IT-λ-400 meter (Russia, measurement accuracy ±10 %). The instrument was calibrated with the aid of sample measures of thermal conductivity made of fused quartz and copper. Measurements were carried out using a monotonous regime method. The specific thermal conductivity of the sample, λ, was calculated as

$$\lambda = h / R_s, \tag{7.2}$$

h is the sample thickness, R_s is thermal resistance.

7.2.2.3 X-Ray Scattering

The structure of composite at a small spatial scale was investigated using a wide-angle X-ray scattering (WAXS) instrument DRON-2 (Bourevestnik, Inc., St. Petersburg, Russia.) with CuK_α source of emission at a wavelength 0.154 nm.

The size of correlation zone, L_c, was estimated from Scherrer equation [46]:

$$L_c = \alpha\lambda/\beta\cos\theta_m, \tag{7.3}$$

where α is the coefficient accounting for the form of correlation zone (it is close to unity, $\alpha \approx 0.9$), β is the width of a peak at half-height expressed in radians of 2θ (width measured in 2θ degrees and then multiplied by $\pi/180$), θ_m is half of the diffraction angle corresponding to position of the scattering peak. For calculation of L_c, the most intensive maximum in the WAXS pattern was used.

WAXS patterns were used also for derivation of the fractional crystallinity of PEG-1000 based composites, χ_{XRD}, defined as a ratio [47]:

$$\chi_{XRD} = Q_c/Q_t. \tag{7.4}$$

Here, Q_c is the area fraction under the diffraction peak that corresponds to the crystalline structure and Q_t is the total area under diffraction peak.

The interlayer distance of the nanoplatelets, d_i, (MMT, MMTO, Lap, LapO) was calculated from Bragg formula [46]:

$$d_i = \lambda/2\sin\theta_m. \tag{7.5}$$

Small-angle X-ray scattering (SAXS) investigations were done in the diapason of the scattering angles, θ, between 0.05° and 4° using a Kratky-camera system. The K_α radiation was selected with a filter and data, measured by step scanning with a scintillation counter, were obtained.

7.2.2.4 Electrical Conductivity

The concentration dependences of DC-electrical conductivity, σ_{dc}, were estimated by impedance spectroscopy method using Z-2000 (Elins, Inc., Moscow, Russia) instrument in the frequency range within 1 Hz–2 MHz. The real, Z', and imaginary, Z'' parts of the sample impedance were measured using a plate geometry cell with 0.11 mm distance between the electrodes. The complex impedance plane plots (Z'' vs. Z') were used to separate the bulk material and the electrode surface polarization phenomena. The frequency-dependent value of complex impedance was evaluated as:

$$Z^*(\omega) = Z' - jZ'' = \frac{R}{1+(\omega CR)^2} - j\frac{\omega CR^2}{1+(\omega CR)^2}, \tag{7.6}$$

where $\omega = 2\pi f$ is the angular frequency, C and R are the capacitances and the equivalent parallel resistance of the cell with sample, respectively.

The direct current electrical conductivity, σ_{dc},was estimated from analysis of a complex impedance plane plots (Z'' vs. Z') in the limit of very small frequencies [48].

The electrical conductivity, σ, was also estimated by the inductance, capacitance and resistance (LCR) meter 819 Instek, 12 Hz–100 kHz (Instek America

Corp. Chino, CA, USA) in a cell equipped with two horizontal platinum electrodes (diameter of 12 mm, inter-electrode space of 0.5 mm). The applied external voltage was $U = 1.275$ V. The measuring frequency of 10 kHz was chosen in order to avoid polarization effects on the electrodes and the electric field-induced asymmetric redistribution of MWCNTs between the electrodes [49].

7.2.2.5 Microstructure

The optical microscopy images were obtained using an OI-3 UHL 4.2 microscope (LOMO, Russia). The microscope detector unit was interfaced with a digital camera and a personal computer.

The digital images were captured with magnification of 10–100, the layer thickness was 50 μm. In order to obtain 2D projections of the MWCNT aggregates, the focal plane was located, approximately, on the middle plane of the aggregates. The binary images were analysed using the box-counting method with the help of the image analysis software ImageJ v1.42q. The fractal dimension, D_f, was obtained from dependence of the number of boxes necessary to cover the boundary of an aggregate, N_b, versus the box size, L_b [50]:

$$N_b \propto L_b^{D_f}. \tag{7.7}$$

7.2.2.6 Statistical Analysis

All experiments and measurements of characteristics were repeated using, at least, five replicates. One-way analysis of variance was used for statistical analysis of the data using the Statgraphics plus (version 5.1, Statpoint Technologies Inc., Warrenton, VA). For each analysis, significance level of 5 % was assumed. The error bars presented on the figures correspond to the standard deviations.

7.3 Polyglycols Doped by Carbon Nanotubes

The polymers doped by carbon nanotubes can display high electrical conductivity even at very low content of nanotubes. The dielectric-conductor percolation threshold can be smaller than 0.01 wt% and is dependent on aspect ratio of nanotubes a, type of polymer and dispersion method [51].

7.3.1 PEG

Nanocomposites on the base of PEG doped with MWCNTs display many intriguing physical properties. For example, such systems have unique percolation behaviour

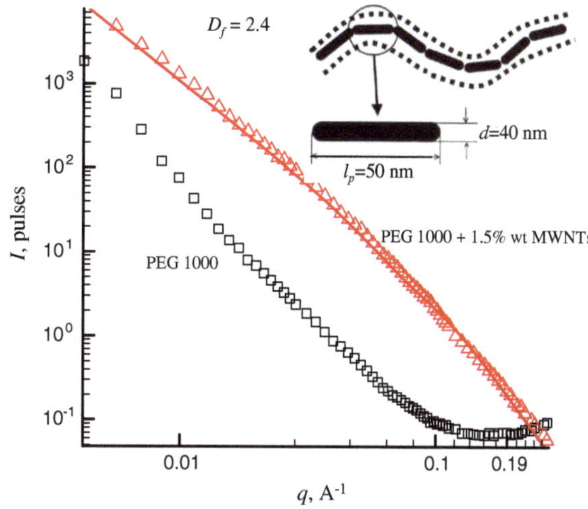

Fig. 7.3 SAXS profiles for pure PEG-1000 and for MWCNTs (1.5 wt%) dispersed in PEG-1000. *Solid line* corresponds to the model of fractal flexible rods [60]. T = 298 K

of electrical and thermal parameters. One of the reasons of such unique percolation behaviour may be related to the impact of nanofiller on the structure of the polymer matrix. Different studies were devoted to the behaviour of PEGs + MWCNT systems [39–41, 52–59].

SAXS method was applied to study the samples of PEG-1000 doped with MWCNTs [39]. SAXS profiles for pure PEG-1000 and for MWCNTs (1.5 wt%) dispersed in PEG-1000 are presented in Fig. 7.3. Doping with nanotubes resulted in a noticeable increase in scattering intensity, I. The estimated mean external diameter of MWCNTs was 40 nm. However, other parameters of MWCNTs were different for the free species and solvated species inside PEG matrix. For example, the persistence length, l_p, was ≈100 and ≈ 50 nm for the free and solvated species, respectively. From other hand, the fractal dimension, D_f, was ≈2.1 and ≈2.4 for the free and solvated species, respectively. These changes in l_p and D_f reflected the more flexible state for solvated species.

The decrease in persistence length upon transferring the MWCNTs inside PEG-1000 can reflect the effect of sonication that used for preparation of composite. It can be speculated that two effects are responsible for such behaviour. Sonication can provoke the decrease of the length of nanotubes (Fig. 7.4a) and the increase in their flexibility due to the disentanglement of aggregates (Fig. 7.4b) [39].

WAXS method was applied to study the samples of PEG-300 + MWCNTs [61]. The half-width of diffraction peak increased systematically from 7.7° to 9.8° with increasing of MWCNT content.

That corresponded to a decrease in the size of a correlation zone, L_c, (a zone with the long range order) up to 0.87 nm. This fact is an evidence of the presence of supplementary disorder of amorphous polymer matrix under the influence of nanofiller particles. This disorder reflects the restriction of conformational sets of polymer chains. WAXS data for the samples PEG-1000 + MWCNTs evidenced that the degree of crystallinity, χ_{XRD}, decreases from 82 to 62.5 % with increasing

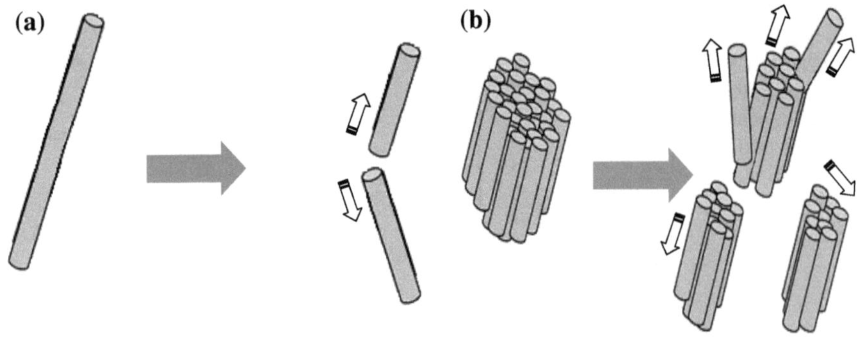

Fig. 7.4 Impact of sonication on the persistence length of MWCNTs. Sonication can provoke the decrease of the length of nanotubes (**a**) and the increase in their flexibility due to the disentanglement of aggregates (**b**)

Table 7.1 Thermal characteristics (melting temperature, T_m, and enthalpy, ΔH_m) and the degree of crystallinity, χ_{DSC}, obtained for the samples PEG-10000 + MWCNTs from the DSC measurements [58]

C_n (wt%)	T_m (K)	ΔH_m (J/g)	χ_{DSC} (%)
0	334.6	128.1	77.6
0.2	335.7	132.7	80.4
0.3	335.8	134.0	81.2
0.35	335.5	132.9	80.5
0.4	336.8	133.7	81.0
0.45	335.2	137.8	83.5
0.5	336.0	142.6	86.4
0.55	335.3	130.3	78.9
0.6	336.3	122.3	74.1
0.8	336.6	124.6	75.5
1	337.0	126.2	76.5
2	338.5	130.7	79.2
5	340.0	128.9	78.1

MWCNT content [39]. These data for χ_{XRD} were good agreement with the estimation of the degree of crystallinity, χ_{DSC}, obtained from the DSC method [40].

The examples of thermal characteristics (melting T_m, and enthalpy of melting, ΔH_m) and the degree of crystallinity, χ_{DSC}, for the samples PEG-10000 + MWCNTs estimated from the DSC data are presented in the Table 7.1 [58].

A rather complex dependence of the degree of crystallinity, χ_{DSC}, versus the concentration of nanotubes, C_n, was observed (Fig. 7.5). The value of χ_{DSC} increased monotonically, reaching a maximum at $C_n = 0.5$ %, and then decreased to a minimum at $C_n = 0.6$ %, and then started to increase again [58].

The increase of crystallinity of PEG in the concentration range $C_n = 0$–0.5 wt% was explained by the fact that the nanotubes could serve as the centres of heterogeneous nucleation of crystalline phase. With increase of the concentration of these centres the crystallization rate and degree of crystallinity of the polymer matrix are

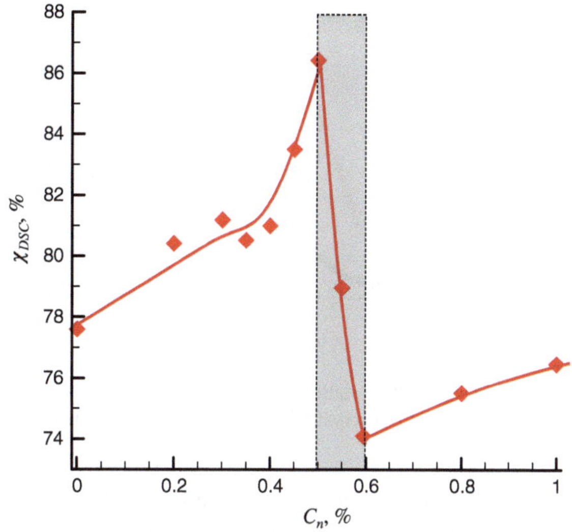

Fig. 7.5 Degree of crystallinity of PEG-10000, χ_{DSC}, versus the concentration of MWCNTs, C_n

also increasing. As the result a more perfect structure of the polymer with a higher degree of crystallinity was realized. Sharp reduction in crystallinity in the concentration diapason 0.5–0.6 wt% (Fig. 7.5, dashed area) can be explained by the effect of surface of the nanotubes on the studied thermal characteristics. At $C_n \approx 0.6$ wt% the processes of thermal motions of macromolecular chains are restricted owing to the presence of well-developed surface of nanotubes. The presence of steric barriers for macromolecules of PEG results in losing of their mobility and thus in ability to form a crystalline structure. However, at higher concentrations ($C_n > 0.6$ wt%) the strong aggregation of carbon nanotubes inside PEG matrix was observed. This aggregation resulted in reducing the surface of the nanotubes, available to interact with the PEG matrix.

It is interesting to note that practically for all samples at different C_n the single-peak DSC crystallization thermograms were observed. It reflects the mechanism of volume crystallization of PEG. However, for the concentration $C_n \approx 0.5$ wt% the additional crystallization peak was observed. The presence of this peak reflected the crystallization of PEG macromolecules that are in close contact with the surface of MWCNTs.

The degree of crystallinity, χ_{DSC}, was dependent also on the cooling rate, V_c, during the experiment. Figure 7.6 presents the typical dependences of the degree of crystallinity, χ_{DSC}, of PEG-10000 (a) and of PEG-10000 + 0.5 wt% MWCNTs (b) versus the time of crystallization, t, at different rates of cooling [41].

To determine the kinetic parameters of the non-isothermal crystallization the modified Avrami equation was used [62]:

$$\chi_{DSC}(t)/\chi_{DSC}(\infty) = 1 - \exp(-Kt^n), \tag{7.8}$$

where K is a rate constant and $n(1-4)$ is an Avrami exponent.

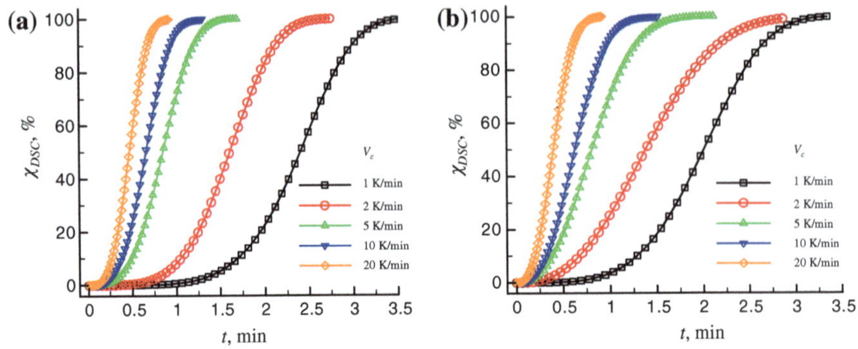

Fig. 7.6 Degree of crystallinity, χ_{DSC}, of PEG-10000 (**a**) and of PEG-10000 + 0.5 wt% MWCNTs (**b**) versus the time of crystallization, *t*, at different rates of cooling [41]

The Avrami exponent can indicate the mechanism of crystallization. For example, the value $n = 4$ corresponds to the three dimensions of growth. It may be also related with the fractal dimension of the growing crystallites, D_f, [63]:

$$D_f = n - 1. \tag{7.9}$$

Figure 7.7 presents the Avrami exponent, *n*, versus the concentration of nanotubes, C_n, for the samples PEG-10000 + MWCNTs at different rates of cooling [41].

It is interesting that the Avrami exponent, *n*, goes through a minimum in the same concentration range C_n = 0.5–0.6 wt% where the minimum of a degree of crystallinity, χ_{DSC}, is observed (Fig. 7.5). The similarity in behaviour of $n(C_n)$ and $\chi_{DSC}(C_n)$ suggests the strong development of aggregation at C_n = 0.5–0.6 wt%, resulting in a decrease of the surface contact between PEG and MWCNTs [41].

The examples of aggregates for PEG-1000 doped by MWCNTs at different concentrations, C_n, are presented in Fig. 7.8 [40]. The similar aggregation was observed also for PEG-300, PEG-400 and PEG-10000.

These aggregates have a fractal structure. The fractal dimension, D_f, was estimated from analysis of 2D projections (7.7). The value D_f depicts morphology of the aggregates and it varies between 1 (corresponding to a linear aggregate) and 2 (corresponding to a compact aggregate). It was an increasing function of C_n, which is typical for behaviour of the percolating MWCNT networks (Fig. 7.9).

At small concentration of nanotubes ($C_n < 0.01$ wt%) the aggregates were rather small (<10 μm) and isolated. The increase of C_n results in increase of aggregate sizes and at the certain critical concentration they coalesce into the infinite percolation cluster that spans through the system. This concentration dependence of the microstructure of the composites is rather typical for different systems doped by nanotubes [40, 56, 58, 59, 64]. The microstructure can be affected by the functionalization of the surface of nanotubes [59] or by the changing the method applied for their dispersion inside matrix [64].

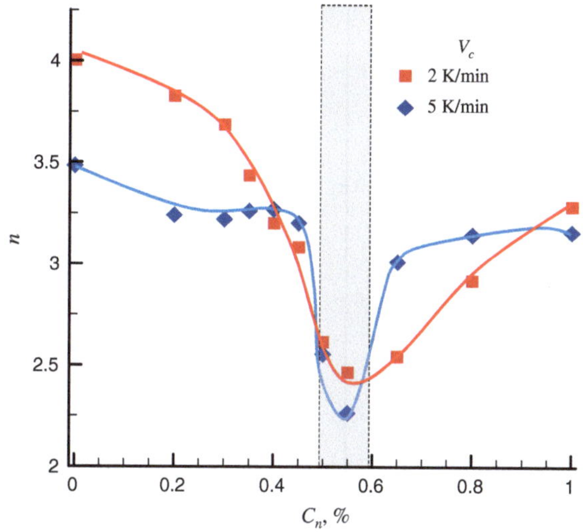

Fig. 7.7 Avrami exponent, n, versus the concentration of MWCNTs, C_n, in PEG-10000 at different rates of cooling, V_c [41]

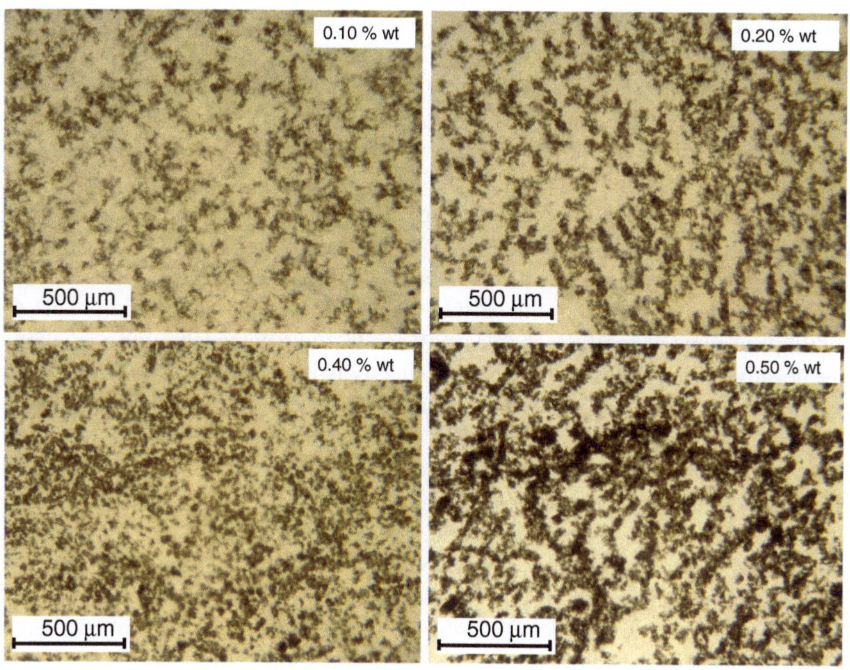

Fig. 7.8 Microscopic images of PEG-1000 doped by with MWCNTs at different concentrations C_n, T = 310 K (From [40]. With permission)

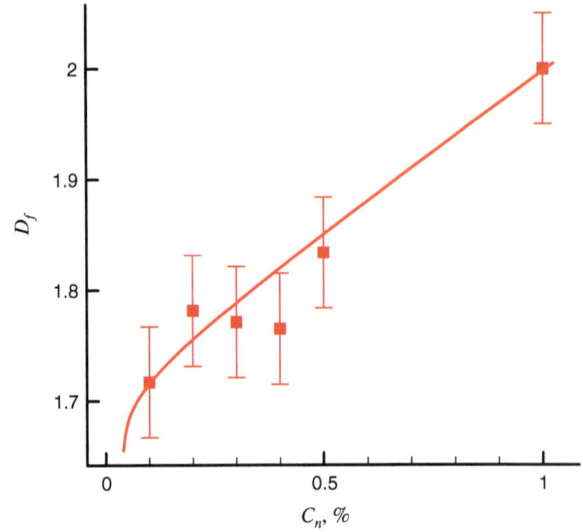

Fig. 7.9 Fractal dimension, D_f, versus concentration of MWCNTs, C_n. $T = 310$ K (From [40]. With permission)

Doping of PEG by nanotubes noticeably affect the system electro-physical behaviour. Figure 7.10 presents temperature dependencies of electrical conductivity of PEG with different molecular weight undoped (dashed lines) and doped (solid lines) with 0.1 wt% MWCNTs. For PEG-400 sample no significant hysteresis in the heating-cooling cycle was observed. However, for PEG-1000 and PEG-10000 samples the hysteresis loop in the heating-cooling cycle in the vicinity of the melting-freezing temperatures was observed.

In general, with increasing molecular weight of the polymer, M_w, and increasing degree of its crystallinity, a significant decrease in electrical conductivity was observed. This effect can be attributed to a decrease in electrical charge mobility with increase of the crystallinity.

The rather unexpected effect was observed in the temperature dependence of electrical conductivity for the PEG-10000 sample. In the vicinity of the melting-freezing temperatures the electrical conductivity passed through a minimum. The negative temperature coefficient of conductivity in the vicinity of 290–325 K can be explained by the effect of the temperature on the conductive pathways in a polymer matrix in this temperature interval. PEG-10000 has a rather high degree of crystallinity, and its structure can be represented as a mixture of crystalline grains with small electrical conductivity covered by amorphous films with large electrical conductivity. The observed changes in electrical conductivity near the melting/freezing points can reflect the thermal expansion of the polymer crystalline grains [65].

Introduction of 0.1 wt% of MWCNTs have resulted in a significant increase in the electrical conductivity. For PEG-400 doping by MWCNTs did not resulted in significant hysteresis behaviour of electrical conductivity. However, for PEG-1000 and PEG-10000 hysteresis loops became large and were observed in a wider temperature range (Fig. 7.10).

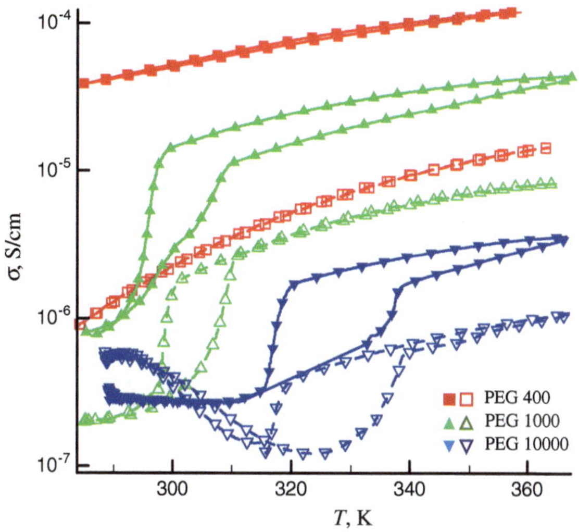

Fig. 7.10 Electrical conductivity, σ, versus the temperature, T, for different undoped PEG (*dashed lines, open symbols*) and PEG doped (*solid lines, closed symbols*) with 0.1 wt% MWCNTs

The noticeable increase in electrical conductivity, σ, was observed for MWCNT concentrations, C_n, above the percolation threshold (≥0.1 wt%). The percolation theory predicts the following scaling law for the concentration dependence of electrical conductivity [66]:

$$\sigma \propto (C_n - C_n^c)^{t_c} \text{ at } C_n > C_n^c, \tag{7.10}$$

where t_c is a critical conductivity exponent, and C_n^c is a percolation concentration of the conductive filler.

The critical exponent, t_c, depends on the system dimensionality and type of percolation. For the random percolation problem the estimated values are $t_c \approx 1.33$ in two and $t_c \approx 2$ in three dimensions [66]. A value of $t_c \approx 2.5$ has been derived for a continuum "Swiss cheese" model with distributed bond strengths or contact resistances [66].

For homogeneously distributed particles (statistical percolation) the theoretically predicted dependence of the percolation threshold on the aspect ratio ($a = l/d$) is usually fulfilled, $2C_n^c \sim \phi_n^c \sim 0.5/a$ [67], here ϕ_n^c is a volume percolation concentration. Accounting for the mean aspect ratio $a \approx 100$–300 of MWCNTs this estimation gives $C_n^c \sim 0.08$–0.25 wt%.

However, the numerous studies have found that for polymers doped by nanotubes the value of C_n^c can reflect the interactions between fillers and polymers, the homogeneity of distribution of nanotube inside polymer matrices and many other factors [51]. The experimental data have suggested that the percolation threshold concentration is inversely proportional to the degree of crystallinity of polymer matrix, i.e., $C_n^c \propto 1/\chi$ [54, 55]. For example, MWCNT doped systems on the base of the high crystalline PEG-10000 demonstrated the smaller percolation concentration

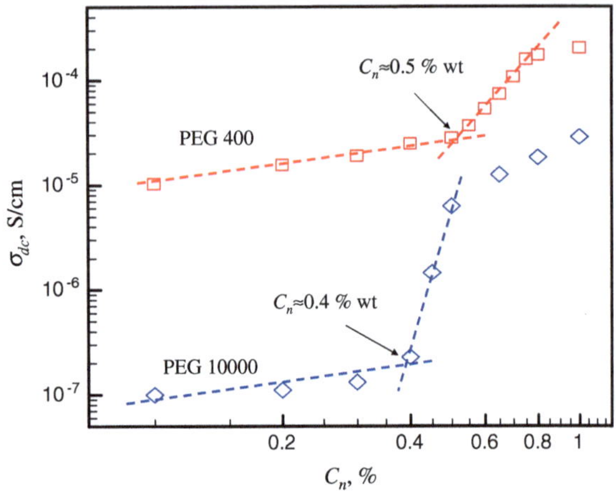

Fig. 7.11 Direct current electrical conductivity, σ_{dc}, versus the concentration of MWCNTs, C_n. $T = 293$ K

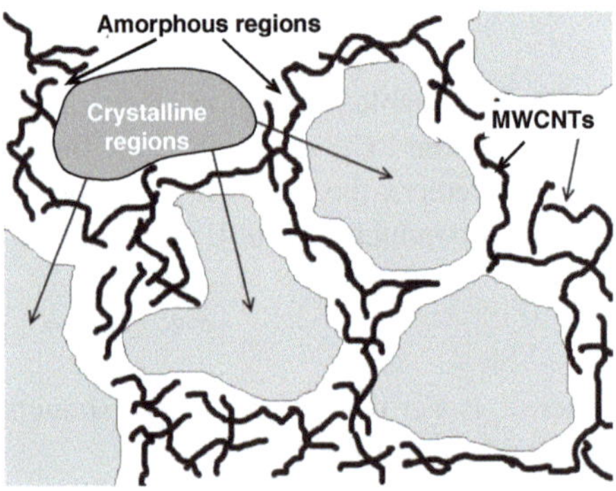

Fig. 7.12 Model of the structure of semi-crystalline polymer doped with MWCNTs [55]

(≈0.4 wt%) than the systems on the base of high amorphous PEG-400(≈0.5 wt%) (Fig. 7.11) [55].

This observation can be explained accounting for the structure of the semi-crystalline polymers doped with MWCNTs (Fig. 7.12). The integration of nanotubes inside the crystalline regions of polymers is restricted due to the compact packing of the polymer chains inside these regions. On the other hand, MWCNTs-polymer interaction can immobilize the nanotubes inside the amorphous regions. During the growth of crystallites in the polymer MWCNTs are expulsed into in amorphous regions and covers the crystalline grains [55]. The situation is rather similar to that observed in segregated composites where percolation can be observed at a very small content of filler [65, 68].

The percolation behaviour and properties of theses composites can be influenced by the spatial distribution and alignment of MWCNTs inside PEG matrix. The different characteristics of the samples PEG-1000 doped with MWCNTs prepared in the presence and absence of direct current electric field (10 kV/cm) were compared [57]. The electric field noticeably affected the melting temperature and

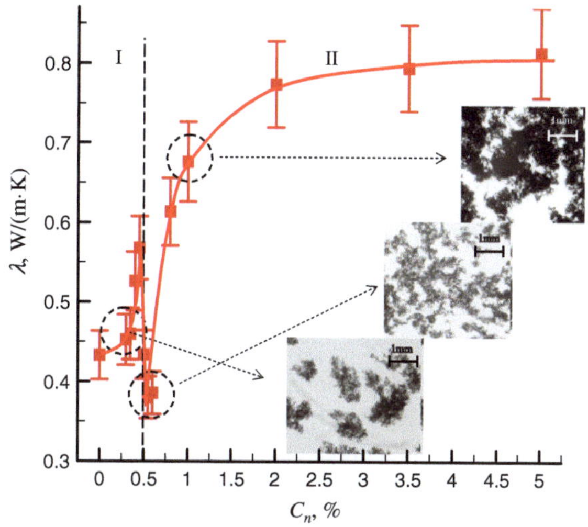

Fig. 7.13 Thermal conductivity, λ, versus the concentration of MWCNTs, C_n, inside PEG-10000. *Insets* show microscopic images of composites at appropriate concentrations. $T = 313$ K [58]

enthalpy for all concentrations of MWCNTs in composite (0–1 wt%) and its application resulted in increase of the size of crystallite grains and degree of crystallinity in the samples. The percolation threshold concentrations were $C_n^c \approx 0.4$ wt% and $C_n^c \approx 0.1$ wt% for the samples obtained without and with curing in the electric field, respectively.

The percolation can be also important in the behaviour of thermal conductivity, λ, of PEGs doped by MWCNTs. The thermal conductivity strongly depends upon the interactions between fillers and polymers and the homogeneity of distribution of nanotubes [58]. Figure 7.13 presents example of the thermal conductivity, λ, versus the concentration of MWCNTs, C_n, inside PEG-10000 [58].

The concentration dependence of λ was rather complex and it was in correspondence with dependence of the degree of crystallinity, χ_{DSC}, (Fig. 7.5) and Avramy exponent, n, (Fig. 7.7) for same system. Initial increase of λ (region I) can be explained by the impact of nanotube on the increase of the degree of crystallinity. In the region II ($C > C_n^c$) the increase of λ can be explained by the contribution of the MWCNT networks to the thermal conductivity. The intrinsic thermal conductivity of the individual MWCNTs may be rather high, $\lambda \approx 2000$–3000 W/(m K) [69, 70] as compared with those for the pure PEG-1000, $\lambda \approx 0.4$ W/(m K). However, in the vicinity of the percolation threshold the jump in the electrical conductivity was noticeably larger (Fig. 7.11) than the changes in the thermal conductivity (Fig. 7.13). This effect reflects the thermal resistance of contacts between MWCNTs and presence of well-developed PEG-MWCNT interfaces.

The different models based on approximations of Maxwell-Eucken [71], Bruggeman [72], Lewis-Nielsen [73], Pal [74], Meredith and Tobias [75], the percolation theory [76] were developed for theoretical description of concentration dependence of the thermal conductivity, λ. The percolation theory utilises the scaling dependence that is in full analogy with dependence for electrical conductivity (7.10) [58]:

$$\lambda \sim (C_n - C_n^c)^k \text{ at } C_n > C_n^c, \tag{7.11}$$

where k is the thermal conductivity exponent.

7.3.2 PPG

The polyethylene and polypropylene glycols are quite similar in chemical structure and properties. However, the PEG has no side branches, whereas PPG has one side group in each monomer unit. In general, PEG + MWCNT and PPG + MWCNT composites have many similar properties.

Figure 7.14 presents microscopic images of the PPG-400 doped by MWCNTs at different concentration of nanotubes [77]. Below the percolation threshold, nanotubes form many individual non-spanning clusters. In the vicinity of percolation concentration, $C_n \approx C_n^c \approx 0.4$ wt% (Fig. 7.14d, e), the spanning cluster appeared. At higher concentration the clusters become more developed and compacted (Fig. 7.14f).

This behaviour was in correspondence with the data on the electrical conductivity for these systems [78–80]. Figure 7.15 presents direct current electrical conductivity, σ_{dc}, versus the concentration of MWCNTs, C_n, for PPG-400 + MWCNT systems [80]. The sharp jump in the value of σ_{dc} was observed at $C_n \approx 0.5$ wt% and it was attributed to the formation of the percolation cluster of MWCNTs.

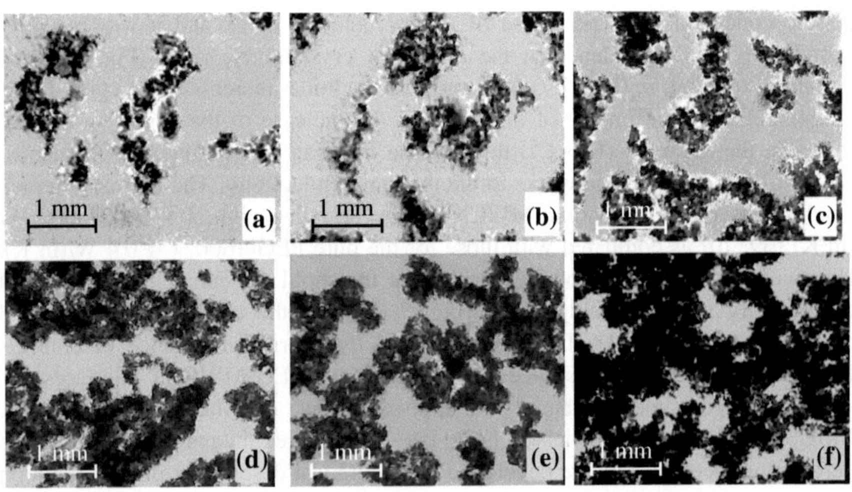

Fig. 7.14 Microscopic images of the PPG-400 doped by MWCNTs at different concentration of nanotubes, C_n (wt%) = 0.075 (**a**), 0.1 (**b**), 0.03 (**c**), 0.4 (**d**), 0.5 (**e**), 0.75 (**f**) [77]

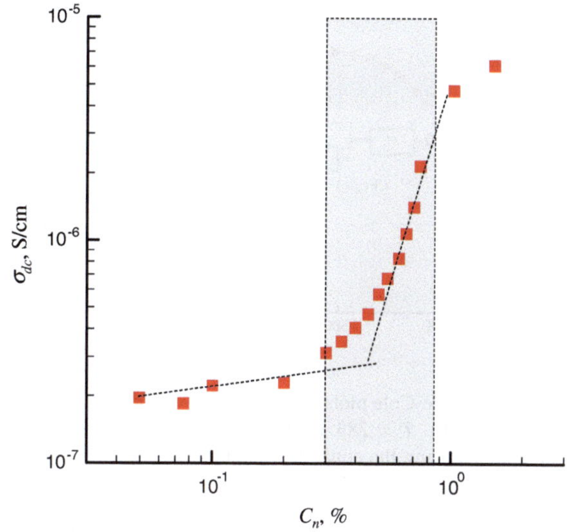

Fig. 7.15 Direct current electrical conductivity, σ_{dc}, versus the concentration of MWCNTs, C_n, in the PPG-400 [80]. $T = 293$ K. *Dashed area* corresponds to the percolation transition

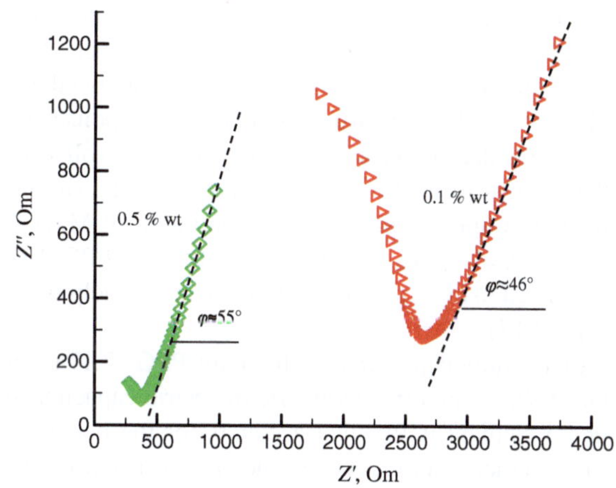

Fig. 7.16 Cole-Cole plots at different concentrations of MWCNTs in PPG 400. $T = 283$ K [79]

The electro-physical properties of PPG 400 + MWCNT systems were also studied using the impedance spectroscopy [79]. Figure 7.16 shows the real impedance Z' versus the imaginary impedance Z'' (Cole-Cole plots) at different concentrations of MWCNTs in PPG-400 and temperature $T = 283$ K. Some parts of Cole-Cole plots for these systems show linear behaviour in the complex plane with an angle of φ to the real axis (Fig. 7.16).

The value of φ reflects the mechanism of the charge transport in the system [81]. The classical Cole-Cole response with $\varphi = 45°$ corresponds to a singular charge transfer mechanism in the system, e.g., diffusion mechanism. The deviation from 45° can be explained by the presence of several mechanisms. For concentration of

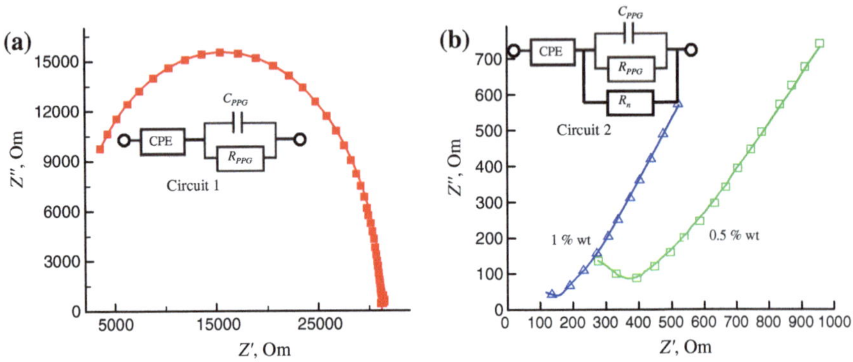

Fig. 7.17 Cole-Cole plots and equivalent circuits for pure PPG-400 (**a**) and PPG-400 doped with MWCNTs (**b**). T = 283 K. *Symbols* are the experimental data, *solid lines* correspond to the simulations using the equivalent circuits [78, 79]

MWCNTs C_n = 0.1 wt% below percolation threshold the value of $\varphi \approx 46°$ was close to the classical response with the singular mechanism.

For pure PPG-400 this mechanism corresponds to the diffusion of the H^+ ions limited by segmental motions of the polymer chains. From the other hand, for concentration of MWCNTs C_n = 0.5 wt% above the percolation threshold, the value of $\varphi \approx 55°$ deviated from the classical response. It indicates the occurrence of a supplementary conductivity mechanism controlled by the charge transport through the clusters formed by carbon nanotubes. The similar behaviour was observed also for the samples with higher concentration of MWCNTs (1.0 and 1.5 wt%) [79].

Cole-Cole plots for pure PPG-400 and PPG-400 doped with MWCNTs were simulated using equivalent circuits with the help of EIS Spectrum Analyser [78, 79] (Fig. 7.17).

The equivalent circuit 1 for pure PPG (7.17a) includes the total volume resistance, R_{PPG}, and the geometric dielectric capacitance, C_{PPG}, of PPG, and constant phase element (CPE) that accounts for the behaviour of the imperfect dielectrics. This element can reflect exponential distribution of parameters of electrochemical reaction, impedance behaviour related with fractal structure of a sample surface, etc.

The frequency dependence of CPE impedance was simulated using the following relationship:

$$Z_{CPE} = A(j\omega)^{-g}, \tag{7.12}$$

where A and g ($0 \le g \le 1$) are frequency independent constants, ω is a cyclic frequency.

The equivalent circuit 1 in Fig. 7.17a allowed a good fitting of the experimental data for pure PPG-400 and PPG-400 doped with small quantity of MWCNTs (C_n = 0.1 wt%). From other hand, the equivalent circuit 1 failed to simulate the experimental data at high concentration of MWCNTs above percolation threshold ($C_n \ge 0.5$ wt%).

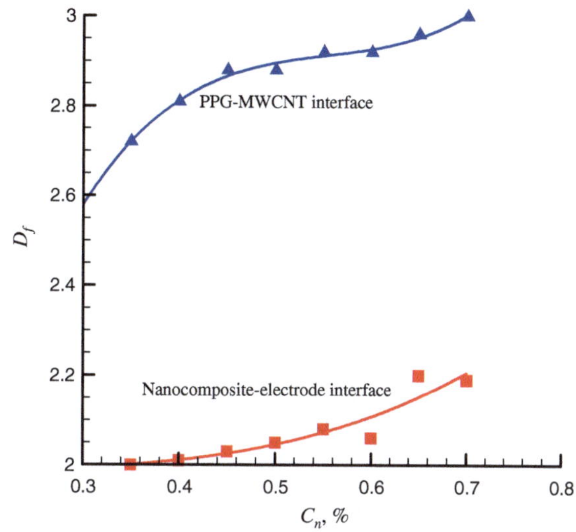

Fig. 7.18 Fractal dimensions of the nanocomposite-electrode and PPG-MWCNT interfaces, D_f, versus the concentration of MWCNTs, C_n, in PPG-400. $T = 298$ K [80]

To account for the charge transport through the clusters formed by carbon nanotubes the equivalent circuit 2 (Fig. 7.17b) was used. In this circuit the element R_n correspond to the total volume resistance of MWCNTs. The simulation for circuit 2 allowed obtaining a good fitting of the experimental data at high concentration of MWCNTs near and above percolation transition (Fig. 7.17b).

The more complicated equivalent circuits were also tested in order to account for the processes at the composite-electrodes and PPG-MWCNT interfaces [80]. The fractal dimensions of the nanocomposite-electrode, D_f^e, and PPG-MWCNT, D_f^c, interfaces increase with increase the concentration of MWCNTs (Fig. 7.18).

At small concentration of nanotubes the value of D_f^e is close to 2 and it means that nanocomposite-electrode interface is smooth. However, this interface becomes rougher near the percolation threshold. From other hand, the fractal dimension of PPG-MWCNT interface, D_f^c, is noticeably higher than D_f^e and it approaches to 3 at $C_n \approx 0.7$ wt%. It corresponds to the formation of infinitely rough PPG-MWCNT interface above percolation threshold.

7.4 Nanoplatelets (Montmorillonite, Laponite)

The structure and properties of nanocomposites based on PEG-1000 doped with MMTO and LapO were studied in detail using small- and wide-angle X-ray diffraction, differential scanning calorimetry, and impedance spectroscopy [82]. Figure 7.19 shows WAXS patterns of PEG-1000 doped with 5 wt% MMTO (a) and LapO (b) [82].

The two diffraction reflexes were observed at small angles (at $2\theta < 10°$) for PEG-1000 + MMTO composite (Fig. 7.19a) and they corresponded to the different

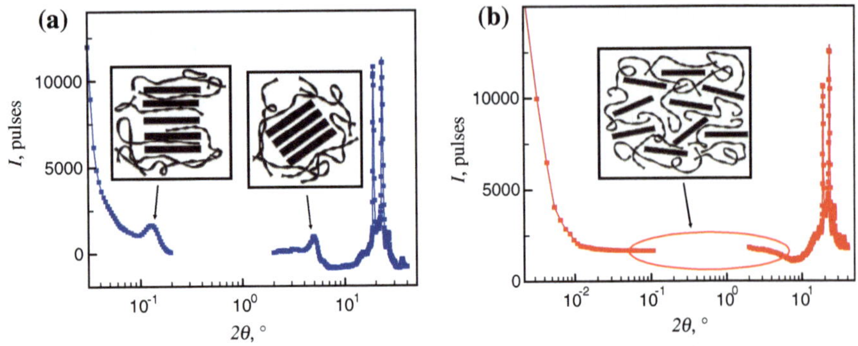

Fig. 7.19 WAXS patterns of PEG-1000 doped with 5 wt% MMTO (**a**) and LapO (**b**). *Insets* show the presentations of structure of the composites. T = 298 K [82]

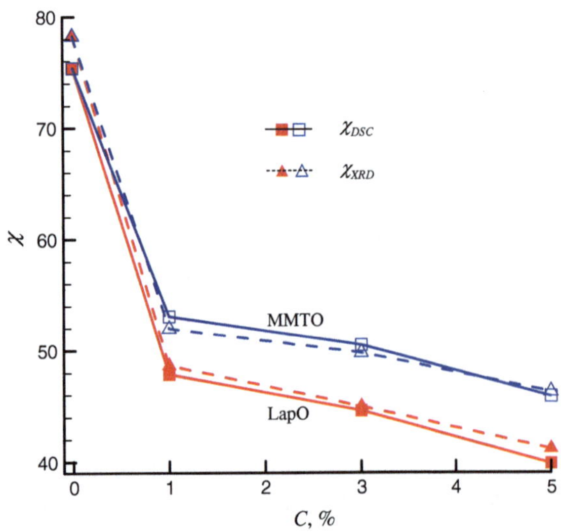

Fig. 7.20 Degree of crystallinity χ_{DSC} and χ_{XRD} of PEG-1000 doped with MMTO and LapO versus the concentration of nanoplatelets, C [82]

interlayer distances, d_i. The first peak (d_i = 3.7 nm) corresponds to the partial intercalation of macromolecules inside interlayer space of MMTO and second peak (d_i = 1.9 nm) evidences the presence of undamaged tactoids (stacks of parallel clay platelets). From other hand, these crystalline reflexes were absent for PEG-1000 + LapO composite (Fig. 7.19b). It evidences the complete exfoliation of LapO platelets inside PEG-1000 matrix.

Figure 7.20 presents degree of crystallinity determined by different methods (χ_{DSC} and χ_{XRD}) of PEG-1000 doped with MMTO and LapO versus the concentration of nanoplatelets [82]. The both DSC and X-ray methods gave the rather results and the introduction of filler resulted in noticeable decrease of χ_{DSC} and χ_{XRD}. It can be explained by the steric restrictions created by a layered silicate surface. The more significant disordering was produced by LapO than MMTO.

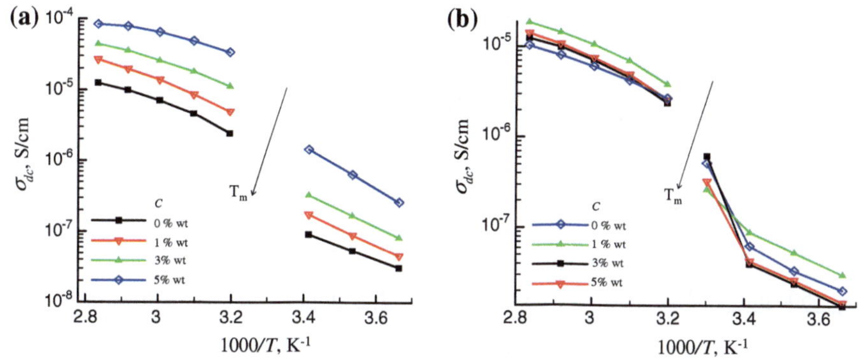

Fig. 7.21 Temperature dependencies of direct current electrical conductivity, σ_{dc}, for PEG-1000 doped with platelets of MMTO (**a**) and LapO (**b**). The *arrows* divides Arrhenius behaviour below the melting temperature, at $T < T_m$, and Vogel–Fulcher–Tammann behaviour above it, at $T > T_m$ [82]

It can be explained by the more deep integration and exfoliation inside PEG of LapO than MMTO.

Figure 7.21 presents temperature dependences of direct current electrical conductivity, σ_{dc}, for PEG-1000 doped with platelets of MMTO (a) and LapO (b) [82]. For these composites the Arrhenius behaviour of $\sigma_{dc}(T)$ was observed below the melting temperature, at $T < T_m$:

$$\sigma_{dc}(T) = \sigma_0 \exp\left(-\frac{E_a}{kT}\right), \qquad (7.13)$$

where E_a is the activation energy, σ_o is the electrical conductivity at infinite temperature.

In the crystalline phase the charge transport is dominated by the hopping mechanism and value of E_a/q_i (q_i is a charge of an ion) corresponds to the blocking potential between two potential wells.

However, above the melting temperature the significant increase in electrical conductivity was observed. The transport of ions at $T > T_m$ is controlled by segmental relaxation of the polymer chains. It was observed that the temperature dependence of the electrical conductivity at $T > T_m$ can be fitted with high precision with a Vogel–Tamman–Fulcher (VTF) type equation [40, 53, 78, 79, 82, 83]:

$$\sigma_{dc} = \sigma_0 \exp\left(-\frac{B}{T - T_0}\right), \qquad (7.14)$$

where B is the parameter and T_0 is a Vogel temperature.

It is interesting that increase of the filler content resulted in a noticeable increase in the electrical conductivity, σ_{dc}, for MMTO, while insignificant changes were observed for LapO (Fig. 7.21). Such differences between behaviour of MMTO and

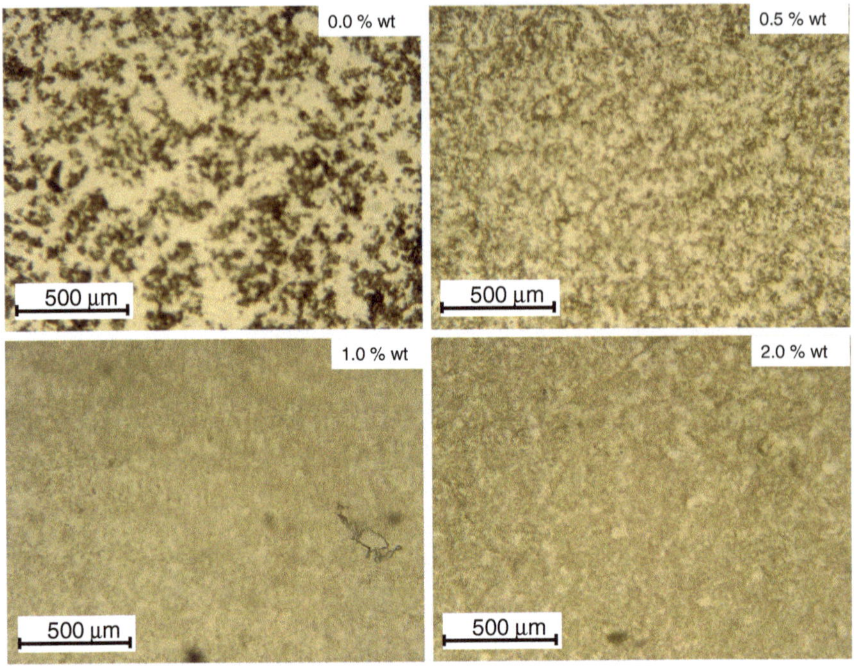

Fig. 7.22 Microscopic images of PEG-1000 doped by with MWCNTs (0.5 wt%) at different concentrations of MMTO, C_p. T = 310 K (From [40]. With permission)

LapO fillers were explained by the different effects of these fillers on the free volume in the polymer matrix [82].

The phase behaviour, microstructure and percolation of PEG-1000 doped by MWCNTs, MMTO, and MWCNTs + MMTO hybrids were studied in [40]. The percolation threshold transitions were observed at ≈0.1 wt%, and at ≈0.5–1.0 wt% for filling of PEG by MWCNTs and MMTO, respectively. It is interesting that adding of MMTO allowed facilitating the dispersion of MWCNTs inside PEG-1000 matrix (Fig. 7.22).

The increase in electrical conductivity was observed with addition of MMTO (0–5 wt%) to the PEG-1000 + MWCNTs (0.5 wt%) composition (Fig. 7.23). It is remarkable because the noticeable improvement of electrical conductivity was observed above the percolation threshold and at high loadings by both MWCNTs (=0.5 wt%) and MMTO (5 wt%).

The percolation behavior of polypropylene glycol (PPG-400) doped by with MWCNTs and Lap or LapO was studied by using different experimental methods [84].

The Lap and LapO displayed different affinity to PPG-400. The data have evidenced finite PPG integration inside Lap and complete exfoliation of LapO stacks in a PPG matrix. The percolation was observed at C_n ≈ 0.4 wt% for

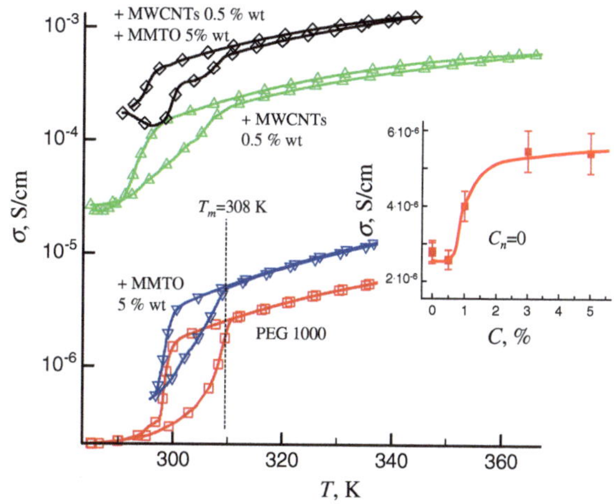

Fig. 7.23 Electrical conductivity, σ, versus the temperature for PEG-1000 doped with MMTO, MWCNTs, and their mixtures. *Inset* shows σ versus the concentration MMTO, *C*, in absence of MWCNTs (at $C_n = 0$ wt%) (From [40]. With permission)

PPG-400 + MWCNT and PPG-400 + MWCNT + Lap systems and at $C_n \approx 0.2$ wt% for PPG-400 + MWCNT + LapO systems (Fig. 7.24). The observed behaviour was attributed to the effects exerted by LapO on the size of MWCNT aggregates, state of their dispersion and homogeneity of their spatial distribution [84].

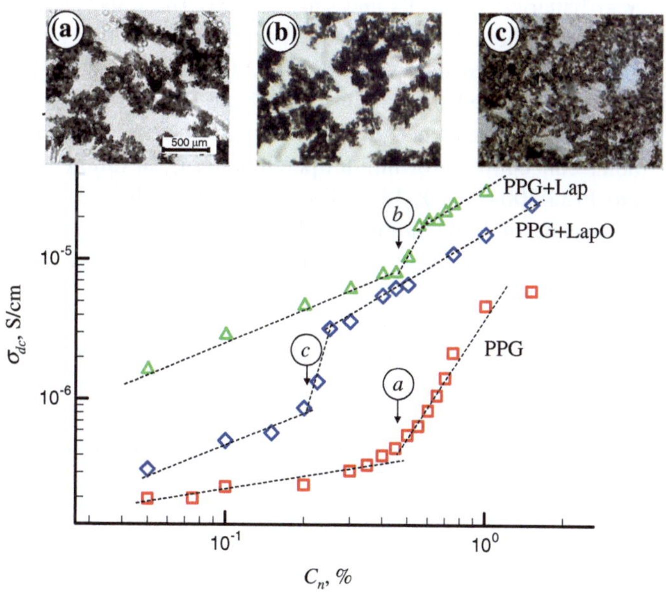

Fig. 7.24 Direct current electrical conductivity, σ_{dc}, versus concentration of MWCNTs, C_n, in PPG-400, PPG-400 + Lap (0.1 wt%) and PPG-400 + LapO (0.1 wt%) systems. Microscopic images of composites with MWCNT concentrations in the vicinity of percolation thresholds are also shown (From [84]. With permission)

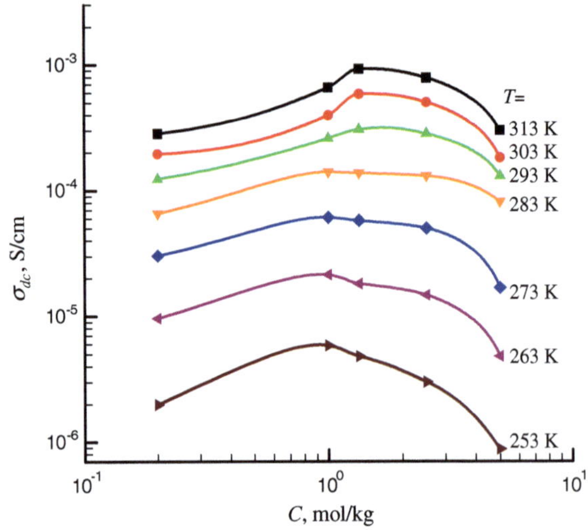

Fig. 7.25 Direct current electrical conductivity, σ_{dc}, versus the molar concentration of lithium perchlorate, C, at different temperatures for the system PEG-300–$LiClO_4$

7.5 Effects of Inorganic Salts

Composite polymer electrolytes on base of polyglycols with added lithium salts were studied in many works in relation with their wide range of applications in lithium polymer batteries, fuel cells and sensors [85]. The important characteristic of these composites is their electrical conductivity. The studies on high (e.g., PEO, $M_w \approx 10^6$ [86]) or low molecular (e.g., PEG-400 [87], PEG-300 [88], PEG-1000) weights polymers complexed with lithium salts have been done.

Figure 7.25 shows dependence of direct current electrical conductivity, σ_{dc}, versus the molar concentration of lithium perchlorate, C, at different temperatures for the system PEG-300–$LiClO_4$ [88]. The electrical conductivity grows significantly with increasing of temperature. It reflects an increase in the mobility of the polymer chain segments. The electrical conductivity passes through a maximum $C = C_m$ with increase of C. The value of C_m of was ≈1 mol/kg and increases with increase of temperature.

The increase of σ_{dc} in the range $C < C_m$ reflects the increase of concentration of charge carriers. The increase of σ_{dc} in the range $C > C_m$ reflects the formation of ionic associates. The association degree of ions in lithium electrolytes can be estimated using a Fuoss-Kraus theory [89]. For lithium perchlorate the following ion-ion transformation are allowed:

$$Li^+ + ClO_4^- \overset{K_I}{\Leftrightarrow} LiClO_4,$$

$$LiClO_4 + Li^+ \overset{K_T}{\Leftrightarrow} Li_2ClO_4^+, \tag{7.15}$$

$$\mathrm{LiClO_4} + \mathrm{ClO_4^-} \overset{K_T}{\Leftrightarrow} \mathrm{Li(ClO_4)_2^-}.$$

Here, K_I and K_T are the association constants of individual ions and triplets, respectively.

The formation of triple ions in solution is also accompanies with formation of ion pairs and free ions. The molar concentration of different ionic species can be calculated as:

$$\begin{aligned} C_{Li^+} &= C_{ClO_4^-} = \alpha_I \cdot C, \\ C_{Li_2(ClO_4)^+} &= C_{Li(ClO_4)_2^-} = \alpha_T \cdot C, \\ C_{LiClO_4} &= C\alpha_P = C(1-\alpha_I - 3\alpha_T), \end{aligned} \tag{7.16}$$

where α_I, α_P, α_T are the fractions of individual ions, pairs and triplets, respectively.

The concentration of non-dissociated molecules of $\mathrm{LiClO_4}$ decreases due to the formation of individual ions and triplets.

The association constants of individual ions and triplets can be calculated from the following equations:

$$\begin{aligned} K_I &= \frac{\mathrm{LiClO_4}}{\mathrm{Li^+} \cdot \mathrm{ClO_4^-}} = \frac{(1-\alpha_I)}{\alpha_I^2 \cdot C}, \\ K_T &= \frac{\mathrm{Li_2(ClO_4)^+}}{\mathrm{Li^+} \cdot \mathrm{LiClO_4^-}} = \frac{\mathrm{Li(ClO_4)_2^-}}{\mathrm{LiClO_4} \cdot \mathrm{ClO_4^-}} = \frac{\alpha_T}{\alpha_I(1-\alpha_I-3\alpha_T)C}. \end{aligned} \tag{7.17a}$$

In these derivations it was assumed that the species they are spherical and have the same radii.

The inverse transformations give:

$$\begin{aligned} \alpha_I &= (-1 + \sqrt{1+4K_IC}))/(2K_IC), \\ \alpha_T &= K_T\alpha_I(1-\alpha_I)C/(1+3K_I\alpha_IC). \end{aligned} \tag{7.17b}$$

So, this approach allows estimation the fractions of individual ions, α_I, pairs, $\alpha_P = 1 - \alpha_I - 3\alpha_T$), and triplets, α_T, from the known values of K_I and K_T.

The constants K_I and K_T for the system PEG-300–$\mathrm{LiClO_4}$ were experimentally estimated from analysis of $\sigma_{dc}(C)$ dependences [88]. It was shown that value of α_I decreases with increase of concentration, C, and increases with increase of temperature, T. The tendency of the formation of ion pairs, triplets and higher order aggregates with increasing of salt concentration was observed. This tendency may be explained using the following mechanism. The interactions between polyglycols and lithium salts result in formation of inter- or intra-molecular short-cross-links between polymer chains. These cross-links have limited flexibility and can be

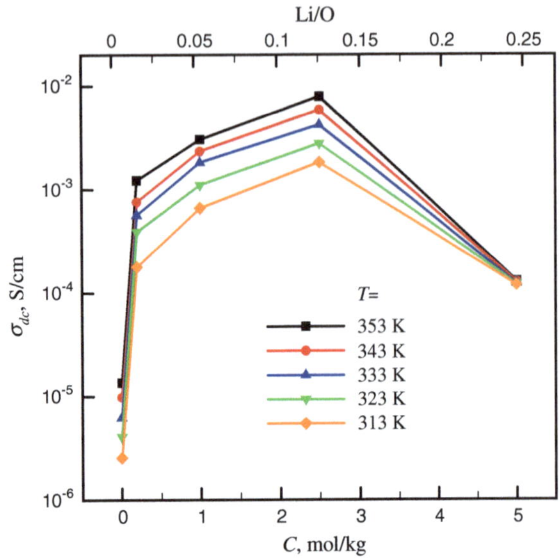

Fig. 7.26 Direct current electrical conductivity, σ_{dc}, versus the molar concentration of lithium perchlorate, C, at different temperatures for the system PEG-1000–$LiClO_4$

formed across the Li^+ ions or $Li_2ClO_4^+$ triplets. The doping of PEG-300–$LiClO_4$ electrolyte by MWCNTs resulted in increase of the fractions of individual ions, α_I, and decrease of the values of α_P and α_T [88]. It can be speculated that MWCNTs prevents formation of cross-links between Li^+ ions and ether oxygens, or formation of ionic associates in polymeric electrolytes.

The concentration dependences $\sigma_{dc}(C)$ for the system PEG-1000–$LiClO_4$ also demonstrated the similar tendencies and the presence of maximums (Fig. 7.26) [90]. The maximum of electrical conductivity for this polymeric electrolyte was observed at $C \approx 2.5$ mol/kg that corresponds to the ratio of Li/O = 1/8. The data of WAXS and DSC evidence about complete amorphization of this polymeric electrolyte at Li/O = 1/8. Ion transfer is carried out mainly in the amorphous phase, and therefore the maximum conductivity can be explained by high mobility of ionic species the amorphous phase. A further increase of C leads to a decrease of ionic conductivity that can be explained by the same reasons as for the system PEG-300–$LiClO_4$.

It is interesting that at high concentration of lithium perchlorate ($C = 5$ mol/kg, Li/O = ¼) the electrical conductivity become independent of temperature. This behaviour is atypical for polymeric electrolyte and can be explained by the high degree of ionic association and formation of high order ionic aggregates. The existence of such aggregates was confirmed by the data of SAXS. The theoretical "core-shell" model gives 1.9 and 3.3 nm for the radii of ionic core and ionic core with shell, respectively.

The similar maximum of electrical conductivity for the system PPG-400 + $LiClO_4$ was also observed at $C \approx 1$ mol/kg (at the ratio of Li/O = 1/20) [83]. The size of correlation zone, L_c, (of the amorphous clusters) was ≈0.75 nm at this concentration of electrolyte.

7.6 Conclusions

The experimental data evidence that doping of polyglycols (polyethylene glycol and polypropylene glycol) by nanoparticles with anisotropic shape (carbon nanotubes and inorganic nanoplatelets) leads to additional disorder in amorphous polymer matrix. Impact of nanofillers on the structure in these polymer composites at rather low concentrations (<0.3–0.5 wt%) was observed. Doping of polyglycols by MWCNTs changes the polymer crystallinity degree. Moreover, the flexibility of MWCNTs inside the polymer matrix becomes higher. The MWCNTs noticeably affect the processes of nucleation and growth of crystals inside polymer matrix. Increase of concentration results in formation of clusters of MWCNTs, which, in the vicinity of percolation threshold, can merge into the large cluster spanning through the system. Anomalous behavior of most parameters that characterize the crystallization kinetics of polyglycols doped with MWCNTs are associated with percolation threshold in the systems. These behaviors are also typical for the concentration dependencies of Avramy exponent, fractal dimension of crystallites and many other properties.

Doping of polyglycols by nanoplatelets of MMTO or LapO resulted in intercalation of polymers inside the interlayer space. Moreover, it was established that partial intercalation was observed for MMTO, whereas the complete exfoliation was observed of LapO. The more developed amorphization was observed for doping by LapO than for MMTO. It was demonstrated that variation of nanofiller loading allows the noticeable changes in the structure and properties of the polymeric electrolytes on the base of polyglycols and additives of inorganic salts. The introduction of small quantity of nanofiller ($\sim$5 wt%) results in increase of the crystallinity degree of polymer electrolyte. The different mechanisms of charge transfer in the studied systems were realized: activation Arrhenius mechanism and mechanism evolving the free volume.

Finally, the impact of the temperature and concentration of salt and MWCNTs on the ion association in PEG + $LiClO_4$ + MWCNTs was analyzed. The increase of temperature resulted in increase of concentration of individual ions and in decrease concentration of ionic pars and triplets. With increase of salt concentration the electrical conductivity growths initially, goes through the maximum at the certain concentration and then decreases. It can be explained by the ionic association. Introduction of nanofiller results in the increase of the concentration of the individual ions, the decrease of ionic aggregates and give the increase of the electrical conductivity.

References

1. E.A. Stefanescu, C. Daranga, C. Stefanescu, Materials **2**, 2095 (2009)
2. S. Nuriel, L. Liu, A.H. Barber, H.D. Wagner, Chem. Phys. Lett. **404**, 263 (2005)
3. L. Vaisman, G. Marom, H.D. Wagner, Adv. Funct. Mater. **16**, 357 (2006)

4. J. Amiran, V. Nicolosi, S.D. Bergin, U. Khan, P.E. Lyons, J.N. Coleman, J. Phys. Chem. C **112**, 3519 (2008)
5. S. Kubota, T. Maruyama, H. Nishikiori, N. Tanaka, M. Endo, T. Fujii, Chem. Lett. **38**, 890 (2009)
6. J. Chattopadhyay, F. De Jesus Cortez, S. Chakraborty, N.K.H. Slater, W.E. Billups, Chem. Mater. **18**, 5864 (2006)
7. Y. Wen, H. Wu, S. Chen, Y. Lu, H. Shen, N. Jia, Electrochim. Acta **54**, 7078 (2009)
8. J. Lee, E.J. Park, J. Choi, J. Hong, S.E. Shim, Synth. Met. **160**, 566 (2010)
9. L. Niu, Y. Luo, Z. Li, Sens. Actuators, B: Chem. **126**, 361 (2007)
10. T. Hirata, S. Amiya, M. Akiya, O. Takei, T. Sakai, T. Nakamura, J. Kawamura-Tsuzuku, T. Yamamoto, R. Hatakeyama, Jpn. J. Appl. Phys. **47**, 2068 (2008)
11. K. Takagi, T. Hirata, M. Akiya, IEEJ Trans. Sens. Micromachines **129**, 333 (2009)
12. A. Sarafraz-Yazdi, A. Amiri, G. Rounaghi, H. Eshtiagh-Hosseini, J. Chromatogr., B: Anal. Technol. Biomed. Life Sci. **908**, 67 (2012)
13. Z. Jin, X. Sun, G. Xu, S.H. Goh, W. Ji, Chem. Phys. Lett. **318**, 505 (2000)
14. Q.-J. Meng, X.-X. Zhang, S.-H. Bai, J. Appl. Polym. Sci. **106**, 2018 (2007)
15. Q.-J. Meng, X.-X. Zhang, X.-C. Wang, High Perform. Polym. **19**, 451 (2007)
16. C.-L. Huang, M.-J. He, M. Huo, L. Du, C. Zhan, C.-J. Fan, K.-K. Yang, I.-J. Chin, Y.-Z. Wang, Polym. Chem. **4**, 3987 (2013)
17. S.M. Kang, S.H. Kwon, J.H. Park, B.K. Kim, Polym. Bull. **70**, 885 (2013)
18. M. Zenkiewicz, J. Richert, A. Różański, Polym. Testing **29**, 251 (2010)
19. A. Mohaddespour, H. Abolghasemi, S.J. Ahmadi, J. Compos. Mater. **42**, 2163 (2008)
20. H.-F. Lei, P. Wang, Y.-M. Zhang, W.-B. Yuan, Cailiao Kexue Yu Gongyi/Mater. Sci. Technol. **17**, 457 (2009)
21. G. Ozkoc, S. Kemaloglu, J. Appl. Polym. Sci. **114**, 2481 (2009)
22. J. Gai, H. Li, J. Appl. Polym. Sci. **105**, 1200 (2007)
23. Y. Lu, S.-T. Kong, H.-J. Deiseroth, W. Mormann, Macromol. Mater. Eng. **293**, 900 (2008)
24. B. Chen, J.R.G. Evans, J. Phys. Chem. B **108**, 14986 (2004)
25. R.J. Sengwa, S. Choudhary, S. Sankhla, Colloids Surf., A **336**, 79 (2009)
26. S. Zhu, H. Peng, J. Chen, H. Li, Y. Cao, Y. Yang, Z. Feng, Appl. Surf. Sci. **276**, 502 (2013)
27. S. Zhu, J. Chen, H. Li, Polym. Bull. **63**, 245 (2009)
28. S.-P. Zhu, J.-Y. Chen, H.-L. Li, Y. Cao, Gaofenzi Cailiao Kexue Yu Gongcheng/Polymeric Mater. Sci. Eng. **26**, 27 (2010)
29. S. Zhu, J. Chen, H. Li, Y. Cao, J. Appl. Polym. Sci. **128**, 3876 (2012)
30. R. De Lisi, M. Gradzielski, G. Lazzara, S. Milioto, N. Muratore, S. Prévost, J. Phys. Chem. B **112**, 9328 (2008)
31. F.M. Gray, J.A. Connor (ed.), *Polymer Electrolytes* (The Royal Society of Chemistry (RSC Materials Monographs), Cambridge, UK, 1997)
32. F.M. Gray, M. Armand, in *Handbook of Battery Materials*, ed. by C. Daniel and J. Besenhard (Wiley-VCH, Weinheim, 2011), pp. 627–656
33. M.R. Johan, O.H. Shy, S. Ibrahim, S.M.M. Yassin, T.Y. Hui, Solid State Ionics **196**, 41 (2011)
34. F. Croce, G.B. Appetecchi, L. Persi, B. Scrosati, Nature **394**, 456 (1998)
35. S. Kim, E.-J. Hwang, Y. Jung, M. Han, S.-J. Park, Colloids Surf., A **313–314**, 216 (2008)
36. J.-H. Ahn, Y.-J. Kim, G.X. Wang, Met. Mater. Int. **12**, 69 (2006)
37. C. Branca, S. Magazù, G. Maisano, F. Migliardo, P. Migliardo, G. Romeo, J. Phys. Chem. B **106**, 10272 (2002)
38. A.V. Melezhyk, Y.I. Sementsov, V.V. Yanchenko, Appl. Chem. **78**, 938 (2005). (in Russian)
39. E.A. Lysenkov, Y.P. Gomza, V.V. Klepko, Y.A. Kunitsky, Phys. Chem. Solid State (Fizika i Himiya Tverdogo Tila, Ivano-Frankivsk, Ukraine) **11**(2), 361 (2010)
40. N.I. Lebovka, E.A. Lysenkov, A.I. Goncharuk, Y.P. Gomza, V.V. Klepko, Y.P. Boiko, J. Compos. Mater. **45**(24), 2555 (2011)
41. E.A. Lysenkov, V.V. Klepko, V.M. Golovanets, Polym. J. (Polimerniy Zhurnal, Kiev, Ukraine) **36**, 269 (2014)

42. W. Wieczorek, A. Zalewska, D. Raducha, Z. Florjańczyk, J.R. Stevens, J. Phys. Chem. B **102**, 352 (1998)
43. H.-W. Chen, F.-C. Chang, Polymer **42**, 9763 (2001)
44. Y. Kong, J.N. Hay, Eur. Polym. J. **39**, 1721 (2003)
45. K. Pielichowski, K. Flejtuch, Polym. Adv. Technol. **13**, 690 (2002)
46. A. Guinier, *Theorie Et Technique De La Radiocristallographie* (Dunod, Paris, 1964)
47. J.L. Matthews, H.S. Peiser, R. R.B., Acta Crystallographica **2**, 85 (1949)
48. A. Kyritsis, P. Pissis, J. Grammatikakis, J. Polym. Sci., Part B: Polym. Phys. **33**, 1737 (1995)
49. L. Liu, Y. Yang, Y. Zhang, Physica E **24**, 343 (2004)
50. J. Feder, *Fractals* (Plenum Press, New York, 1988)
51. W. Bauhofer, J.Z. Kovacs, Compos. Sci. Technol. **69**, 1486 (2009)
52. V.V. Klepko, B.B. Kolupaev, E.A. Lysenkov, M.O. Voloshyn, Mater. Sci. **47**, 14 (2011)
53. E.A. Lysenkov, Y.P. Gomza, M. Minenko, V.V. Klepko, Polym. J. (Polimerniy Zhurnal, Kiev, Ukraine) **32**(1), 17 (2010)
54. E.A. Lysenkov, V.V. Klepko, J. Nano- and Electron. Phys. **5**, 03052(6 pp) (2013)
55. A.A. Lysenkov, V.V. Klepko, Y.V. Yakovlev, Nanostructural Mater. Sci. (Nanostrukturnoe Materialovedenie, Kiev, Ukraine) **3–4**, 46 (2013)
56. E.A. Lysenkov, V.V. Klepko, Y.V. Yakovlev, J. Nano- Electron. Phys. (Sumy, Ukraine) **7**, 01031 (6 pp) (2015)
57. E.A. Lysenkov, V.V. Klepko, V.M. Golovanets, V.L. Demchenko, Ukrainian J. Phys. **59**(9), 906 (2014)
58. E.A. Lysenkov, V.V. Klepko, J. Eng. Phys. Thermophys. (Inzhenerno-Fizicheskii Zhurnal, Minsk, Belarus) **88**(4), 973 (2015)
59. E.A. Lysenkov, Y.V. Yakovlev, V.V. Klepko, Phys. Surf. Eng. (Kharkiv, Ukraine) **12(1)**, 31 (2014)
60. J. Teixeira, J. Appl. Crystallogr. **21**, 781 (1988)
61. E.A. Lysenkov, Y.P. Gomza, V.V. Davydenko, V.V. Klepko, Polym. J. (Polimerniy Zhurnal, Kiev, Ukraine) **32**(2), 99 (2010)
62. M. Avrami, J. Chem. Phys. **9**, 177 (1941)
63. M.K. Hassan, J. Kurths, arXiv:cond-mat/0407715v1 [cond-mat.stat-mech] (2004)
64. E.A. Lysenkov, Y.V. Yakovlev, V.V. Klepko, Phys. Surf. Eng. (Kharkiv, Ukraine) **12**(2), 223 (2014)
65. M.O. Lisunova, Y.P. Mamunya, N.I. Lebovka, A.V. Melezhyk, Eur. Polym. J. **43**, 949 (2007)
66. D. Stauffer, A. Aharony, *Introduction to Percolation Theory* (Taylor and Francis, London, 1992)
67. I. Balberg, C.H. Anderson, S. Alexander, N. Wagner, Phys. Rev. B **30**, 3933 (1984)
68. N. Lebovka, M. Lisunova, Y.P. Mamunya, N. Vygornitskii, J. Phys. D Appl. Phys. **39**, 2264 (2006)
69. P. Kim, L. Shi, A. Majumdar, P.L. McEuen, Phys. Rev. Lett. **87**, 215502 (2001)
70. M. Fujii, X. Zhang, H. Xie, H. Ago, K. Takahashi, T. Ikuta, H. Abe, T. Shimizu, Phys. Rev. Lett. **95**, 65502 (2005)
71. A. Eucken, Forschung Auf Dem Gebiete Des Ingenieurwesens **11**, 6 (1940)
72. D.A.G. Bruggeman, Annalen Der Physik. Leipzig **24**, 636 (1935)
73. T.B. Lewis, L.E. Nielsen, J. Appl. Polym. Sci. **14**, 1449 (1970)
74. R. Pal, J. Reinf. Plast. Compos. **26**, 643 (2007)
75. R.E. Meredith, C.W. Tobias, J. Appl. Phys. **31**, 1270 (1969)
76. W. Tian, R. Yang, Comput. Model. Eng. Sci. **24**, 123 (2008)
77. E.A. Lysenkov, Y.V. Yakovlev, V.V. Klepko, Ukrainian J. Phys. **58**, 378 (2013)
78. E.A. Lysenkov, Y.P. Gomza, V.V. Klepko, M.A. Rehteta, Polym. J. (Polimerniy Zhurnal, Kiev, Ukraine) **32**(5), 429 (2010)
79. E.A. Lysenkov, V.V. Klepko, Ukrainian J. Phys. **56**, 484 (2011)
80. E.A. Lysenkov, Y.V. Yakovlev, V.V. Klepko, Polym. J. (Polimerniy Zhurnal, Kiev, Ukraine) **34**(4), 345 (2012)

81. E. Barsoukov, J.R. Macdonald (eds.), *Impedance Spectroscopy Theory, Experiment, and Applications* (A John Wiley & Sons Inc, Publication, 2005)
82. E.A. Lysenkov, Y.P. Gomza, V.V. Klepko, Y.A. Kunitsky, L.A. Kunitska, Nanosystems, Nanomaterials, Nanotechnol. (Kiev, Ukraine) **8**(3), 677 (2010)
83. E.A. Lysenkov, Y.P. Gomza, V.V. Klepko, M.A. Rehteta, Y.A. Kunitsky, I.M. Shabelnik, Nanosystems, Nanomaterials, Nanotechnol. (Kiev, Ukraine) **8**(3), 693 (2010)
84. E.A. Lysenkov, N.I. Lebovka, Y.V. Yakovlev, V.V. Klepko, N.S. Pivovarova, Compos. Sci. Technol. **72**, 1191 (2012)
85. M.B. Armand, P.G. Bruce, M. Forsyth, B. Scrosati, W. Wieczorek, in *Energy Materials*, ed. by D.W. Bruce, D.O'Hare, R.I. Walton (John Wiley & Sons, Ltd, 2011), pp. 1–31
86. J.H. Shin, Y.T. Lim, K.W. Kim, H.J. Ahn, J.H. Ahn, J. Power Sources **107**, 103 (2002)
87. D. Shanmukaraj, R. Murugan, J. Power Sources **149**, 90 (2005)
88. E.A. Lysenkov, V.V. Klepko, Research Bulletin of National Technical University of Ukraine "Kyiv Polytechnic Institute" (Naukovi Visti) **N3**, 88 (2010)
89. R.M. Fuoss, C.A. Kraus, J. Am. Chem. Soc. **55**, 2387 (1933)
90. E.A. Lysenkov, Y.P. Gomza, V.V. Klepko, Polym. J. (Polimerniy Zhurnal, Kiev, Ukraine) **32** (3), 223 (2010)

Chapter 8
Colloidal Solution of 3 nm Bucky Diamond: Primary Particles of Detonation Nanodiamond

N.O. Mchedlov-Petrossyan, N.N. Kamneva, E. Ōsawa, A.I. Marynin, S.T. Goga, V.V. Tkachenko and A.P. Kryshtal

Abstract The nanodiamond (ND) hydrosol with positively charged 2.7 ± 0.3 nm primary particles behaves as a hydrophobic colloidal dispersion. The coagulation by inorganic electrolytes with anion charges of −1, −2, −3, and −4 occurs in line with the Schulze–Hardy rule for "positive" sols. The single-charged anions are arranged according to their coagulating ability in the lyotropic series. The sole exception is the hydrophilic HO^- ion, which displays much stronger coagulation impact than those of Cl^- and BF_4^- ions. This particularizes the acidic nature of the positive charge of the colloidal species. The last-named readily adsorb anionic dyes, which results in batochromic shifts of their absorption bands. Application of an acid-base indicator bromocresol green allowed estimating the value of the interfacial electrical potential of the nanodiamond particles $\Psi = +89$ to $+123$ mV, depending on the concentration of the hydrosol. These values are higher as compared with those of the zeta-potential, $\varsigma = +43$ to $+62$ mV. The size distribution of the dispersed system is strongly concentration-dependent. The dilution of the initial 5.0 wt/vol% ND hydrosol by water results in gradual increase in the average particle size, up to

N.O. Mchedlov-Petrossyan (✉) · N.N. Kamneva · S.T. Goga · V.V. Tkachenko
Department of Physical Chemistry, Kharkov V. N. Karazin National University, 4, Svobody Square, Kharkiv 61022, Ukraine
e-mail: mchedlov@yandex.ru

E. Ōsawa
Faculty of Textile Science and Technology, Nano Carbon Research Institute Ltd, AREC (Asama Research Extension Center), Shinshu University, 3-15-1 Tokida, Ueda, Nagano 386-8567, Japan
e-mail: osawa@nano-carbon.jp

A.I. Marynin
National University of Food Technologies, 68, Volodymyrskaya, Kyiv 01601, Ukraine
e-mail: a_marinin@ukr.net

A.P. Kryshtal
Department of Physics and Technology, Kharkov V. N. Karazin National University, 4, Svobody Square, Kharkiv 61022, Ukraine
e-mail: kryshtal@mail.ru

L. Bulavin and N. Lebovka (eds.), *Physics of Liquid Matter: Modern Problems*, Springer Proceedings in Physics 171, DOI 10.1007/978-3-319-20875-6_8

ca. 30 nm in 0.01 % colloidal solution. These results of dynamic light scattering were confirmed by transmission electron microscopy. Accordingly, the viscosity of the hydrosol decreases along with dilution. This phenomenon was explained in terms of the periodic colloidal structures, or colloidal crystals, formed in concentrated solutions.

8.1 Introduction

This report is devoted to colloid properties of the hydrosol formed by the primary 3 nm-sized detonation diamond species recently produced in the NanoCarbon Research Institute, Japan.

The interests in nanodiamonds (ND) increase rapidly from a variety of different fields of science and technology [1–3]. In liquid media, the dispersions of ND are actually a kind of either suspensions or sols. Hence, it is necessary to enlarge our knowledge of these colloid systems. Though the most common way to obtain the ND is treating the detonation products, a recently reported new approach should be mentioned [4]. Among a variety of parameters characterizing a colloid system, the size of particles is of special importance. The detonation ND contains the smallest diamond particles available, and generally offered by many vendors as the source of primary particles 4–10 nm in diameter. However, the commercial powder in fact consists of rather large hard agglomerates. The last-named appear as a result of strong electrostatic interactions, which are comparable in bonding energies to covalent C–C bonds [5]. Attrition milling of aqueous suspension of the crude detonation ND provides, when performed under optimum conditions, a black colloidal solution of perfectly disintegrated primary particles [6]. This aqueous solution of colloidal particles from nanocrystals is kinetically stable, but dilution often leads to aggregation. The present study was undertaken in order to gain insight into the colloidal characteristics of the hydrosol formed by single-nano diamond particles mono-dispersed in water.

8.2 Experimental

8.2.1 Preparation of the ND

The hydrosol consisting of the primary particles of dispersed detonation ND was prepared by NanoCarbon Research Institute. The commercial crude grey powder of detonation ND (manufactured by FMD Nano Tech Co., Guangzhou, China) was disintegrated in portions in a 150 ml attrition mill (constructed by Kotobuki Industries Co., Tokyo) using 30 μm zirconia beads in distilled water. Milling conditions have been locally optimized by applying Taguchi's Method of quality engineering [7] and will be disclosed shortly elsewhere [8]. The product of beads milling is a black but translucent and smooth colloidal solution with 4–5 % concentration. The size of the primary particles of detonation ND normally equals to 3.0 ± 0.5 nm, as determined via the Dynamic Light Scattering (DLS) method. The colloidal solution was used in this work virtually as it is except for light centrifugal separation at 5000 rpm for 1 h. As an inevitable consequence of attrition milling, the solid fraction of the product contained 0.4 wt% of zirconia, which is so far difficult to remove [9].

8.2.2 Materials

Sodium chloride, sulfate, tetrafluoroborate, and tetraphenylborate, potassium hexacyanoferrate (III) and hexacyanoferrate (II), and hydrochloric acid were of reagent grade. The aqueous solution of sodium hydroxide was prepared from saturated stock solution using CO_2-free water and kept protected from the atmosphere. Sodium *n*-dodecylsulfate (99 %), fluorescein isothiocyanate and 5′-aminofluorescein were from Sigma-Aldrich, bromocresol green was used in our previous studies [10]. Toluene, *n*-hexane, 1-octanol, and trichloromethane were purified by standard methods. Double-distilled water was used in all the procedures.

8.2.3 Apparatus

Electrical conductance measurements were carried out using platinized platin electrodes in molybdenum-glass cells with the Precision LCR Meter GW Instek LCR-817 (Taiwan) apparatus operating at a frequency of 1 kHz. The cells were calibrated using 12 standard aqueous solutions of potassium chloride, within the concentration range 1×10^{-4} to 0.01 M (1 M ≡ 1 mol L^{-1}). All the measurements were done at 25.00 ± 0.05 °C. The pH determinations were performed by using an R 37-01 potentiometer and pH-121 pH-meter with an ESL-43-07 glass electrode (Labtech, Russia) in a cell with liquid junction, calibrated using standard buffers at 25 °C. The Cl-selective electrodes ELIS-131Cl (NV Lab, Russia) and Orion Research Incorporated (USA), Model 93-17 and the NO_3-selective electrode ELIS-121NO_3 (NV Lab, Russia) were used for estimating the Cl^- and NO_3^- concentrations. For the viscosity measurements, the apparatus VPZh-2 (Labtech, Russia) with capillary diameter of 0.56 mm was used at 25.0 ± 0.2 °C. The uncertainty of time measurements was ±0.1 s. The particle size distribution and zeta-potentials were determined using Zetasizer Nano ZS Malvern Instruments (Great Britain) apparatus at 25 °C, scattering angle 173° in the National University of Food Technologies, Kyiv, Ukraine. Some preliminary experiments have been made in the Laboratory of Professor Paavo Kinnunen, Department of Biomedical Engineering and computational science, Aalto University, Espoo, Finland. Absorption spectra were run with Hitachi U-2000 and SF-46 spectrophotometer against solvent blanks. For the electron microscopy studies, the Selmi TEM-125 K microscope (Sumy, Ukraine) was used. The procedure was as follows. In a vacuum vessel VUP-5 M (Selmi, Sumy, Ukraine), a 10–20 nm carbon film was deposited from the Volta arc on fresh cleavages of KCl monocrystals at the pressure of residual gases around 10^{-5} Torr. After floating off in distilled water, the carbon films were picked up on copper electron-microscopy grids. The portions of the examined solutions were deposited on the films and studied after drying in the bright-field and diffraction modes of the TEM at accelerating voltage of 100 kV. The images were registered using the CCD camera or photographic plates. For registering the chemical elements in the sample by the energy-dispersive X-ray spectroscopy (EDX), a droplet of concentrated ND solution was placed on silicon wafer and after drying EDX analysis was performed on scanning electron microscope (SEM) JSM-840 (Jeol, Japan) fitted with Selmi EDS-1 spectrometer (Sumy, Ukraine). The spectrometer is capable to register elements with $Z > 5$.

8.3 Results and Discussion

8.3.1 Characterization of the Samples

The concentration of the initial ND hydrosol was 5.00 wt/vol%; in some experiments, 3.81 wt/vol% solutions have been used. The pH value of the sols was 4.75–5.00. The light absorption increases gradually from 700 to 350 nm. The solutions do not obey the Bouguer–Lambert–Beer law. The primary species were positively charged. Together with the non-diamond layer, the size of the colloidal particles in the initial sol was 2.7 ± 0.3 nm as found via DLS, while the entire diamond size was around 2 nm. The numerical concentration of the primary particles was about 1.1×10^{21} L^{-1}. In the solid phase, the EDX spectra reveal, besides carbon, 0.28 Cl and 0.28 Zr (wt%, ±0.14 %); no sulfur was detected above 0.02 %. The nitrogen content was around 2 wt%, as determined by elemental analysis. The IR spectrum of the dried sample exhibited the bands normally reported for detonation ND; the presence of adsorbed water, OH, C=O, C–O–C, and (probably) NH groups may be supposed.

The ND species are sparingly extractable into 1-octanol, chloroform, *n*-hexane, and toluene. In the last case, the layer of the very stable emulsion, probably stabilized via the ND particles, is formed between the water and toluene phases (Fig. 8.1).

Interestingly, the average particle size of 3 nm was registered only in concentrated ND solutions. Along with dilution, the particles became larger; this effect was repeated several times and reproduced in three different laboratories (in Japan,

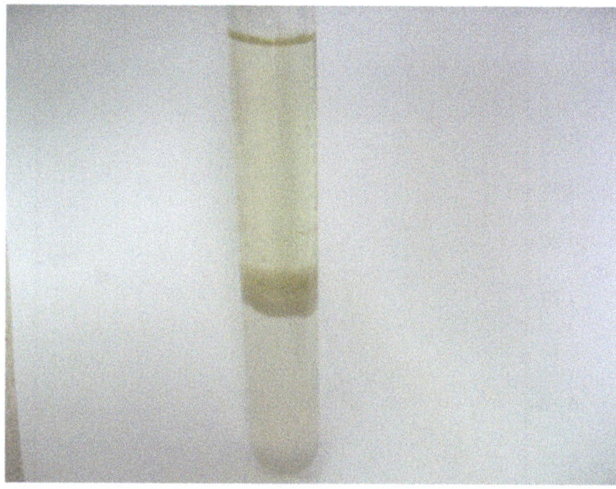

Fig. 8.1 Photo of the distribution of the nanodiamond particles between water (*bottom layer*) and toluene (*upper layer*) phases

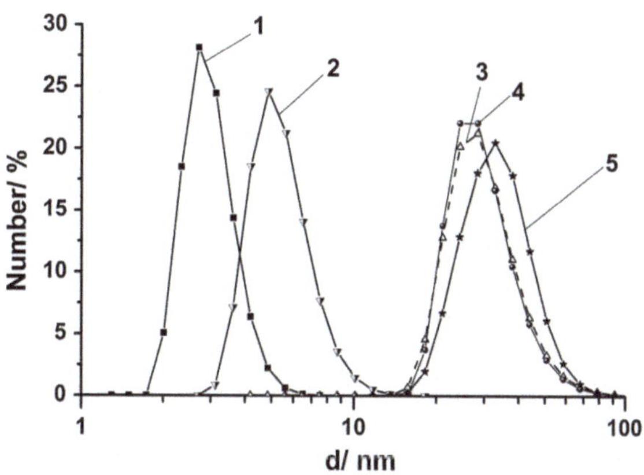

Fig. 8.2 The size distribution for the ND hydrosols of different concentrations, wt/vol%: 5.00 (*1*); 1.00 (*2*); 0.19 (*3*); 0.036 (*4*); and 0.010 (*5*)

Finland, and Ukraine). The results are exemplified in Fig. 8.2. These size changes are to some extent reversible.

Taking into account the significance of the size of the ND colloidal species for their versatile applications, this phenomenon became a matter of detailed consideration in the present research. To better understand this and other properties of the system under study, a comprehensive examining was undertaken.

8.3.2 Charge and Interfacial Potential of the Hydrosol Particles

The zeta potential of the ND colloidal species varies within the range of ca. +(40–60) mV depending on the ND concentration. So, for hydrosols of different wt/vol%

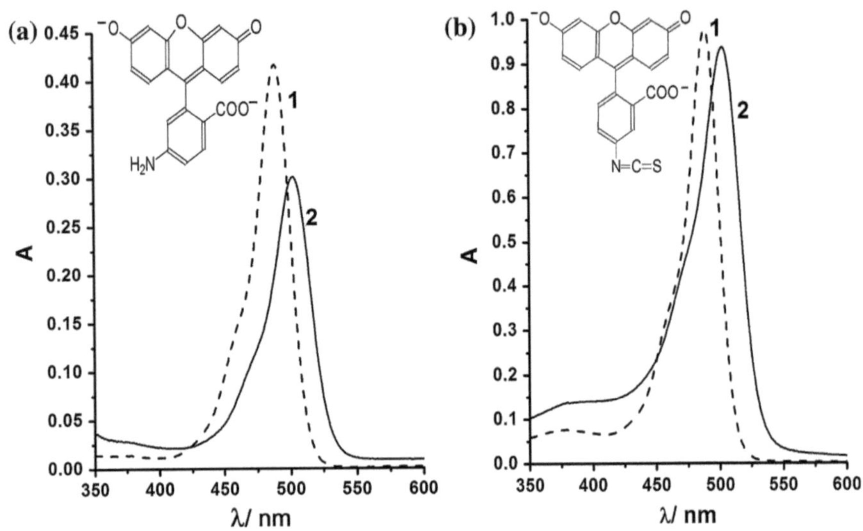

Fig. 8.3 Absorption spectra of 5′-aminofluorescein (**a**) and fluorescein isothiocyanate (**b**) dianions in water, λ_{max} = 489 nm (*1*) and in ND 0.048 wt/vol% solution, λ_{max} = 502 nm (*2*), with subtracting the blank

concentrations, we obtained the following data: ς/mV = +51.9 (2.90 %); +50.1 (2.5 %); +43.3 (0.19 %); and +62.2 (0.036 %).

The confidence range was 0.8–2.7 mV. The specific conductance of 1.00 % ND solution at 25 °C corresponds to that of 4.84×10^{-4} M KCl; the values are 6.874×10^{-5} and 7.797×10^{-5} S cm^{-1} respectively. The Cl^- content in the same ND solution was $(5.5 \pm 1.3) \times 10^{-4}$ M, as found using the Cl-selective electrode, whereas the concentration of the NO_3^- ions was two orders of magnitude lower. This chlorine content is in line with the rough estimate by EDX (see above).

The negatively charged dyes are readily adsorbed on the nanoparticles thus confirming the positive charge of the latter. These results are typified in Fig. 8.3.

Another proof of the positive charge of the species consists in the coagulation of ND hydrosols by electrolytes. Indeed, the ratio of the reciprocal critical coagulation concentration, CCC, by NaCl, Na_2SO_4, $K_3Fe(CN)_6$, and $K_4Fe(CN)_6$ is 1:16:175:538. This is in accordance with the famous Schulze–Hardy rule and is typical for "positive" sols. The CCC value for NaCl equals 2.8 mM (1 mM = 1×10^{-3} M). The coagulation by electrolytes will be considered and discussed below in a more detailed way.

The above-mentioned ς values correspond to the slipping area, or surface of shear. In order to estimate the value of the interfacial potential, Ψ, we made attempt to use measurement with an acid-base indicator. This approach has been adapted for micelles of ionic surfactants and well documented for these systems and other dispersed phases of lyophilic colloids [10–12]. In the case of lyophobic dispersions, the electrolytes necessary for creation of varying pH values may cause the coagulation of the sols. However, one of the peculiarities of the system under consideration consists in relatively high CCC value by HCl. This allows utilization of the hydrochloric acid for creating the pHs within the range of 3–4; the buffer capacity at lower HCl concentrations is insufficiently small. As a suitable acid-base indicator, the sulfonephthalein dye bromocresol green was chosen.

If an acid-base indicator is located on the charged interface, the indices of the so-called apparent ionization constant may be determined by spectrophotometry:

$$\mathrm{p}K_{\mathrm{a}}^{\mathrm{app}} = \mathrm{pH} + log\frac{[\text{acidic form}]}{[\text{basic form}]}. \tag{8.1}$$

The pH value refers to the bulk phase and can be measured using the glass electrode in a cell with liquid junction, whereas the concentration ratio of the equilibrated acid-base couple is available via the absorption spectra. Thus obtained $\mathrm{p}K_{\mathrm{a}}^{\mathrm{app}}$ value is connected with the interfacial potential or the Stern layer potential, Ψ, through (8.2) [10–12]:

$$\mathrm{p}K_{\mathrm{a}}^{\mathrm{app}} = \mathrm{p}K_{\mathrm{a}}^{\mathrm{i}} - \frac{\Psi F}{2.303\,RT}. \tag{8.2}$$

Here F is the Faraday constant, R is the gas constant, T is the absolute temperature, and $K_{\mathrm{a}}^{\mathrm{i}}$ is the so-called intrinsic ionization constant. Then the Ψ value at 25 °C is equal to:

$$\Psi/\mathrm{mV} = 59.16\,(\mathrm{p}K_{\mathrm{a}}^{\mathrm{i}} - \mathrm{p}K_{\mathrm{a}}^{\mathrm{app}}). \tag{8.3}$$

The dye bromocresol green displays the color transition from yellow (HR^-) to green-blue (R^{2-}) (Fig. 8.4). Though in the presence of the ND species, the HCl concentrations over 0.006 M lead to coagulation, the absorption in ND-free water

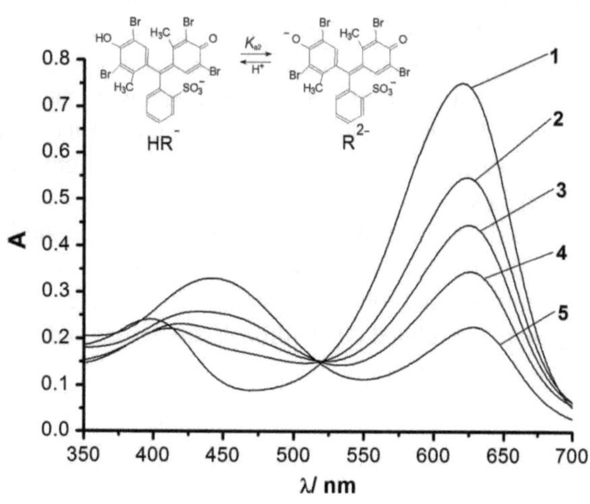

Fig. 8.4 The spectra of the indicator bromocresol green at different pH in 0.024 wt/vol % ND: the spectrum of R^{2-} without HCl additions, pH = 6.2 (*1*); with HCl additives, pH = 3.78 (*2*), 3.59 (*3*), 3.39 (*4*), 3.04 (*5*)

Table 8.1 The pK_a^{app} values of bromocresol green and the estimates of the Ψ values, at 25 °C

w(ND) (wt/vol %)	pK_a^{app}	Ψ/mV
0	4.90	–
0.024	3.40 ± 0.03	+89
0.13	2.82 ± 0.12	+123
0.67	2.99 ± 0.09	+113

solutions at pH 1–2 (not shown in Fig. 8.4) in the region around 600–650 nm is negligible. Thus, this indicator may be considered as a one-color one in this portion of the spectrum. Similar data for this and other indicator dyes in cationic surfactant micelles have been considered in a recent paper [13]. The limiting absorption of the basic form may be measured just in the colloid solution without any additives. The pK_a^{app} values are gathered in Table 8.1; the pK_a value in water was taken from the literature [10].

The next problem in the Ψ estimation was the uncertainty of the pK_a^i value in (8.3). In the case of ionic surfactant micelles, the pK_a^i value of the given indicator is usually equated to its pK_a^{app} in non-ionic micelles, where $\Psi = 0$; otherwise, some more complicated algorithms may be used [12]. However, our experiments with two solvatochromic dyes, methyl orange and Reichardt's betaine (not shown here), demonstrated but small band shifts as compared with their spectra in water. This observation allows concluding that the ND surface being, in fact, the graphene patches, is well-hydrated. Therefore, the pK_a^i value may be equated to the value in water. Thus determined Ψ values are presented in Table 8.1.

The Ψ values are rough estimates only, but they look out reasonable, because the experimentally measured ς values are 1.5–2 times lower, in agreement with the theory of the double electrical layer. However, there are two points to be taken into account. First, as the locus of the adsorbed indicator is probably the Stern layer, the Ψ value should be more close to that of ς. Such conclusions have been made by Mukerjee and Banerjee for the dyes bromocresol green and bromophenol blue associated with micelles of cationic surfactants [14]. Second, taking into account that the TEM images (Fig. 8.5) give no evidence for tight spherical shape of the ND secondary aggregates, which are probably porous, the determined Ψ values may reflect rather the local electrostatic potential in the region where the indicator dye is situated. In turn, the local positive charges of the primary particles, as revealed via electrophoresis measurements, repel the tight association in the secondary aggregates.

The decrease in the pK_a^{app} values and hence the increase in the Ψ values along with the rise of ND concentrations may be attributed to the alterations of the structure of the colloidal species. For instance, the larger secondary aggregates registered in diluted ND hydrosols are porous and thus the interfacial charge density is probably lower.

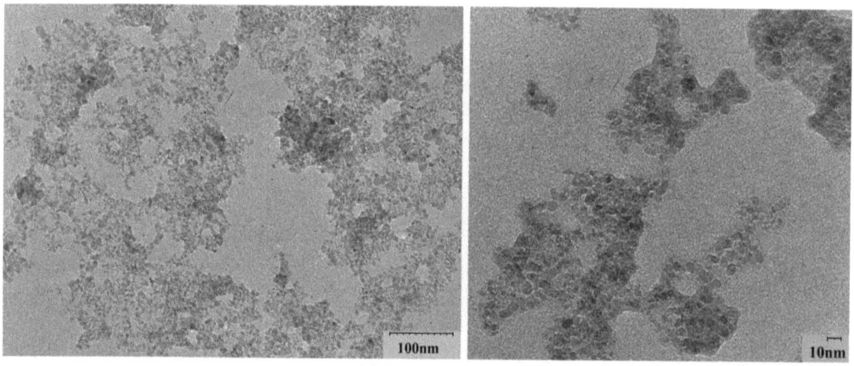

Fig. 8.5 The TEM images of the 0.036 wt/vol% ND hydrosol after evaporation of water (representative examples from several series of experiments)

Fig. 8.6 The time dependence of the absorbance of 0.024 wt/vol% ND hydrosol at NaCl concentrations of 2.0 mM (*1*), 2.5 (*2*), 3.2 (*3*), and 3.5 mM (*4*)

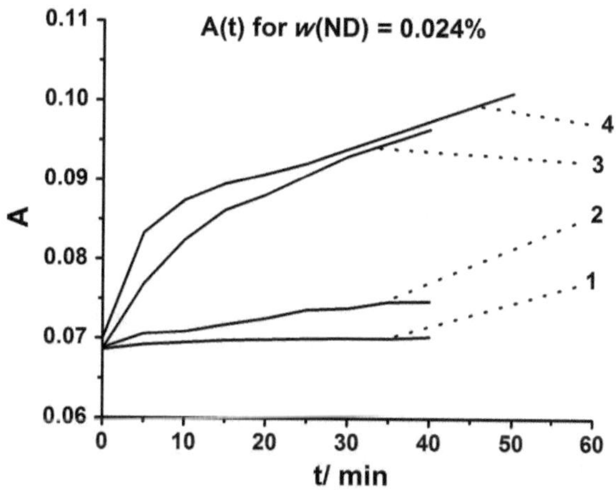

8.3.3 *Coagulation of the ND Hydrosols by Electrolytes*

The coagulation of the ND hydrosols has been studied at 25 °C using two procedures. In both cases, the criterion of the rapid coagulation was the distinct increase in the absorbance (mainly at 525 nm) caused by the turbidity of the systems under study.

First, the absorption spectra of the solutions with constant ND content and varying electrolyte concentrations have been measured immediately and within a period of time (Fig. 8.6).

Otherwise, the CCC values have been determined using the procedure of spectrophotometric titration. In this case, both the (slight, not more than 1.5-fold) dilution of the initial solution by the electrolyte and alteration of the blank

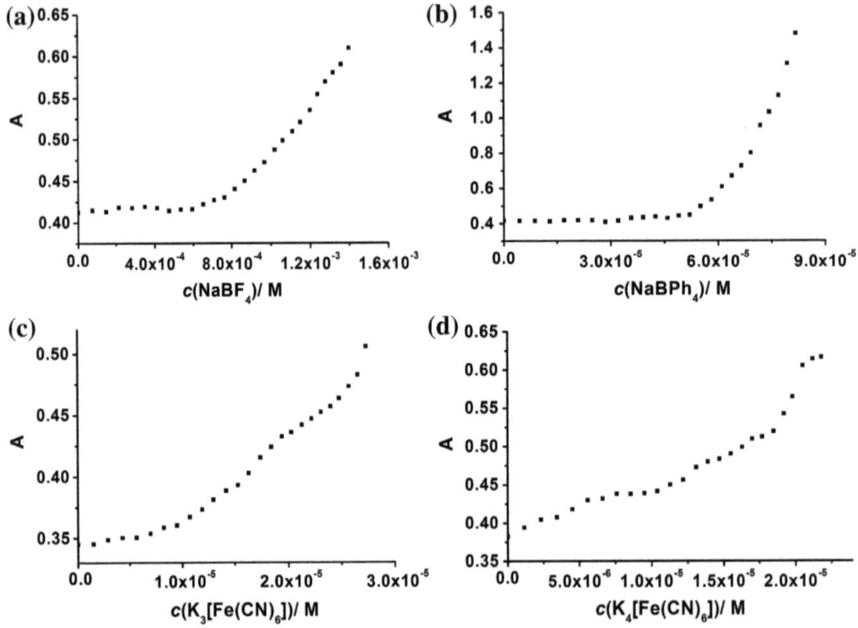

Fig. 8.7 The typical titration curves of the 0.19 wt/vol% ND hydrosol by solutions of electrolytes: $NaBF_4$ (**a**); $NaB(C_6H_5)_4$ (**b**); $K_3Fe(CN)_6$ (**c**); and $K_4Fe(CN)_6$ (**d**)

(electrolyte-free) solvent as a result of dilution with corresponding amounts of pure water (including the deviations from the Bouguer–Lambert–Beer law) were taken into account (Fig. 8.7).

The increase in the absorption by 15 % due to the turbidity was chosen as a criterion of the rapid coagulation. The solutions have been prepared and titrated in such a manner that allowed avoiding the high local concentration of the electrolytes.

Both methods lead to similar results, and the variation of the CCC values along with the ND concentration is negligible. For instance, the data in Fig. 8.8 demonstrate that the jump of the size, which causes the rise in turbidity, occurs always at 3 mM NaCl.

Figure 8.8 reflects the peculiar of particle size distribution during the coagulation process. Note that the lower the ND concentration is, the larger the aggregates are not only before, but even more after the coagulation. The coagulation by NaCl is reversible; the peptization of the freshly coagulated hydrosols by dilution is quite possible. This is typical for the case of coagulation caused by condensing the diffuse electrical layer around the colloidal species: dilution restores the pre-threshold salt concentration. After several hours, however, this procedure of the restoration becomes invalid.

The primary 3 nm particles are evidently too small to be described by the DLVO theory in its classical version [15]. Even the secondary species, i.e., aggregates, are

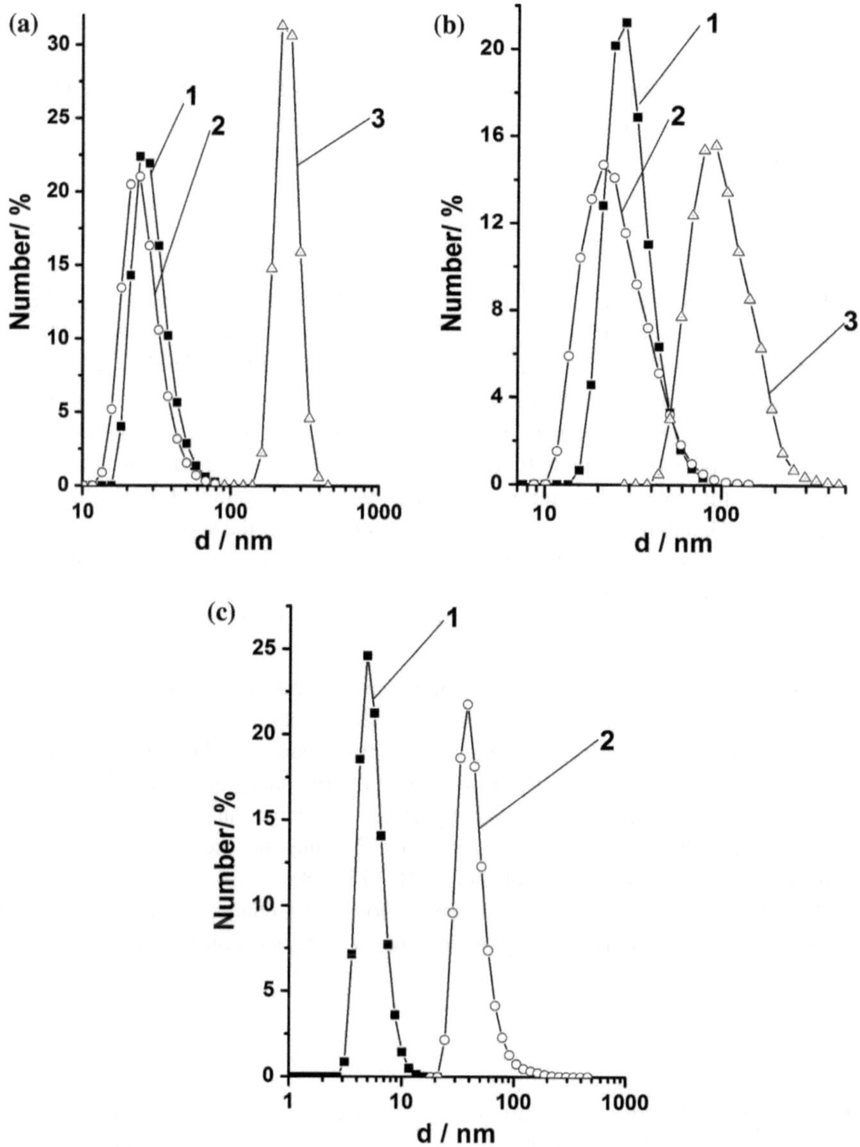

Fig. 8.8 The size distribution in the ND hydrosols of various wt/vol% concentrations. **a** 0.036 % ND (*1*); 0.024 % ND with 2 mM (*2*) and 3 mM NaCl (*3*); **b** 0.19 % ND (*1*); 0.13 % ND with 2 mM (*2*) and 3 mM NaCl (*3*); and **c** 1.00 % ND (*1*); and 0.67 % ND with 3 mM NaCl (*2*)

well below the limit of ca. 100 nm [16] and probably non-spherical, as follows from the TEM data. For approximate estimates, the old criterion of the stability of small colloidal particles proposed by Derjaguin may be used [15, 17]:

$$4\pi\varepsilon_r\varepsilon_0 r\Psi^2 > A^*. \quad (8.4)$$

In the case under consideration, the value of the Hamaker constant, $A^* = 1.19 \times 10^{-19}$ J, originates from the values for diamond, $A_{DD} = 28.9 \times 10^{-20}$ J, and water, $A_{WW} = 3.7 \times 10^{-20}$ J, selected from the literature data [18, 19]:

$$A^* = \left(A_{DD}^{1/2} - A_{WW}^{1/2}\right)^2. \quad (8.5)$$

For the mean r value of 15 nm, the value $\Psi > 30$ mV meets the stability condition. Indeed, the 0.036 wt/vol% ND hydrosol without NaCl additives exhibits the value of $\varsigma = +62$ mV and is quite stable. The Ψ value is evidently even higher. The additional condition of stability, i.e., $4\pi\varepsilon_r\varepsilon_0 r\Psi^2 > 20kT$ [15], also holds because in our case A^* equals to $29kT$. However, at NaCl concentrations between 2 and 3 mM (Fig. 8.6), ς equals to +(32–33) mV, while the coagulation takes place therein. So, the colloidal system under study is less stable than follows from criterion (8.4). This may originate from the loose structure of the secondary aggregates existing under the considered conditions. Another possible reason is the interaction between the facets possessing negative and positive electrostatic potential [20, 21]. More detailed consideration of the applicability of the DLVO theory is given elsewhere [22].

The CCC values for different electrolytes are collected in Table 8.2. There are three important conclusions to be made from these data.

Firstly, the CCCs are in line with the positive charge of the colloidal species. Indeed, the classical Schulze–Hardy rule predicts the sharp increase in the coagulation power of multi-charged well-hydrated inorganic anions in the case of "positive" sols. The ratio of the reciprocal CCC values for the anions Cl^-, SO_4^{2-}, $Fe(CN)_6^{3-}$, and $Fe(CN)_6^{4-}$ is as follows: 1:16:175:538.

The DLVO theory in its classical version for highly charged surfaces of colloid species predicts the following dependence of CCC versus counter ion charge, z:

$$\mathrm{CCC} = \mathrm{const} \times z^{-n}, \quad (8.6)$$

Table 8.2 The coagulation points, CCC/mM, of the 0.19 % ND hydrosol

Electrolyte	CCC	z (anion)	CCC_{NaCl}:CCC
NaCl	2.8	−1	1.00
HCl	5.7	−1	0.5
$NaBF_4$	1.9	−1	1.5
NaOH	0.27	−1	10
$C_{12}H_{25}OSO_3Na$	0.039	−1	72
$NaB(C_6H_5)_4$	0.027	−1	104
Na_2SO_4	0.17	−2	16
$K_3Fe(CN)_6$	0.016	−3	175
$K_4Fe(CN)_6$	0.0052	−4	538

where n equals 6. Lower n values, such as 2; 2.5; 3.5; or 4 have been usually explained either in terms of coagulation in the "distant minimum", i.e., through the water layers, or by adsorption of counter ions and low interfacial charge, and also by taking into account the retardation of the molecular forces [23, 24]. In our case, the average n value is 4.4. For the multi-charged anions $Fe(CN)_6^{3-}$ and $Fe(CN)_6^{4-}$, neither compressing of the diffuse part of the double electrical layer nor the neutralization via adsorption alone may lead to the coagulation. Probably, these factors are acting in concert. Indeed, the significance of the account of specific adsorption for understanding the Schulze–Hardy rule has been underlined recently by Lyklema [25]. This may be the reason of the sweep and less distinct shape of the titration curves in the case of highly charged $Fe(CN)_6^{3-}$ and $Fe(CN)_6^{4-}$ (Fig. 8.7).

Secondly, another universal regularity follows from comparing the coagulation power of single-charged anions with increasing hydrophobicity. The sequence $Cl^- < BF_4^- < C_{12}H_{25}OSO_3^- < B(C_6H_5)_4^-$ is in accordance with the so-called lyotropic row (Hofmeister series).

Thirdly, the strong coagulating action of the most hydrophilic anion HO^- explores the acidic nature of the positive charge of ND species. This statement is in line with the elevated CCC value for HCl: the addition of the acid results not only in the condensing of the diffuse part of the double electrical layer and screening of the interfacial charge of the ND particles, but also in protonation of some residual basic centers. This increases the surface charge and thus additionally stabilizes the hydrosol.

Figure 8.9 reflects the alterations of the size distribution on adding HCl and NaOH. After coagulation by NaCl around the CCC value, the colloidal system is easily restored just by immediate slight dilution. However, if the coagulation is accomplished by the alkali, the addition of a large amount of water displays no changes in the deposited coagulate.

But the complete neutralization of the alkali by HCl does restore the hydrosol (Fig. 8.10). Of course, after the neutralization the final electrolyte concentration, in fact NaCl, should not reach the corresponding CCC value (2.8 mM). This gives evidence for the occurrence of the acid-base neutralization of the interfacial positively charged groups by the HO^- ions, which is impossible to invert by dilution with water.

These groups may be protonated alcohols or ethers, pyrones, ammonium groups or lyonium ions attached to the negatively charged [111] facets of the diamond [26–31] (Fig. 8.11).

In the last case, the local positive potential of the interfacial [100] facets may cause the positive zeta-potential of the surface in the case if the negative [111] graphitized facets are neutralized via protonation.

Interestingly, after about 1 year after preparation of the ND hydrosol, the CCC values began to rapidly decrease. For instance, for NaCl CCC = 0.33–0.46 mM as determined for the 0.19 % colloidal solution. The CCC value for HCl, however, stays unaffected (5.7 mM). Moreover, small addition of the hydrochloric acid to the sol, 0.16 mM, restores the CCC value for NaCl. Also, the pK_a^{app} value of the indicator dye as determined in fresh and aged ND solutions practically coincide.

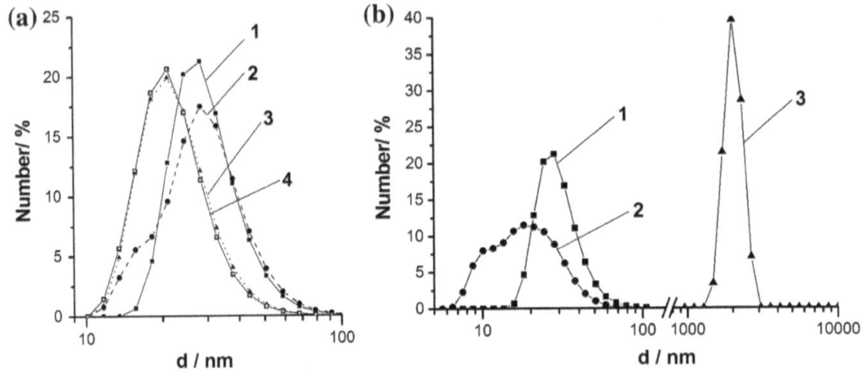

Fig. 8.9 The alterations of the particle size distribution along with adding HCl and NaOH. **a** 0.19 % ND (*solid line*, *1*); 0.13 % ND with: 0.1 mM HCl (*dashed*, *2*); 0.3 mM HCl (*dotted*, *3*); 1 mM HCl (*solid*, *4*); **b** 0.19 % ND (1); 0.13 % ND with: 0.08 mM NaOH (*2*); 1 mM NaOH (*3*)

Fig. 8.10 Photo of the ND hydrosol coagulated by NaOH (*right*) and the same hydrosol after restoration via HCl (*left*). The final NaCl concentration formed after neutralization of the alkali was below the CCC value of NaCl

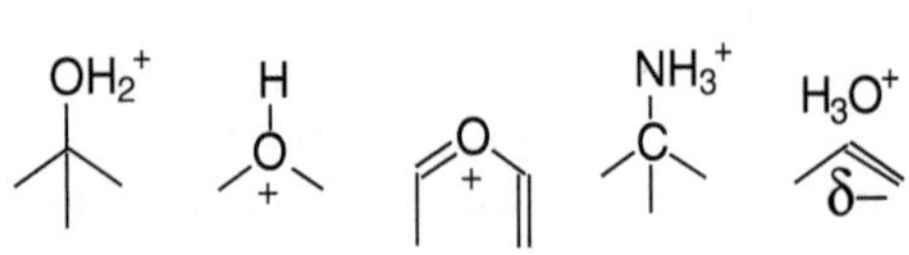

Fig. 8.11 Examples of the (possible) positively charged interfacial groups

Probably, the acid protonates some basic interfacial centers of the colloidal species thus enhancing the stability of the sol. But the ς-potential values of the freshly prepared, aged, and restored via HCl admixtures ND hydrosol differ only within the error limits.

8.3.4 Concentration-Dependent Size Distribution in the ND Hydrosols

The mean size of the colloid particles increases along with dilution; the results of DLS typified in Fig. 8.2 are representative examples from a much larger body of data. The TEM data for the dried 0.036 and 0.0036 wt/vol% ND solutions (Figs. 8.5 and 8.12) give support to this observation.

At the same time, the viscosity drops (Table 8.3). Similar viscosity data have been already reported for ND hydrosols [32].

We propose the following explanation. In concentrated solutions, the small particles interact through the surrounding water layers and form a kind of the so-called periodic colloidal structures [15, 33, 34], or colloidal crystals [35, 36] that leads to some structuring of the sol and viscosity enhancement. After dilution, the distances between the particles became larger, and it becomes more preferable to form secondary aggregates, much more separated from each other in aqueous medium.

For such kind of colloidal systems, some peculiarities have been noticed long ago, mostly for much larger particles of the dispersed phase [33, 34]. The most significant one is as follows: the "secondary" or "distant" minimum, resulting from the cooperative interactions, may be deeper as compared with that calculated from the interaction of two particles only. Namely, the fixation of particles without direct contact may take place even if the repulsive forces overcome the attractive ones

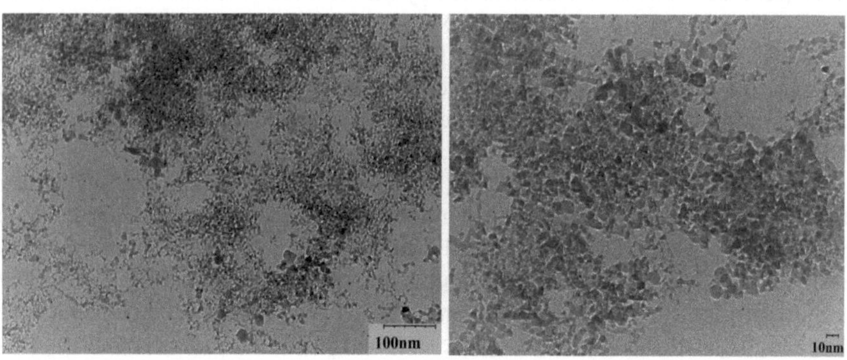

Fig. 8.12 Some typical TEM images of the 0.0036 wt/vol% ND hydrosol after evaporation of water. The initial solution was the same as in Fig. 8.5, but was additionally diluted tenfold

Table 8.3 The DLS and viscosity data of the ND hydrosol at different concentrations; 25 °C

ND conc. (wt/vol%)	Average size (nm)	Polydispersity index	$\eta \times 10^4$ (Pa s)
5.00	2.7 ± 0.3	0.58 ± 0.03	17.01
3.40	3.0 ± 0.7	0.585 ± 0.006	14.27
2.90	–	–	13.40
2.50	–	–	12.58
1.74	–	–	11.40
1.00	5.1 ± 1.2	0.532 ± 0.017	10.16
0.50	–	–	9.498
0.30	30.9 ± 1.4	0.158 ± 0.006	–
0.19	31 ± 3	0.241 ± 0.007	9.146
0.10	33 ± 4	0.18 ± 0.03	–
0.036	30.2 ± 0.3	0.188 ± 0.002	8.910
0.03	32.0 ± 1.8	0.171 ± 0.008	–
0.027	30.3 ± 0.8	0.180 ± 0.006	–
0.01	34 ± 5	0.19 ± 0.04	–
0	–	–	8.900

[33]. The structural forces may also hinder the aggregation [33, 34]. Small concentrations of the electrolyte may favor the fixation in the distant minimum, whereas elevated concentrations result in adherence of the species. Note, that at least ca. 3 mM of Cl^- ions are already present in the entire 5.0 % ND hydrosol, whereas adding NaCl results in coagulation of the more diluted sol at the same CCC value of ≈3 mM.

The 5.0 wt/vol% ND hydrosol may be considered as a colloidal system being on the way to gelation. But the zones of periodic colloidal structure do not necessarily cover the whole solution. Indeed, the polydispersity index, PDI, gradually increases along with dilution from PDI = 0.58 for 5.00 % to 0.19 for 0.036 % hydrosol. Probably, some numbers of very large islands of associated species are also present in concentrated sols, as is demonstrated by the DLS measurements in the volume–size and intensity–size distributions. Somewhat different interpretation has been proposed basing on the SANS data for similar systems [37, 38]. Namely, the presence of fractal aggregates in concentrated ND colloidal solutions has been deduced following from the dependence of the forward scattering intensity on the ND volume fraction [37, 38].

On dilution accompanied by stirring, the cooperative interactions become weaker; the entropy factor ($\Delta S > 0$) favors the uniform distribution of species in the enlarged volume. Instead of the network of small particles, larger aggregates are formed as a result of tight interaction of several neighboring primary particles (Fig. 8.13). Here, the driving forces are hydrophobic and van der Waals (dispersive) interactions. Further coagulation is hindered by the decrease in the ionic concentration in the bulk and thus expansion of the diffuse parts of the double electrical layer and also by the Brownian motion and the lower probability of particle collision.

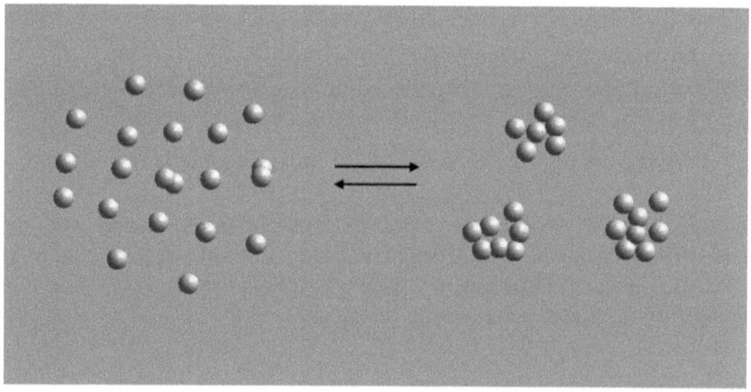

Fig. 8.13 The schematic picture of the dilution of the colloidal periodic structure

It should be noted that the entire concentrated ND sols are inclined to very slow aging. As a result, the mean size is somewhat increasing and the viscosity decreases.

8.4 Conclusions

The nanodiamond hydrosol with positively charged colloidal particles behaves as a typical hydrophobic dispersion. The addition of different electrolytes leads to coagulation; the latter occurs in accordance with the Schulze–Hardy rule and the lyotropic (Hofmeister) series. The anomalous coagulation power of the hydrophilic HO^- ion underlines the acidic nature of the positive interfacial charge of the colloidal particles.

The batochromic shift of the absorption bands of the anionic dyes gives evidence for their adsorption on the nanodiamond/water interface. The indices of the apparent ionization constant of the adsorbed acid-base indicator bromocresol green, pK_a^{app}, has been determined. This, in turn, allowed estimating the interfacial electrical potential of the nanodiamond particles, Ψ. The last value depends on the concentration of the hydrosol, but is always substantially higher as compared with the zeta-potential.

The colloidal particles in the initial 5.0 wt/vol% hydrosol have the size of 2.7 ± 0.3 nm, whereas the dilution leads to the ca. ten-fold increase in the particle size. In 0.3–0.01 wt/vol% hydrosol, the average size is around 30 nm. Along with the growth of colloidal species, the polydispersity index drops from 0.58 to 0.19. These results have been obtained by dynamic light scattering and confirmed by transmission electron microscopy images. Taking into account the high viscosity of the concentrated hydrosol, the results have been explained in terms of structuring of the system and formation of the so-called periodic colloidal system. After dilution, the distances between the primary particles became larger, and it becomes more preferable to form secondary aggregates, much more separated from each other in aqueous medium.

The further research of the ND colloids is now in progress.

Acknowledgments We are grateful to Ekaterina Vus for measurements with the Zetasizer Nano ZS Malvern Instruments in Aalto University, Espoo, Finland. We also express our gratitude to Professor Paavo Kinnunen, Department of Biomedical Engineering and computational science, Aalto University, Espoo, Finland, for putting to our disposal the above-mentioned apparatus.

References

1. A. Krueger, J. Mater. Chem. **21**, 12571 (2011)
2. V.N. Mochalin, O. Shenderova, D. Ho, Y. Gogotsi, Nat. Nano. **7**, 11 (2012)
3. O.A. Shenderova, D.M. Gruen, Ultrananocrystalline diamond: synthesis, properties and applications, 2nd edn. William Andrew Publishing, Norwich (2012)
4. A. Kumar, P.A. Lin, A. Xue, B. Hao, Y.K. Yap, R.M. Sankaran, Nature Commun **4**, 2618 (2013)
5. L.Y. Chang, E. Osawa, A.S. Barnard, Nanoscale **3**, 958 (2011)
6. E. Osawa, in ed. by S. Somiya, Academic Press, Oxford (2013)
7. G. Taguchi, S. Chowdhury, Y. Wu, *Quality Engineering Handbook* (Wiley, New York, 2007) pp. 56–123
8. E. Osawa et al., Under preparation
9. S. Sasaki, R. Yamanoi, E. Ōsawa, Dispersion of Detonation Nanodiamond: A Progress Report, presented before the 46th Fullerenes-Nanotubes-Graphenes General Symposium. University of Tokyo (2013)
10. N.O. Mchedlov-Petrossyan, N.A. Vodolazkaya, N.N. Kamneva, in Micelles: Structural Biochemistry, Formation and Functions and Usage, eds. by D. Bradburn, T. Bittinger, Nova Science Pub Inc, New York, (2013), Chap. 1
11. F. Grieser, C.J. Drummond, J. Phys. Chem. **92**, 5580 (1988)
12. N.O. Mchedlov-Petrossyan, Pure Appl. Chem. **80**, 1459 (2008)
13. N.N. Kamneva, A.Y. Kharchenko, O.S. Bykova, A.V. Sundenko, N.O. Mchedlov-Petrossyan, J. Mol. Liq. **199**, 376 (2014)
14. P. Mukerjee, K. Banerjee, J. Phys. Chem. **68**, 3567 (1964)
15. B.V. Deryagin, Russ. Chem. Rev. **48**, 363 (1979)
16. L.B. Boinovich, Russ. Chem. Rev. **76**, 471 (2007)
17. B. Derjaguin, Trans. Faraday Soc. **36**, 730 (1940)
18. J.N. Israelachvili, *Intermolecular and Surface Forces: Revised*, 3rd edn. (Academic press, Amsterdam 2011)
19. J. Visser, Adv. Colloid Interface Sci. **3**, 331 (1972)
20. A.S. Barnard, J. Mater. Chem. **18**, 4038 (2008)
21. A.S. Barnard, E. Osawa, Nanoscale **6**, 1188 (2014)
22. N. O. Mchedlov-Petrossyan, N. N. Kamneva, A. I. Marynin, A. P.Kryshtal, E. Ōsawa. Phys. Chem. Chem. Phys. **17**, 16186 (2015)
23. B.V. Derjaguin, *The Theory of Stability of Colloids and Thin Films* (Nauka, Moscow, 1986)
24. B.V. Derjaguin, N.V. Churaev, V.M. Muller, *Interfacial Forces* (Nauka, Moscow, 1985)
25. J. Lyklema, J. Colloid Interface Sci. **392**, 102 (2013)
26. J. Hees, A. Kriele, O.A. Williams, Chem. Phys. Lett. **509**, 12 (2011)

27. J.T. Paci, H.B. Man, B. Saha, D. Ho, G.C. Schatz, J. Phys. Chem. C **117**, 17256 (2013)
28. O.A. Williams, J. Hees, C. Dieker, W. Jäger, L. Kirste, C.E. Nebel, ACS Nano **4**, 4824 (2010)
29. C.A. Leon y Leon, J.M. Solar, V. Calemma, L.R. Radovic, Carbon **30**, 797 (1992)
30. X. Xu, Z. Yu, Y. Zhu, B. Wang, Diam. Relat. Mater. **14**, 206 (2005)
31. A.N. Zhukov, F.R. Gareeva, A.E. Aleksenskii, Colloid J. **74**, 463 (2012)
32. A.Y. Vul, E.D. Eydelman, M. Inakuma, E. Ōsawa, Diam. Relat. Mater. **16**, 2023 (2007)
33. I.F. Efremov, *Periodic Colloidal Structures* (Khimiya, Leningrad, USSR, 1971)
34. I.F. Efremov, O.G. Us'yarov, Russ. Chem. Rev. **45**, 435 (1976)
35. P. Bartlett, R.H. Ottewill, J. Chem. Phys. **96**, 3306 (1992)
36. A. Rugge, W.T. Ford, S.H. Tolbert, Langmuir **19**, 7852 (2003)
37. M.V. Avdeev, N.N. Rozhkova,V.L. Aksenov, V.M. Garamus, R. Willumeit, E. Ōsawa, J. Phys. Chem. C **113**, 9473 (2009)
38. M.V. Avdeev, V.L. Aksenov, O.V. Tomchuk, L.A. Bulavin, V.M. Garamus, E. Ōsawa, J. Phys.: Condens. Matter. **25**, 445001 (2013)

Chapter 9
Kaolin Suspensions with Negative First Normal Stress Difference

M. Bombrowski, H.-J. Mögel, M. Wahab and P. Schiller

Abstract It is well known that kaolin suspensions can have a negative first normal stress difference. We analyze rheological models that relate the flow alignment of platelets and normal stress differences. Using a freezing technique for arresting the platelets of sheared kaolin suspensions, X-ray data of the frozen material reveal that the flow alignment angle and a scalar orientational order parameter depend on the shear rate. A mesoscopic theoretical approach predicts that a negative value of the first normal stress difference is not compatible with a stationary flow alignment regime. If the first normal stress difference is negative, the model suggests a tumbling or wagging motion of the platelet normals in analogy to the director dynamics in polymeric liquid crystals under shear.

9.1 Introduction

The rheological properties of aqueous clay suspensions depend on many physical parameters such as volume density of the dispersed particles, the ionic strength of the electrolyte solution and the particle shape. The microstructure of clay dispersions has a determining influence on the stress arising in shear flows. Kaolinite particles resemble plates of moderate thickness. Typical aspect ratios in many

M. Bombrowski · H.-J. Mögel (✉) · M. Wahab · P. Schiller
Department of Physical Chemistry, TU Bergakademie Freiberg,
Leipziger Straße 29, 09599 Freiberg, Germany
e-mail: hans-joerg.moegel@chemie.tu-freiberg.de

M. Bombrowski
e-mail: martin.bombrowski@chemie.tu-freiberg.de

M. Wahab
e-mail: mirco.wahab@chemie.tu-freiberg.de

P. Schiller
e-mail: peter.schiller@chemie.tu-freiberg.de

L. Bulavin and N. Lebovka (eds.), *Physics of Liquid Matter: Modern Problems*,
Springer Proceedings in Physics 171, DOI 10.1007/978-3-319-20875-6_9

polydisperse natural dispersions of kaolinite platelets vary between about ten and twenty, the platelet diameter may range between a few hundred nanometers and a few micrometers. Anisometric particles with negligible pair interaction in diluted suspensions do not have a stable orientation in a shear flow. Platelets lying in the shear plane are supposed to rotate around the vorticity axis [1]. In concentrated colloidal suspensions excluded volume effects are very important. Strong particle-particle interactions may suppress the flow-induced particle rotation. The oblate shape of the particles in dense kaolin suspensions favors ordered domains with parallel platelets. This alignment can be improved in a shear flow [2–5]. Similar to nematic liquid crystals [6], where molecular axes of long molecules align in a shear flow, the platelet normals have a preferred orientation. In this case a certain flow alignment angle between the flow gradient vector and the plate normals appears. Investigating orientational order of the platelets in sheared kaolin dispersions by neutron scattering, Brown et al. [3] found a pronounced alignment of the platelet normals. But in contrast to nematic non-polymeric liquid crystals, where the flow alignment angle is independent of the shear rate, the angle between kaolinite platelet normals and the shear gradient vector was found to decrease with increasing shear rate. The flow alignment angle is a macroscopic quantity defined by the orientation of domains containing many platelets in a flow field. Orientational fluctuations of single platelets may be described by an orientational distribution function, which is a probability density for the orientation of the platelet normal. A narrowing of the orientational distribution is accompanied with an increase of the scalar Maier-Saupe order parameter [6]. This parameter describes microscopically the degree of nematic ordering. It was found that the orientational distribution becomes narrower with increasing shear rate [3]. Hence, a shear flow in kaolin suspensions may lead to macroscopic flow alignment and an increase of the nematic order parameter. Both aspects of ordering are incorporated in the de Gennes tensor order parameter for nematic liquid crystals. Most papers on flow of anisotropic fluids, which use this order parameter tensor for describing orientational order, mainly consider flow alignment instabilities leading to tumbling and wagging of molecular or particle axes [7–12].

The influence of a shear flow on molecular orientation has been studied both experimentally and theoretically for polymeric nematic liquid crystals [1, 13–15]. Compared to non-polymer nematic liquid crystals a fixed flow alignment angle does not occur at low and moderate shear rates. At low shear rates the long molecular axes tumble in the shear flow, i.e. the long molecules that lie in the shear plane rotate around an axis parallel to the shear plane normal. At higher shear rates the tumbling changes into a wagging motion and a further increase of the shear rate to very large values eventually leads to flow alignment of the long molecules along a certain direction in the shear plane. The unsteady motion of the molecular axes causes a non-monotonic dependence of the first normal stress difference on the shear rate $\mathring{\gamma}$ as depicted schematically in Fig. 9.1. The flow alignment regime starts close to the minimum of the first normal stress difference T_I. Compared to the vast

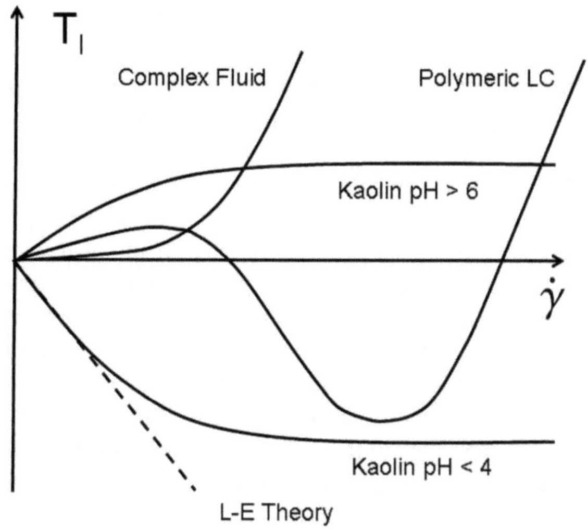

Fig. 9.1 First normal stress difference T_I versus the shear rate $\dot{\gamma}$ for several materials. The *dashed line* refers to the result of the macroscopic Leslie-Erickson theory in combination with the microscopic Baalss-Hess theory for nematic liquid crystals

majority of complex fluid materials, the first normal stress difference of polymeric nematic liquid crystals behaves exceptionally. Most complex fluids have a positive first normal stress difference satisfying the asymptotic law $T_I \propto \mathring{\gamma}^2$ at low shear rates $\mathring{\gamma}$. Theoretical considerations for nematic liquid crystals lead to a different result (Fig. 9.1). The generally accepted Leslie-Erickson theory for non-polymeric nematic liquid crystals is compatible with the appearance of negative and positive first normal stress differences. This macroscopic theory predicts a linear dependence of the first normal stress difference on $\mathring{\gamma}$, where $T_I \propto |\mathring{\gamma}|$ is a non-analytic function at $|\mathring{\gamma}| = 0$.

Investigating concentrated kaolin suspensions, Moan et al. [5] found that the values of the first normal stress difference were negative and they changed linearly with $\dot{\gamma}$ at low and moderate shear rates. Hence, the first normal stress difference T_I of kaolin can be described by the Leslie-Erickson theory as long as the shear rate is relatively small. At higher shear rates, however, the first normal stress difference approaches a constant value in a plateau region as illustrated in Fig. 9.1. At still higher shear rates $\mathring{\gamma}$, the value of T_I was no longer constant but increased gradually. Thus, T_I increased linearly from −800 Pa to about −500 Pa after changing $\mathring{\gamma}$ from 10 to 100 s^{-1}. In our measurements [16] for kaolin suspensions the first normal stress difference was found to be approximately constant at high shear rates. There was only a weak minimum of T_I and a slight increase above the plateau value, even at shear rates as high as 100 s^{-1}. It should be noted, however, that the suspensions used in our measurement have lower *pH* values and the kaolin platelets are considerably larger than in [5]. Furthermore, we only found negative values of the first normal stress difference T_I for *pH* values below about 5. At higher *pH* the value of T_I was positive as shown in Fig. 9.1.

The observation that the Leslie-Erickson theory can give an acceptable description of the first normal stress difference at low shear rates (Fig. 9.1) supports the assumption that the shear flow aligns the kaolinite platelets in analogy to the flow alignment of non-polymer nematic liquid crystals. However, the approximately constant value of the first normal stress difference at moderate and high shear gradients (Fig. 9.1) indicates that a purely macroscopic theory is not sufficient to describe the rheological data. In this paper we propose a rheological model for kaolin suspensions which correlates rheological with structural data. For this purpose, we modify a model for polymeric liquid crystals proposed by Farhoudi and Rey [7, 8]. In order to avoid an excessive large number of unknown viscosity coefficients, an order parameter expansion of an appropriately defined Rayleighian is truncated after lowest order terms. This procedure is justified, since the nematic order of kaolin suspensions is weak. Most material parameters of the rheological model are related to the orientational order of the platelets. The consequences of the model are compared with experimental results on structural and rheological data.

9.2 Orientational Distribution and Order Parameter

The orientation of a disk-like platelet is described by the normal **m** to its face. Since platelets have two equivalent parallel faces with normals **m** and –**m**, any orientational distribution function f for the orientation of platelet normals **m** should remain unchanged if some or all platelet normals **m** are replaced by –**m**. The inversion transformation $(x, y, z) \rightarrow (-x, -y, -z)$ of Cartesian space coordinates turns the normals of each platelet into the opposite direction (**m** → –**m**), but the orientational distribution cannot be influenced by this operation. Using spherical coordinates, where $m_x = \sin\theta\cos\phi, m_y = \sin\theta\sin\phi$ and $m_z = \cos\theta (0 \leq \theta \leq \pi, 0 \leq \phi < 2\pi)$, the inversion is accompanied with the transformation $(\theta, \phi) \rightarrow (\pi - \theta, \phi + \pi)$. Hence, an orientational distribution function $f(\theta, \phi)$ obtained by averaging over the orientation of many platelet normals in a nematic domain obeys the condition $f(\theta, \phi) = f(\pi - \theta, \phi + \pi)$. A nematic phase has cylindrical symmetry. If the z-axis of the Cartesian coordinate system is chosen to be parallel to the symmetry axis, the orientational distribution is independent of the azimuthal angle ϕ and the condition $f(\theta) = f(\pi - \theta)$ restricts the number of allowed terms in an expansion to Legendre polynomials. This expansion may be expressed as

$$f(\theta) = \frac{1}{4\pi} + \sum_{l=2,4,6,\ldots} c_l P_l(\cos\theta), \qquad (9.1)$$

where $P_l(\cos\theta)$ is a Legendre polynomial [17] and the coefficients c_l describe the degree of nematic orientational order. If the nematic order is weak, it is sufficient to

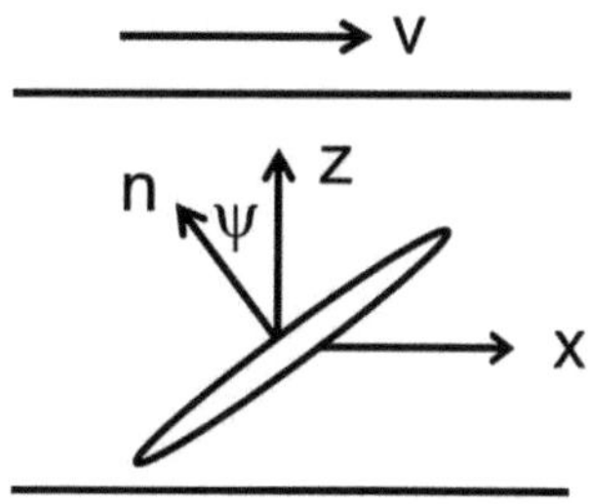

Fig. 9.2 Flow alignment angle ψ in the shear plane (*x–z—plane*). The director **n** is a unit vector parallel to the favored alignment of the platelet normals in the shear flow (**v**, velocity of the upper rheometer plate)

consider only lowest order terms. Using the notation $c_2 = 5S/4\pi$, the lowest order terms of the series for the orientational distribution (9.1) are written as

$$f(\theta) = \frac{1}{4\pi} + \frac{5}{4\pi} SP_2(\cos\theta) \tag{9.2}$$

for $l < 4$. The scalar order parameter S describes the nematic ordering. S is equal to zero for the isotropic phase and equal to one for perfectly aligned platelet normals along the z-axis.

If the preferred direction of the platelet normals is turned away from the z-axis by an angle $\theta_{\max}$, which is caused in our experiments by flow alignment (Fig. 9.2), the orientational distribution (9.2) of nematic order is transformed into

$$f(\theta, \phi) = \frac{1}{4\pi} + \frac{5}{4\pi} SP_2(\cos\omega), \tag{9.3}$$

where $\cos\omega$ is obtained from the addition theorem [17] $\cos\omega = \cos\theta \cos\theta_{max} + \sin\theta \sin\theta_{max} \cos(\phi - \phi_{max})$. The orientational distribution function $f(\theta, \phi)$ can be obtained experimentally by using techniques based on X-ray or neutron scattering [3], since the intensity $I(\theta, \phi)$ of scattered neutrons or X-rays is proportional to the orientational distribution function ($f(\theta, \phi) \propto I(\theta, \phi)$). Figure 9.8 (Appendix) gives a brief sketch of our experimental setup, which uses X-ray scattering on kaolin suspensions frozen after the application of shear stress (see explanation in the Appendix). Figure 9.3 shows X-ray scattering results for the sheared kaolin suspension.

The data for the intensity distribution $I(\theta, \phi)$ refer to the shear plane ($\phi = 0$ and $\phi = \pi$) and perpendicular to it ($\phi = \pi/2$ and $\phi = 3\pi/2$). Using these data the scalar order parameter S can be obtained from the relation

$$S = \int_0^{2\pi} d\phi \int_0^{\pi} P_2(\cos\omega) f(\theta, \phi) \sin\theta d\theta. \tag{9.4}$$

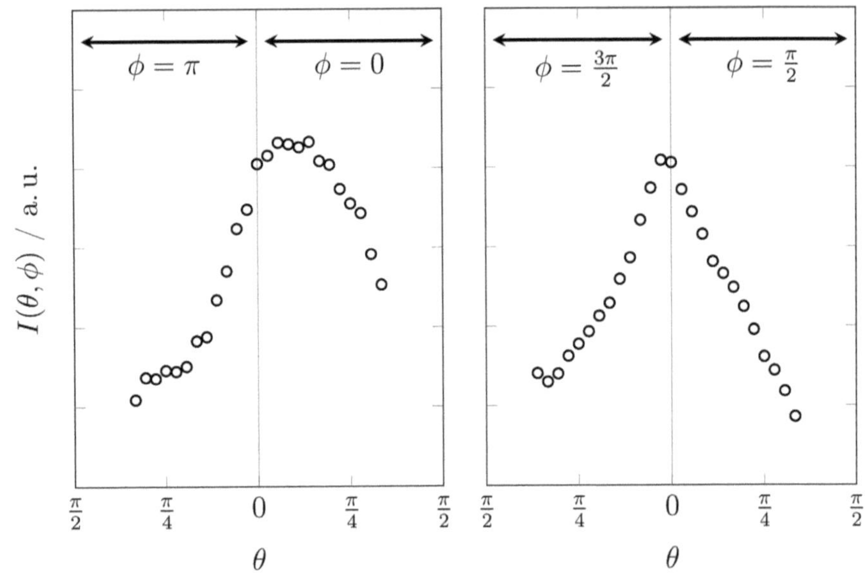

Fig. 9.3 Intensity distribution $I(\theta, \phi)$ of the scattered X-ray beam (Fig. 9.8, Appendix) for $\phi = 0, \pi, \pi/2$ and $3\pi/2$. The kaolin suspension ($pH = 3.5, \Phi = 0.2$) was sheared with a rate of $\dot{\gamma} = 25\,\text{s}^{-1}$

At higher shear rates, the narrowing of peaks of the orientational distribution function indicates an increase of the order parameter S. Furthermore, the maxima of the functions $f(\theta, \phi = 0)$,$f(\theta, \phi = \pi)$,$f(\theta, \phi = \pi/2)$ and $f(\theta, \phi = 3\pi/2)$ indicate the flow alignment direction of the platelet normals, which is found to lie in the shear plane as illustrated in Fig. 9.2. It should be noted that the flow alignment angle ψ and the angle θ_{max} accompanied with a maximum value of $f(\theta, \phi)$ are related by

$$\psi = \begin{cases} \theta_{\text{max}}, & \text{if} \quad \phi = \pi, \\ -\theta_{\text{max}}, & \text{if} \quad \phi = 0. \end{cases} \tag{9.5}$$

9.3 Flow Alignment of a Nematic Phase: Macroscopic Theory

9.3.1 Flow Alignment far away from Interfaces

The macroscopic Leslie-Erickson theory [1, 6] of nematic liquid flow is based on the introduction of a director $\mathbf{n} = (n_x, n_y, n_z)$, which is a unit vector with orientation parallel to the symmetry axis of the nematic phase. In the case of disk-like platelets the director indicates the preferred direction of the platelet normals. According to the Leslie-Erickson theory, the effective shear viscosity of a nematic phase depends on the orientation of the director. It is convenient to use the notation $x_1 = x$, $x_2 = y$ and $x_3 = z$ for a Cartesian coordinate system. In expressions with repeating indices Einstein's sum convention is applied. Let us consider a nematic phase with director $\mathbf{n} = (n_1, n_2, n_3)$ subjected to a flow with a velocity field $\mathbf{v} = (v_1, v_2, v_3)$ that may vary in space. We introduce the symmetric part of the velocity gradient tensor

$$A_{lm} = \frac{1}{2}\left(\frac{\partial v_l}{\partial x_m} + \frac{\partial v_m}{\partial x_l}\right), \tag{9.6}$$

and the vector $\omega = \frac{1}{2}\text{curl } \mathbf{v}$ with components

$$\omega_m = \frac{1}{2}\varepsilon_{mks}\frac{\partial v_s}{\partial x_k}, \tag{9.7}$$

where ε_{mks} denotes the totally antisymmetric isotropic tensor of rank three. The vector $\mathbf{N} = (N_1, N_2, N_3)$ with components

$$N_m = \frac{\partial n_m}{\partial t} + v_k\frac{\partial n_m}{\partial x_k} - \varepsilon_{mks}\omega_k n_s \tag{9.8}$$

is the corotational time derivative of director $\mathbf{n}$. Then the viscous part of the Leslie stress tensor is defined as [6]

$$t_{lm} = \alpha_1 n_l n_m n_k n_s A_{ks} + \alpha_2 n_l N_m + \alpha_3 N_l n_m + \alpha_4 A_{lm} + \alpha_5 n_l n_k A_{km} + \alpha_6 n_m n_k A_{kl}. \tag{9.9}$$

The Onsager relation $\alpha_2 + \alpha_3 = \alpha_6 - \alpha_5$ [6] reduces the number of independent kinetic coefficients. A flow with non-zero velocity gradients produces a torque that affects the director alignment. In the balance equation for this torque $\mathbf{n} \times \mathbf{\Gamma} = 0$ the components of vector $\mathbf{\Gamma}$ are obtained from

$$\Gamma_m = -(\alpha_3 - \alpha_2)N_m - (\alpha_3 + \alpha_2)[n_k A_{km} - n_m(n_k n_s A_{ks})]. \tag{9.10}$$

It is useful to introduce the flow alignment parameter

$$\lambda = -\frac{\alpha_3 + \alpha_2}{\alpha_3 - \alpha_2}. \tag{9.11}$$

For a simple stationary shear flow of a nematic liquid crystal with oblate particles the velocity and the director are expressed as $\mathbf{v} = (\mathring{\gamma} x_3, 0, 0)$ and $\mathbf{n} = (-\sin\psi, 0, \cos\psi)$. A positive angle ψ corresponds to an anticlockwise rotation about the y-axis (Fig. 9.2). Inserting (9.6)–(9.8) into the torque balance $\mathbf{n} \times \mathbf{\Gamma} = 0$ yields

$$\frac{\partial \psi}{\partial t} = -\frac{\mathring{\gamma}}{2}(1 + \lambda \cos(2\psi)). \tag{9.12}$$

Hence, in the stationary case ($\partial\psi/\partial t = 0$), equation $\cos(2\psi) = -\lambda^{-1}$ must be satisfied for the flow alignment angle ψ. Flow alignment without tumbling of the director requires that condition $|\lambda| > 1$ is satisfied. Experiments on lyotropic polymer and thermotropic non-polymeric nematic liquid crystals with prolate molecules revealed that the value of $|\lambda|$ is always close to 1. Carlsson and Skarp [18] emphasized that for discotic liquid crystals consisting of oblate molecules the condition $\alpha_3 \gg \alpha_2 > 0$ is expected to be satisfied, so that λ is slightly lower than -1 ($-\lambda \gtrsim 1$). Then $\psi = arc\cos(-\lambda^{-1})/2 > 0$ corresponds to a small anticlockwise rotation of the director away from the direction of the z-axis (Fig. 9.2).

9.3.2 Influence of Interfaces on Flow Alignment

Equation (9.12) is no longer valid in the vicinity of solid walls such as rheometer plates. A solid surface aligns the director of particles adjacent to the surface. This alignment propagates into the bulk due to elastic interactions. In the framework of the elasticity theory of nematic liquid crystals, the elastic deformation energy density can be expressed in terms of the spatial gradients of director angles [6]. Using an one constant approximation with equal splay and bend elastic constants ($K_{11} = K_{33} = K$), (9.12) is replaced by

$$\frac{\partial \psi}{\partial t} = \left(\frac{K}{\gamma_1}\right)\frac{\partial^2 \psi}{\partial z^2} - \frac{\mathring{\gamma}}{2}(1 + \lambda \cos(2\psi)), \tag{9.13}$$

where $\gamma_1 = \alpha_3 - \alpha_2$ denotes the rotational viscosity and z is the distance between the solid interface and a point in the bulk. We assume that the shear rate $\mathring{\gamma}$ is independent of z, although a small change of ψ from zero to ψ_{max} produces a slight disturbance of the flow field homogeneity. Kaolinite particles are homeotropically aligned at wall interfaces, i.e. $\psi(z = 0, t) = 0$ [19]. In a stationary flow, when $\partial\psi/\partial t = 0$, the flow alignment angle ψ tends to $\psi_{\max}$ if the distance z from the wall increases. Equation (9.13) has the implicit solution

$$\int_0^{\psi} \frac{d\chi}{\sqrt{f(\psi_{max}, \chi)}} = \frac{z}{\xi}, \tag{9.14}$$

with

$$f(\psi_{max}, \chi) = \frac{\sin(2\psi_{max}) - \sin(2\chi)}{2} - (\psi_{max} - \chi)\cos(2\psi_{max}),$$

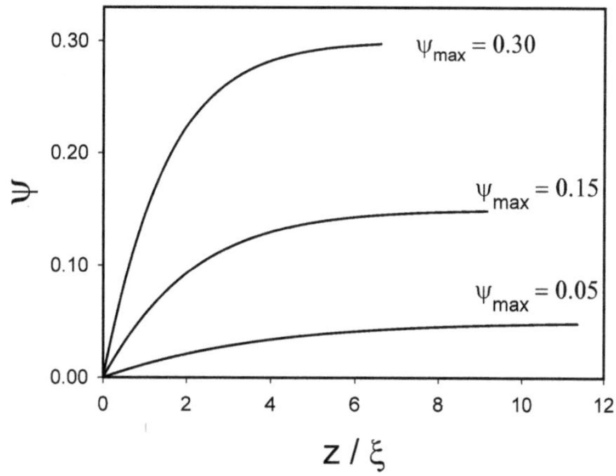

Fig. 9.4 Flow alignment angle ψ versus the reduced distance z/ξ from the sample surface. ψ tends to $\psi_{\max}$ for $z/\xi \to \infty$

and the characteristic length

$$\xi = \sqrt{\frac{K}{|\lambda|\ \gamma_1\ |\mathring{\gamma}|}}. \tag{9.15}$$

Figure 9.4 displays a plot of the director angle ψ in dependence on the reduced distance z/ξ for three different values of $\psi_{\max}$. If $z > 4\xi$ the condition $\psi \simeq \psi_{\max}$ holds fairly good.

However, for $|\mathring{\gamma}| \to 0$ the length ξ diverges to infinity and the angle ψ tends to zero throughout a finite nematic sample. Actually, investigating nematic probes before starting the shear flow in the rheometer, the director was found to be aligned parallel to the normals of the rheometer plates ($\psi = 0$). This homeotropic alignment may also be supported by the squeeze flow appearing when the upper rheometer plate moves down onto the kaolin paste for producing a thin film between the bounding plates. A squeeze flow during sample preparation cannot be avoided in a conventional rheometer. After starting the shear flow for recording viscosity and normal stress data, the flow alignment angle $\psi_{\max} > 0$ should appear first close to the midplane halfway between the plates. The characteristic length falls ($\xi \propto |\mathring{\gamma}|^{-1/2}$) with increasing shear rate $|\mathring{\gamma}|$, and thus the region where the condition $\psi \simeq \psi_{\max}$ is satisfied increases. Finally, at high shear rates almost the whole sample is aligned with the director parallel to the flow alignment direction. In this case the assumption of a constant shear alignment angle $\psi_{\max}$ throughout the sample is an acceptable approximation.

If the shear rate is low, the director alignment is non-homogeneous and an interpretation of rheological data in terms of structural investigations becomes intricate. There is an additional source of error due to the attenuation by X-rays. The penetration depth of an *X*-ray beam into a dense isotropic powder can be estimated by using the approximate equation for the intensity attenuation

$$I(z) = I_0 \exp\left(-\frac{2\mu z}{\sin\beta}\right), \tag{9.16}$$

where β is the angle of incidence, z the distance from the powder surface and μ denotes a material constant, which is proportional to the powder density. The penetration depth $z = \xi_X$ is defined as the distance where $I(z)/I_0 = 1/e$. If β is equal to the Bragg reflection angle for the (002) peak of a dense kaolin suspension ($2\beta \simeq 25°$), the value estimated for kaolinite is about $\xi_X \simeq 10\mu\text{m}$. But considering that the volume density of the solid particles in the suspension ($\Phi \simeq 0.2$) is much smaller than the corresponding density in the dry powder, the penetration depth should be five times larger, i.e. $\xi_X \simeq 50\mu\text{m}$. On the other hand, the solid rheometer walls produce a homeotropic alignment ($\psi = 0$) of the nematic director on the kaolin-wall interface $z = 0$. The thickness of the reorientation region ξ is proportional to the inverse root of the shear rate (9.15). If the shear rate $\mathring{\gamma}$ is zero, the condition $\psi \simeq 0$ is satisfied for small and large values of z. If $|\mathring{\gamma}|$ is small, even very small but not equal to zero, both the macroscopic and the mesoscopic theory of flow alignment predict a finite non-zero flow alignment angle $\psi = \psi_{\max}$ far away from solid walls where $z \gg \xi$ (Fig. 9.4). However, for sufficiently small values $|\mathring{\gamma}|$ the non-homogeneous boundary region with thickness $\xi \propto |\mathring{\gamma}|^{-1/2}$ should be even larger than the X-ray penetration depth ξ_X. In this case X-ray measurements lead to flow alignment angles ψ that are distinctly smaller than the values of ψ in regions where the director alignment is homogeneous ($z \gg \xi$). Unfortunately, the thickness ξ of the region influenced by walls cannot be evaluated, since the material constants K and γ_1 in (9.15) are unknown. Hence, the estimation of experimental errors in the determination of the scalar order parameter S and the flow alignment angle ψ by X-ray diffraction is hardly possible. Nevertheless, experimental X-ray data are suitable for deciding wether the flow alignment angle ψ is negative or positive (rotation in a clockwise or an anti-clockwise direction). Furthermore, it can clearly be distinguished between a monotonous and a non-monotonous dependence of S and ψ on the shear rate $|\mathring{\gamma}|$.

9.3.3 Predictions of a Statistical Theory on Viscosity Coefficients and Normal Stress Differences

Baalss and Hess [20] attempted to derive relations between the Leslie coefficients and the aspect ratio q of anisometric particles. For spheroids q is equal to B/A, where the semi-axis A is the equatorial radius and B is the distance from center to pole of the symmetry axis.

The theory is based on the observation that a fluid of perfectly aligned spheroids can be transformed into a fluid of spheres by applying an affine coordinate transformation $\mathbf{r} \Rightarrow C\,\mathbf{r}$ (Fig. 9.5), where the determinant of matrix C should satisfy the condition $\det(C) = 1$. Then the volume of the system does not change by the

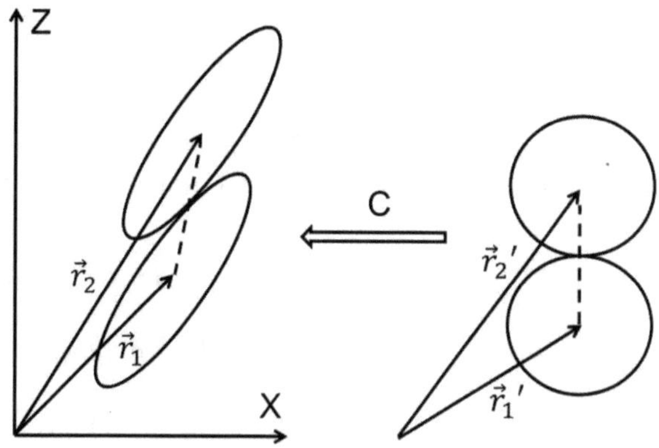

Fig. 9.5 An anisotropic fluid of hard parallel spheroids can be obtained from an isotropic fluid of hard spheres by an affine transformation (matrix C) of space coordinates $r' = (x', y', z')$

transformation. Furthermore, the authors suggest that the pair potential of interacting spheroids is obtained from a spherically symmetric pair potential by

$$\Phi_C(\mathbf{r}) = \Phi(C\mathbf{r}), \tag{9.17}$$

where $\mathbf{r} = \mathbf{r}_2 - \mathbf{r}_1$ is the vector connecting the centers of the spheroids and $\Phi(\mathbf{r}) = \Phi(|\mathbf{r}|)$ is a spherically symmetric pair potential.

Let us check if hard colloidal particle of axially symmetric ellipsoids satisfy (9.17). The hard sphere pair potential is defined as $\Phi_{HS}(\mathbf{r}) = 0$ if $|\mathbf{r}| > \sigma$ and $\Phi_{HS}(\mathbf{r}) = \infty$ if $|\mathbf{r}| < \sigma$, where σ is the diameter of the spheres. If the spheres are transformed into spheroids by an affine coordinate transformation ($\mathbf{r} \Rightarrow Cr$), the pair potential for spheroids $\Phi_C(\mathbf{r}) = \Phi_{HS}(|C\mathbf{r}|)$ is infinity if the corresponding spheres overlap and zero if they do not. Thus the potential for hard colloidal spheroids satisfies the condition (9.17) required for the application of the theory of Baalss and Hess [20]. Velocity and director are again expressed as $\mathbf{v} = (\mathring{\gamma}z, 0, 0)$ and $\mathbf{n} = (-\sin\psi, 0, \cos\psi)$, where ψ is the angle between director and velocity gradient direction in the shear plane (x–z—plane). Prolate molecules of conventional non-polymeric nematic liquid crystals align with their long axis almost parallel to the velocity vector, whereas oblate or plate-like particles are supposed to align with their director perpendicular to this vector ($\psi \simeq 0$). However, in both cases a small non-zero flow-induced director tilt angle should be taken into account. Considering that the flow is parallel to the x-axis and the shear gradient parallel to the z-axis (Fig. 9.2) the first and second normal stress differences are defined as $T_I = T_{11} - T_{33}$ and $T_{II} = T_{33} - T_{22}$, respectively. Using (9.9), these normal stress differences are

$$\begin{aligned} T_I &= \mathring{\gamma} n_1 n_3 \left[\alpha_6 - \alpha_5 + \alpha_1\left(n_3^2 - n_1^2\right)\right], \\ T_{II} &= -\frac{1}{2}\mathring{\gamma} n_1 n_3 \left[\alpha_1 + 2\alpha_6 + \alpha_1\left(n_3^2 - n_1^2\right)\right]. \end{aligned} \tag{9.18}$$

Assuming $\mathring{\gamma} > 0$, the relation $\psi = arc\cos(-\lambda^{-1})/2$ for the flow alignment angle derived in the previous section may also be written as

$$\tan\psi = \sqrt{\frac{\alpha_2}{\alpha_3}}. \tag{9.19}$$

Using the theory of Baalss and Hess [20], the viscosity coefficients of the Leslie-Erickson theory for the fluid of aligned hard spheroids are

$$\begin{aligned}
\alpha_1 &= -\frac{1}{2}\left(q - q^{-1}\right)^2 \eta_{HS},\\
\alpha_2 &= \frac{1}{2}\left(1 - q^2\right)\eta_{HS},\\
\alpha_3 &= \frac{1}{2}\left(q^{-2} - 1\right)\eta_{HS},\\
\alpha_4 &= \eta_{HS},\\
\alpha_5 &= -\alpha_2,\\
\alpha_6 &= \alpha_3,
\end{aligned} \tag{9.20}$$

where in our case η_{HS} is the shear viscosity of a hard sphere fluid with sphere volume equal to the volume of a spheroid particle. The flow alignment angle (9.19) can be expressed as

$$\tan\psi = q\ \mathrm{sgn}(\mathring{\gamma}), \tag{9.21}$$

which corresponds to a counterclockwise tilt of the director with respect to the z-axis if $\mathring{\gamma} > 0$ (Fig. 9.2). Using (9.11) and the results for α_2 and α_3 listed in (9.20), the theoretical value of the flow alignment parameter is found to be $\lambda = (1+q^2)/(q^2-1)$. Combining (9.11) and (9.19), the flow alignment angle ψ and λ are related by

$$\tan\psi = \sqrt{\frac{\lambda+1}{\lambda-1}}. \tag{9.22}$$

Flow alignment requires that condition $|\lambda| > 1$ holds, otherwise (9.22) has no solution and the director tumbles in a shear flow.

Equation (9.21) leads to the director components $n_1 = -q\ \mathrm{sgn}(\mathring{\gamma})/\sqrt{1+q^2}$ and $n_3 = 1/\sqrt{1+q^2}$. Using these expressions for the director components, (9.18) yields the normal stress differences

$$T_I = -\frac{2\eta_{HS} q(1-q^2)|\mathring{\gamma}|}{(q^2+1)^2}, \tag{9.23}$$

$$T_{II} = \frac{\eta_{HS} q(1-q^2)|\mathring{\gamma}|}{(q^2+1)^2}. \tag{9.24}$$

9.4 Mesoscopic Approach for a Weakly Anisotropic Fluid of Oblate Platelets

9.4.1 Rheological Model

Using the scalar order parameter S and the director $\mathbf{n}$, the components of a symmetric second rank tensor

$$Q_{kl} = S\left(n_k n_l - \frac{1}{3}\delta_{kl}\right) \tag{9.25}$$

describe both the degree of nematic order and the preferred direction of the long molecular axes in space [6]. Furthermore, it is useful to introduce the antisymmetric tensor $\Omega_{kl} = (\partial v_k/\partial x_l - \partial v_l/\partial x_k)/2$. The corotational time derivative of this tensor order parameter

$$\mathring{Q}_{kl} = \frac{\partial Q_{kl}}{\partial t} + v_s \frac{\partial Q_{kl}}{\partial x_s} - \Omega_{ks} Q_{sl} + Q_{ks} \Omega_{sl} \tag{9.26}$$

replaces the corotational time derivative $\mathbf{N}$ of the director defined in the framework of the macroscopic theory [9, 10]. We consider experiments with a rheometer that produces an approximately homogeneous flow. Our theoretical approach is based on the Onsager's principle, which allows to derive kinetic equations by minimizing a Rayleighian. The principal way to apply this general method is explained by many examples in the monograph of Doi on soft matter physics [21]. In the present case the Rayleighian depends on the components of the shear gradient tensor A and the corotational time derivatives $\mathring{Q}_{kl}$ (9.26) of the components of tensor Q. The Rayleighian of a flowing nematic fluid can be written as

$$R(A, \mathring{Q}) = \Phi(A, \mathring{Q}) + \mathring{Q}_{kl} \frac{\partial F(Q)}{\partial Q_{kl}}. \tag{9.27}$$

The first term on the right hand side is the dissipation function describing the entropy production by viscous energy dissipation. The second term accounts for the change of free energy arising when the nematic order parameter Q relaxes and accommodates to the flow. $F(Q)$ is the free energy of a nematic liquid crystal in dependence on the order parameter. Onsager's principle states that the system evolves along a path with minimal Rayleighian. Using Onsager's principle, equations of motion are obtained from

$$\frac{\partial R(A,\overset{\circ}{Q})}{\partial \overset{\circ}{Q}_{kl}} = 0, \quad \text{and} \tag{9.28}$$

$$\frac{\partial R(A,\overset{\circ}{Q})}{\partial A_{kl}} = T_{kl}, \tag{9.29}$$

where the components T_{kl} define the stress tensor T, which is assumed to be constant in space. T is produced by the external forces applied to the rheometer plates. The dissipation function $\Phi(A,\overset{\circ}{Q})$ is quadratic in the fluxes A_{kl} and Q_{kl}

$$\begin{aligned}\Phi(A,\overset{\circ}{Q}) = {} & L_{klst}(Q)\overset{\circ}{Q}_{kl}\overset{\circ}{Q}_{st} + M_{klst}(Q)\left(\overset{\circ}{Q}_{kl}A_{st} + A_{kl}\overset{\circ}{Q}_{st}\right) \\ & + N_{klst}(Q)A_{kl}A_{st}.\end{aligned} \tag{9.30}$$

The procedure based on a Rayleighian implies that the kinetic coefficients satisfy Onsager's reciprocal relations. In the present case these relations are

$$M_{(kl)(st)}(Q) = M_{(st)(kl)}(Q). \tag{9.31}$$

Introducing the notation $P_{kl} = \partial R(A,\overset{\circ}{Q})/\partial \overset{\circ}{Q}_{kl}$, (9.28) and (9.29) are written as

$$P_{kl} = 0, \quad \text{and} \tag{9.32}$$

$$T_{kl} = \frac{\partial \Phi(A,\overset{\circ}{Q})}{\partial A_{kl}}. \tag{9.33}$$

Taking into account that the order parameter Q is small, i.e. its matrix norm satisfies the condition $\|Q\| \ll 1$, the kinetic matrices L, M, and N in (9.30) could be expanded into a power series of matrix Q. However, there is a simpler way to obtain the thermodynamic forces utilizing that $\Phi(A,\overset{\circ}{Q})$ is a homogeneous Eulerian function of second order of the tensor components $\overset{\circ}{Q}_{kl}$ and A_{kl} (9.30). Hence, the dissipation function $\Phi(A,\overset{\circ}{Q})$ may be written as a bilinear form in terms of fluxes and forces as

$$\Phi = P_{kl}\overset{\circ}{Q}_{kl} + T_{kl}A_{kl}, \tag{9.34}$$

where the forces P_{kl} and T_{kl} depend linearly on the fluxes $\overset{\circ}{Q}_{kl}$ and A_{kl}. Since the tensors $\overset{\circ}{Q}$ and A are symmetric and traceless, the dissipation function does not change if we add multiple of the unit tensors to the tensors P and T In this case P and T can always be chosen as traceless tensors. Hence, the matrices P_{kl} and T_{kl} may be written in such a way that they are symmetric and traceless without changing the dissipated energy $\Phi(A, \mathring{Q})$. Furthermore, taking into account Onsager's reciprocal relation (9.31) and keeping only terms of lowest order of magnitude, the expansions of the tensors P and T give

$$\begin{aligned} P_{kl} = \mu_0 \mathring{Q}_{kl} - \mu_1 A_{kl} - \mu_2 \left(A_{ks} Q_{sl} + Q_{ks} A_{sl} - \frac{2}{3} A_{st} Q_{ts} \delta_{kl} \right) \\ + \mu_3 \left(\mathring{Q}_{ks} Q_{sl} + Q_{ks} \mathring{Q}_{sl} - \frac{2}{3} \mathring{Q}_{st} Q_{ts} \delta_{kl} \right) + \frac{\partial F(Q)}{\partial Q_{kl}}, \end{aligned} \tag{9.35}$$

and

$$T_{kl} = -\mu_1 \mathring{Q}_{kl} + \upsilon_1 A_{kl} + \upsilon_2 (A_{ks} Q_{sl} + Q_{ks} A_{sl}) - \mu_2 \left(\mathring{Q}_{ks} Q_{sl} + Q_{ks} \mathring{Q}_{sl} \right), \tag{9.36}$$

where only the traceless parts of T_{kl} should be used. If the trace T_{ss} does not vanish, the components of tensors defined by (9.36) should be replaced by the modified expression $\hat{T}_{kl} = T_{kl} - \frac{1}{3} T_{ss} \delta_{kl}$. Using a different mathematical method, (9.35) has already been derived by Farhoudi and Rey [7, 8] for a rheological model of polymeric nematic liquid crystals. The derivative of the free energy $F(Q)$ for the nematic phase

$$\frac{\partial F(Q)}{\partial Q_{kl}} = b Q_{kl} - c Q_{ks} Q_{sl} + d Q_{ks} Q_{st} Q_{tl} \tag{9.37}$$

is obtained from the de Gennes order parameter expansion [6]. In the present stage of experimental accuracy, it is sufficient to assume an uniaxial ordering of the kaolin platelets described by the order parameter tensor (9.25). Possible small corrections due to biaxiality of the nematic order are neglected. The nematic order parameter (9.25) accounts for the order parameter magnitude S and the flow alignment angle ψ. Using again the coordinate system shown in Fig. 9.2, the tensors A and Ω are

$$(A_{kl}) = \begin{pmatrix} 0 & 0 & \frac{1}{2}\mathring{\gamma} \\ 0 & 0 & 0 \\ \frac{1}{2}\mathring{\gamma} & 0 & 0 \end{pmatrix}, \tag{9.38}$$

$$(\Omega_{kl}) = \begin{pmatrix} 0 & 0 & \frac{1}{2}\mathring{\gamma} \\ 0 & 0 & 0 \\ -\frac{1}{2}\mathring{\gamma} & 0 & 0 \end{pmatrix}. \tag{9.39}$$

Taking into account that the director is defined by $\mathbf{n} = (-\sin\psi, 0, \cos\psi)$ the order parameter tensor Q has the representation

$$(Q_{kl}) = \begin{pmatrix} \frac{1}{6}S(1-3\cos(2\psi)) & 0 & -\frac{1}{2}S\sin(2\psi) \\ 0 & -\frac{1}{3}S & 0 \\ -\frac{1}{2}S\sin(2\psi) & 0 & \frac{1}{6}S(1+3\cos(2\psi)) \end{pmatrix}. \tag{9.40}$$

Using (9.36) the shear stress T_{13} and the first ($T_I = T_{11} - T_{33}$) and second ($T_{II} = T_{33} - T_{22}$) normal stress differences are obtained from the relations

$$T_{13} = \frac{1}{6}\mathring{\gamma}[3\upsilon_1 + \upsilon_2 S + (3\mu_1 + \mu_2 S)S\cos(2\psi)], \tag{9.41}$$

$$T_I = -\frac{1}{3}\mathring{\gamma}(3\mu_1 + \mu_2 S)S\sin(2\psi), \tag{9.42}$$

$$T_{II} = \frac{1}{6}\mathring{\gamma}(3\upsilon_2 + 3\mu_1 + \mu_2 S)S\sin(2\psi). \tag{9.43}$$

The flow alignment angle may be evaluated by using the matrix relation $M = PQ - QP = 0$. The only components of the antisymmetric matrix $\boldsymbol{M}$ which do not vanish identically are M_{13} and $M_{31} = -M_{13}$. $M_{13} = 0$ leads to an equation for the flow alignment angle

$$\frac{\partial\psi}{\partial t} = -\frac{1}{2}\mathring{\gamma}\left[1 + \frac{3\mu_1 + \mu_2 S}{S(3\mu_0 + \mu_3 S)}\cos(2\psi)\right] \tag{9.44}$$

which corresponds to the result of Farhoudi and Rey [7, 8]. If the condition $3\mu_0 \gg |\mu_3 S|$ is satisfied the number of adjustable parameters can be reduced. This condition seems to be reasonable, since the order parameter S for the kaolin platelet suspension is not large. Thus, we always assume $\mu_3 = 0$ in the following evaluations. The nematic order parameter magnitude S is obtained from the equation $trace(PQ + QP) = 0$. The trace contains the term $Q_{st}(\partial F(Q)/\partial Q_{st}) + (\partial F(Q)/\partial Q_{st})Q_{st}$, which is a polynomial of degree four in terms of S. This polynomial may be transformed into the product $4dS^2(S - S_b)(S - S_0)/9$, where the roots S_b and S_0 depend on the coefficients c and d in (9.37). S_0 is the magnitude of the nematic order parameter at small shear gradients ($S_0 = S(\mathring{\gamma} \to 0)$). After some rearrangements the equation for the trace of $PQ + QP$ and (9.44) are expressed as

$$\frac{\partial\psi}{\partial t} = -\frac{1}{2}\mathring{\gamma}[1 + \lambda(S)\cos(2\psi)], \tag{9.45}$$

and

$$\frac{\partial S}{\partial t} = -\frac{3}{4}\mathring{\gamma}S[\lambda(S) + \beta]\sin(2\psi) - \chi(S)(S - S_0), \tag{9.46}$$

where the flow alignment parameter $\lambda(S)$ and function $\chi(S)$ are defined by

$$\lambda(S) = \frac{\alpha}{S} + \beta \quad \text{and} \quad \chi(S) = \frac{4d}{9} S(S - S_b) \tag{9.47}$$

with $\alpha = \mu_1/\mu_0$ and $\beta = \mu_2/3\mu_0$. It should be mentioned that a replacement of $\chi(S)$ by the inverse relaxation time $\chi(S_0) > 0$ in (9.46) could lead to a further simplification, which is justified if changes of S remain relatively small as found in our experiments. Equation (9.45) is related to (9.12) of the macroscopic Leslie-Erickson theory for nematic liquid crystals. However, in contrast to the Leslie-Erickson approach the flow alignment parameter in (9.45) is a function of the variable order parameter S, which depends on the shear gradient $\mathring{\gamma}$.

9.4.2 *Theoretical Results for Stationary Flow Alignment*

Let us consider the stationary shear alignment regime appearing for a shear flow if $|\lambda(S)| > 1$. Using (9.46), it can be proven that shear alignment implies that $\psi_\infty = \psi(\mathring{\gamma} \to \infty) = 0$, if the value of order parameter S at high shear rates ($\mathring{\gamma} \to \infty$) does not coincide with the corresponding value S_0 at low shear rates ($\mathring{\gamma} \to 0$). Then, using (9.45), for high shear rates the relation $\lambda(S_\infty) = -1/\cos(2\psi_\infty) = -1$ results. If suitable experimental data are available for low and large shear rates, we can use the order parameter magnitude at low shear rates $S_0 = S(\mathring{\gamma} \to 0)$, the corresponding flow alignment angle $\psi_0 = \psi(\mathring{\gamma} \to 0)$ and the value of S at high shear rates ($S_\infty = S(\mathring{\gamma} \to \infty)$) to evaluate the material constants α and β in the expression for $\lambda(S)$. Thus we obtain

$$\lambda(S) = \frac{\alpha}{S} - \frac{\alpha}{S_0} - \frac{1}{\cos(2\psi_0)}, \tag{9.48}$$

where

$$\alpha = -\frac{S_0 S_\infty (1 - \cos(2\psi_0))}{(S_\infty - S_0)\cos(2\psi_0)}, \tag{9.49}$$

and the parameter β in (9.46) is defined as $\beta = -\alpha/S_0 - 1/\cos(2\psi_0)$, or

$$\beta = -\frac{S_\infty \cos(2\psi_0) - S_0}{(S_\infty - S_0)\cos(2\psi_0)}. \tag{9.50}$$

For solving (9.45) and (9.46) a perturbation approach $S = S_0 + S_1\mathring{\gamma} + S_2\mathring{\gamma}^2 + \ldots$ and $\psi = \psi_0 + \psi_1\mathring{\gamma} + \psi_2\mathring{\gamma}^2 + \ldots$ is useful. Inserting these expansions into (9.45) and (9.46) the lowest order terms of the perturbation expansion are found to be

$$S = S_0 + \frac{3S_0[1 - \beta\cos(2\psi_0)]\tan(2\psi_0)}{4\chi(S_0)}\mathring{\gamma} + O(\mathring{\gamma}^2), \tag{9.51}$$

and

$$\psi = \psi_0 + \frac{3\alpha[1 - \beta\cos(2\psi_0)]\cos(2\psi_0)}{8\chi(S_0)S_0}\mathring{\gamma} + O(\mathring{\gamma}^2), \tag{9.52}$$

where we always assume that $\mathring{\gamma} > 0$. The constant $\chi(S_0) > 0$ may be considered as an inverse relaxation time, which is proportional to the restoring force when the order parameter S relaxes. In the case of stationary flows ($\partial S/\partial t = 0$), relation (9.42) for the first normal stress difference can be simplified. Taking into account that the flow alignment parameter can be expressed as $\lambda(S) = (3\mu_1 + \mu_2 S)/3S\mu_0$ (9.47), (9.42) is transformed into

$$T_I = -\mu_0 S^2 \lambda(S)\mathring{\gamma}\sin(2\psi). \tag{9.53}$$

As the dissipation function (9.34) must be positive definite, the condition $\mu_0 > 0$ holds. If $\mathring{\gamma}$ is small or moderately high, the perturbation approach leads to the first normal stress difference $T_I = -\mu_0 S_0 \lambda(S_0)\mathring{\gamma}\sin(2\psi_0) + O(\mathring{\gamma}^2)$, which is positive, since $\lambda(S_0) < 0$ and the experimentally observed flow alignment angle ψ has a small but positive value ($\sin(2\psi) \simeq 2\psi > 0$). The value of T_I increases with increasing shear gradient $\mathring{\gamma} > 0$. At higher values of $\mathring{\gamma}$, the difference $T_{11} - T_{33} = T_I$ approaches a finite limit. For a stationary flow ($\partial S/\partial t = 0$), (9.46) can be used to replace the product $\mathring{\gamma}\sin(2\psi)$ in (9.53) and we arrive at

$$T_I = \frac{4\mu_0\lambda(S)\chi(S)S(S - S_0)}{3(\lambda(S) + \beta)}. \tag{9.54}$$

In this expression the replacement $\chi(S) \simeq \chi(S_0)$ is a reasonable approximation. Obviously, according to (9.46) the product $\mathring{\gamma}\sin(2\psi)$ remains finite in the shear alignment regime even if $|\mathring{\gamma}| \to \infty$. For high shear rates ($|\mathring{\gamma}| \to \infty$) the insertion of $\lambda(S_\infty) = -1$ into (9.54) leads to

$$T_I = \frac{4\mu_0\chi(S_0)S_\infty(S_\infty - S_0)}{3(1 - \beta)}. \tag{9.55}$$

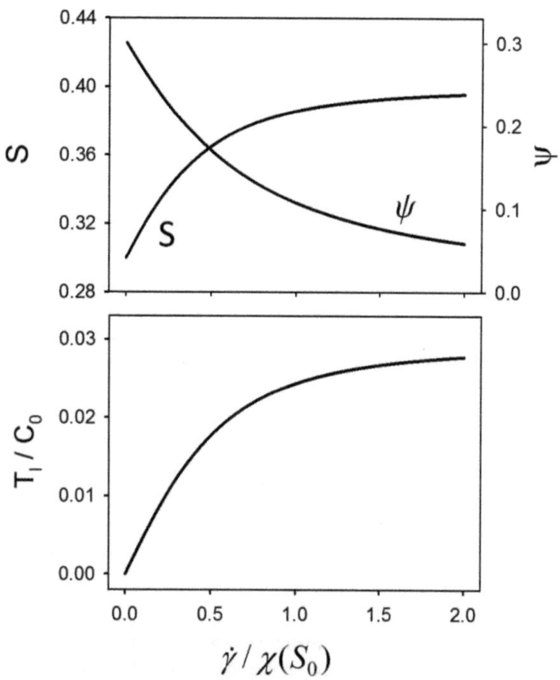

Fig. 9.6 Results of the mesoscopic theory for the scalar order parameter S, the flow alignment angle ψ and the scaled normal stress difference T_I/C_0 versus the scaled shear rate $\dot{\gamma}/\chi(S_0)(S_0 = 0.3, S_\infty = 0.4, \psi_0 = 0.30$ and $C_0 = 4\mu_0\chi(S_0)/3)$

If $S_\infty > S_0$ (9.49) and (9.50) lead to the conditions $\alpha < 0$ and $|\beta| < 1$ for small flow alignment angles ψ_0. In this case (9.54) and (9.55) predict that the first normal stress difference is positive and approaches a constant positive value when $|\mathring{\gamma}|$ tends to infinity. In the whole range of shear rates $\mathring{\gamma}$, the structural parameters S and ψ can be obtained by solving (9.45) and (9.46) numerically. For the stationary state $(\partial S/\partial t = 0$ and $\partial\psi/\partial t = 0)$ and a parameter choice $S_0 = 0.30$, $S_\infty = 0.40$ and $\psi_0 = 0.30$, the first normal stress difference (9.53) can be evaluated.

S, ψ and T_I/C_0 are plotted versus the reduced shear rate $\mathring{\gamma}/\chi(S_0)$ in Fig. 9.6, where $C_0 = 3/4\mu_0\chi(S_0)$ is a constant. We use the simplified version of the model with $\chi(S) \simeq \chi(S_0)$. In stationary flows, the qualitative behavior of the first normal stress difference and the structural parameters S and ψ does not change if $\chi(S_0)$ varies. $\chi(S_0)$ only defines the time scale of dynamic processes. In the other case, if $S_\infty < S_0$, the first normal stress difference may be negative. However, a stability analysis reveals that the director tumbles or waggles if $S_\infty < S_0$. This more involved case will be discussed more thoroughly in another paper.

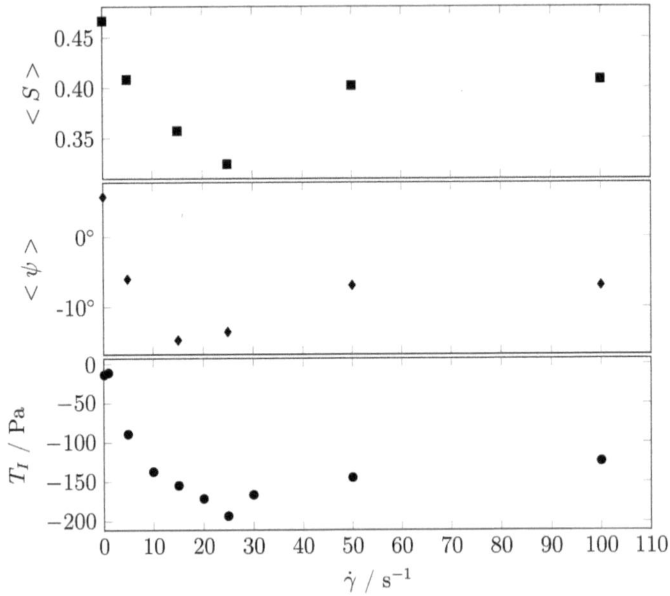

Fig. 9.7 Apparent scalar nematic order parameter ($< S >$), flow alignment angle ($< \psi >$) and first the normal stress difference T_I in dependence on the shear rate $\dot{\gamma}$ for a low $pH-$ value ($pH = 3.5$). As the first normal stress difference is negative, the director is supposed to waggle or tumble. The experimentally observed values of $< S >$ and $< \psi >$ are considered as averages over many domains

9.5 Comparison of Rheological and Structural Data

Kaolin powder obtained from Sigma Aldrich has a broad particle size distribution. Using sedimentation analysis of the kaolin powder, the equivalent spherical diameters $d_{10} = 0.17\,\mu\text{m}$ and $d_{50} = 0.65\,\mu\text{m}$ (10 and 50 % of the particles are smaller than 17 and 0.65 μm, respectively) were found. The average particle size determined by laser diffraction was 3.8 μm (Malvern Instruments Mastersizer 2000). Using SEM, an aspect ratio of roughly $q \simeq 0.1$ was estimated. In all rheological experiments the $pH-$ value was 3.5. The solid particles of the kaolin suspensions had the mass fraction $w = 0.4$, which corresponds to a volume fraction of $\Phi \simeq 0.2$. At low pH-values a concentrated kaolin suspension at rest is supposed to form a gel that impedes Brownian motion of the platelets. In this gel network negatively charged edges are connected to positively charged platelet faces. In a shear flow, however, the weak particle aggregates become disconnected and Brownian motion accompanied with particle reorientation is possible.

A disadvantage of the X-ray scattering method is the relatively low penetration depth as discussed in Sect. 3.2. Nevertheless, general structural results can be compared with the data for T_I. Our experimental results demonstrate that the preferred direction of platelet normals under shear has the azimuthal angle $\phi = 0$ (Fig. 9.3). Thus, according to (9.5), the polar angle $\theta_{\max} = -\psi$ corresponds to a clockwise rotation of the platelet normals in the shear plane ($\psi < 0$). Figure 9.7 shows the apparent scalar order parameter $< S>$, the apparent flow alignment angle $< \psi >$ and the first normal stress difference T_I in dependence on the shear rate for a low pH-value of the suspension ($pH = 3.5$). Both the first normal stress difference T_I and the director tilt angle $< \psi >$ have negative values. If the theory for polymer nematics were applicable to aligned kaolin platelets, a negative first normal stress difference would be the result of a non-stationary flow regime. Then the director, defined as the preferred direction of the platelet normals in a domain, is supposed to waggle or tumble [1, 14, 15]. In this case, the intensities of the X-ray diffraction beams from many differently aligned domains superimpose and the experimental data for S and ψ should be interpreted as averaged values, symbolized by $< S >$ and $< \psi>$. The magnitude of the director tilt angle $< \psi >$ is relatively small (Fig. 9.7). T_I, $< \psi >$ and $< S >$ are non-monotonous functions of the shear rate $\mathring{\gamma}$. The magnitude $|T_I|$ is maximal where both the order parameter $<S>$ and the flow alignment angle $<\psi>$ have a relative extremum. Hence, there is a close relation between platelet alignment and the rheological properties of the suspension.

9.6 Discussion

Let us compare the results of theoretical models with experimental data for concentrated kaolin suspensions which have negative values of T_I. It should be noted that there are also other experimental results for modified kaolin suspensions ($pH > 6$ and $\Phi \geq 0.2$), which always seem to have a positive first normal stress difference as shown in Fig. 9.1 [16]. But in this paper we focus our attention on the case $T_I < 0$. For simplicity, we choose the direction of the flow velocity to be parallel to the positive direction of the x-axis, excluding the antiparallel orientation. In this case we have $\mathring{\gamma} = |\mathring{\gamma}| \geq 0$. The predictions of the macroscopic Leslie-Erickson theory for nematic liquid crystals in combination with the microscopic Baalss-Hess theory [20] for oblate spheroids ($q < 1$) are summarized as follows (9.21, 9.23 and 9.24): (a) A stationary shear flow of spheroids is stable, (b) the flow alignment angle ψ is small and positive (anticlockwise tilt direction) for $q \ll 1$, (c) the first normal stress difference is negative, (d) the second normal stress difference is positive, and (e) both normal stress differences are proportional to $|\mathring{\gamma}|$. The observed negative first normal stress difference T_I combined with a negative value of ψ cannot be explained by the Baalss-Hess theory.

The mesoscopic approach takes into account the nematic order parameter tensor Q, which depends on both the director tilt angle ψ (Fig. 9.2) and the scalar order parameter S (9.4). Equations (9.41)–(9.46) contain several adjustable material parameters. In the case of a stable flow alignment, i.e. if the director does not tumble or waggle, the mesoscopic model provides us rather general predictions (Sect. 4.2): (a) The flow alignment angle ψ is positive (anticlockwise rotation) as depicted in Fig. 9.2, (b) angle ψ decreases and the scalar order parameter S increases with increasing shear rate $\mathring{\gamma}$, (c) the first normal stress difference T_I is positive and increases linearly with the shear rate $\mathring{\gamma}$ at low and moderate values of $\mathring{\gamma}$, and (d) at higher shear rates T_I approaches a plateau value. A stable stationary platelet orientation in combination with a negative first normal stress difference is not compatible with our theoretical model.

Experimentally observed negative values of the first normal stress difference could have the same origin as in the case of nematic polymeric solutions. If $T_I < 0$, the director is supposed to tumble or waggle, at least in a certain interval at low shear rates $\mathring{\gamma}$. Tumbling or wagging must occur if the flow alignment parameter $\lambda(S)$ (9.47) satisfies the condition $|\lambda(S)| < 1$, since in this case (9.22) does not yield a real value for the angle ψ. It is well known that the value of $\lambda(S)$ for disk-like particles is always negative and very close to -1 [1, 18]. Experimental data are mean values averaged over many domains of the kaolin sample. In a similar way as for polymeric liquid crystals, averaging over many different domains could be replaced by time averaging over a single domain [15]. Using angle brackets for symbolizing averaging, (9.42) can be written as

$$T_I = -\mu_0 < S^2 \lambda(S) \mathring{\gamma} \sin(2\psi) > . \tag{9.56}$$

Equation (9.56) relates the flow alignment angle to the first normal stress difference. Considering the experimental data in Fig. 9.7, the apparent flow alignment angle $<\psi>$ is negative in the investigated range of shear rates $0 < \mathring{\gamma} < 100\,\mathrm{s}^{-1}$. Since $\lambda(S) \simeq -1$, $\mu_0 > 0$ and $< \sin(2\psi) > < 0$, the values of the first normal stress difference T_I (9.56) should be negative in agreement with the experimental observation. Further insight into the rheological behavior for the non-stationary regime ($|\lambda(S)| < 1$) can be gained by evaluating $\psi(t)$ and $S(t)$. Solving the coupled differential equations for the dynamics (9.45 and 9.46) requires extensive numerical computations. Results will be published in a forthcoming paper.

In conclusion, a mesoscopic theoretical approach for modelling rheological properties of concentrated kaolin suspensions predicts that a stationary shear alignment is not compatible with a negative first normal stress difference. The observed negative normal stress difference at $pH = 3.5$ is suggested to be a consequence of tumbling or wagging of the platelet normals. The platelet normals are supposed to behave similarly as the director in polymeric liquid crystals under shear.

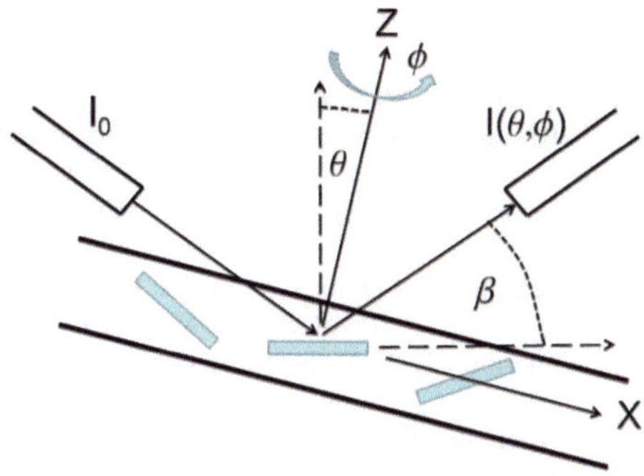

Fig. 9.8 Schematic illustration of the setup for recording the X-ray intensity distribution $I(\theta, \phi)$ for the kaolin platelets in a frozen patch of sheared kaolin

Acknowledgment Financial support by the Deutsche Forschungsgemeinschaft through grant Mo 600/4–1 is gratefully acknowledged.

Appendix

Setup for recording the platelet orientational distribution in kaolin suspensions

After shearing the kaolin suspension has been frozen rapidly to fix the platelet orientation. Figure 9.8 illustrates the experimental setup for recording the intensity $I(\theta, \phi)$ of diffracted X-rays in dependence on the orientation of the rotation table. $I(\theta, \phi)$ is the intensity of the X-ray beam reflected by the crystallographic $(0, 0, 2)$—planes (Bragg angle $\beta \cong 12°$).

If the Bragg condition is satisfied for a platelet, the platelet normal, the incident and the diffracted beam lie in the fixed *x*–*z*—plane of the Cartesian coordinate *x*–*y*–*z*–system shown in Fig. 9.8. The angles θ and ϕ are the polar and the azimuthal angle of the rotation table normal in the fixed *x*–*y*–*z*–system. The orientational distribution function of the platelet normals is obtained from the intensity distribution of diffracted X-rays by using the relation $I(\theta, \phi) \propto f(\theta, \phi)$.

References

1. R.G. Larson, *The Structure and Rheology of Complex Fluids* (Oxford University Press, New York, 1999)
2. S.M. Jogun, C.F. Zukoski, Rheology and microstructure of dense suspensions of plate-shaped colloidal particles. J. Rheol. **43**, 847–871 (1999)
3. A.B.D. Brown, S.M. Clarke, P. Convert, A.R. Rennie, Orientational order in concentrated dispersions of plate-like kaolinite particles under shear. J. Rheol. **44**, 221–233 (2000)

4. S.J.S. Qazi, A.R. Rennie, J.P. Wright, J.K. Cockcroft, Alignment of Plate-Like Particles in a Colloidal Dispersion under Flow in a Uniform Pipe Studied by High-Energy X-ray Diffraction. Langmuir **26**, 18701–18709 (2010)
5. M. Moan, T. Aubry, F. Bossard, Nonlinear behavior of very concentrated suspensions of plate-like kaolin particles in shear flow. J. Rheol. **47**, 1493–1504 (2003)
6. P.G. de Gennes, J. Prost, *The Physics of Liquid Crystals* (Oxford University Press, New York, 1993)
7. Y. Farhoudi, A.D. Rey, Shear flows of nematic polymers: Orienting modes, bifurcations, and steady state rheological predictions. J. Rheol. **37**, 289–313 (1993)
8. Y. Farhoudi, A.D. Rey, Ordering effects in shear flows of discotic polymers. Rheol. Acta **32**, 207–2017 (1993)
9. A.P. Singh, A.D. Rey, Microstructure constitutive equation for discotic nematic liquid crystalline materials-Part I: Selection procedure and shear flow predictions. Rheol. Acta **37**, 30–45 (1998)
10. A.P. Singh, A.D. Rey, Microstructure constitutive equation for discotic nematic liquid crystalline materials-Part II: Microstructure-rheology relations. Rheol. Acta **37**, 374–386 (1998)
11. A.D. Rey, Bifurcational analysis of the isotropic-nematic phase transition of rigid rod polymers subjected to biaxial stretching flow. Macromol. Theory Simul. **4**, 857–872 (1995)
12. D. Greco, A.D. Rey, Transient rheology of discotic mesophases. Rheol. Acta **42**, 590–604 (2003)
13. M. Doi, S.F. Edwards, *The Theory of Polymer Dynamics* (Oxford University Press, Oxford, 1986)
14. G. Marucci, Rheology of Nematic Polymers. in Liquid Crystallinity, in *Polymers, pp 395-422*, ed. by A. Ciferri (VCH Publishers, New York, 1991)
15. R.G. Larson, Arrested tumbling in shearing flows of liquid crystal polymers. Macromolecules **23**, 3983–3992 (1990)
16. M. Bombrowski, P. Schiller,H.-J. Mögel, *Negative Normal Stress Difference in Concentrated Kaolin Suspension*. In: 8th Annual European Rheology Conference, Leuven (2013)
17. G.B. Arfken, H.J. Weber, *Mathematical Methods for Physisists* (Academic Press, San Diego, 1995)
18. T. Carlsson, The possibility of the existence of a positive Leslie viscosity α_2. Proposed flow behavior of disk like Nematic liquid crystals. Mol. Cryst. Liq. Cryst. **89**, 57–66 (1982)
19. P. Schiller, S. Krüger, M. Wahab, H.-J. Mögel, Interactions between Spheroidal Colloidal Particles. Langmuir **27**, 10429–10437 (2011)
20. D. Baalss, S. Hess, The Viscosity Coefficients of Oriented Nematic and Nematic Discotic Liquid Crystals: Affine Transformation Model. Z. Naturforsch. **43**, 662–670 (1988)
21. M. Doi, *Soft Matter Physics* (Oxford University Press, New York, 2013)

Chapter 10
Carbon Nanotubes in Liquid Crystals: Fundamental Properties and Applications

Longin Lisetski, Marat Soskin and Nikolai Lebovka

Abstract The structure and properties of liquid crystalline suspensions filled by carbon nanotubes (CNTs) are critically reviewed. Special attention is paid to interactions between CNTs and molecules of the liquid crystals (LC), which lead to formation of ordered supramolecular structures. These structures, in turn, determine unique physical properties of LC + CNT suspensions, including electrical conductivity, dielectric permittivity, phase transitions, optical transmission, memory effects that can be used in electrooptic and optoelectronic devices, etc. Great variety of LC phases are considered as a host media, such as nematics, cholesterics, smectics of different types (including ferroelectrics), lyotropic, chromonic, ionic and hydrogen-bonded liquid crystals. Alongside multi- and single-walled carbon nanotubes, the suspensions can also contain the platelets of organoclays used for facilitation of CNT dispersing. Recent practical applications of LC + CNT suspensions and nanomaterials based thereon are also outlined.

List of Symbols and Abbreviations

Symbols

a — L/d, aspect ratio

C — Weight concentration of CNTs

C_c — Percolation weight concentration of CNTs

L. Lisetski (✉)
Institute for Scintillation Materials of STC "Institute for Single Crystals", NAS of Ukraine, 60 Lenin Av, Kharkiv 61001, Ukraine
e-mail: lisetski@isma.kharkov.ua

M. Soskin
Institute of Physics, NAS of Ukraine, 46 Nauky Pr, Kyiv 03028, Ukraine
e-mail: marat.soskin@gmail.com

N. Lebovka
Department of Physical Chemistry of Disperse Minerals, F. D. Ovcharenko Biocolloidal Chemistry Institute, 42, Vernadsky Av, Kyiv 03142, Ukraine
e-mail: lebovka@gmail.com

L. Bulavin and N. Lebovka (eds.), *Physics of Liquid Matter: Modern Problems*, Springer Proceedings in Physics 171, DOI 10.1007/978-3-319-20875-6_10

d Outer diameter of MWCNTs
d_i Inner diameter of MWCNTs
d_c Effective size d_c of a coil
d_s Interlayer distance in MWCNT
f Frequency
$f(\theta)$ Angular function
H Heat flow
h Thickness of a sample
K Bending stiffness of CNTs
K_e Frank elastic constant
k Degree of impenetrability
k_BT Thermal energy
L Length of nanotubes
L_p Persistence length
n Number of walls
$\boldsymbol{n}$ Director vector
r Distance between particles
S Specific surface area
s_{LC} Order parameter in the nematic phase
s_n Order parameter of rods (CNTs)
T Temperature
T_{cn} Crystal-nematic transition temperature
T_{ni} Nematic-isotropic transition temperature
T_r Optical transmission
t Critical conductivity exponent
U Interaction potential
u Voltage
W Surface anchoring energy

Greek Letters

δ_T Total Hansen parameters
$\Delta\varepsilon$ $\varepsilon_\parallel - \varepsilon_\perp$, dielectric anisotropy
$\varepsilon_\parallel$ Parallel component of dielectric permittivity
$\varepsilon_\perp$ Perpendicular component of dielectric permittivity
γ Angle between the rods
η Viscosity of LC
φ Volume concentration of CNTs
φ_c Percolation volume concentration of CNTs
Λ Thickness of LC interfacial shells surrounding the MWCNT aggregates
λ Chiral pitch length
θ Angle of the connecting line of the particle centres with the director direction $\boldsymbol{n}$
ρ Density
σ Electrical conductivity

τ	Time of aggregation
ξ	Penetration length

Abbreviations

AC	Alternating current
AFM	Atomic force microscopy
BLB	Beer–Lambert–Bouguer
CNTs	Carbon nanotubes
DC	Direct current
DSC	Differential scanning calorimetry
FLC	Ferroelectric liquid crystal
Lap	Laponite
LapO	Organo-modified laponite
LC	Liquid crystal
LLC	Lyotropic liquid crystal
MMT	Montmorillonite
MMTO	Organo-modified montmorillonite
MWCNTs	Multi-walled carbon nanotubes
NTC	Negative temperature coefficient of conductivity
POM	Polarized optical microscopy
SAXS	Small-angle X-ray scattering
SEM	Scanning electron microscopy
SWCNTs	Single-walled carbon nanotubes
TEM	Transmission electron microscopy
UV-Vis	Ultraviolet and visible

Chemical substances

5CB	4-pentyl-4′-cyanobiphenyl
7OBA	*p*-*n*-heptyloxybenzoic acid
BBBA	4-butoxybenzylidene-4′-butylaniline
$C_{12}E_6$	*n*-dodecyl octaoxyethene monoether
C14mimCl	1-tetradecyl-3-methylimidazolium chloride
CB-15	4-(2-methylbutyl)-4′-cyanobiphenyl, chiral dopant
CC	Cholesteryl chloride
CB	Cholesteryl benzoate
CN	Cholesteryl nonanoate
CCN	Cholesteryl caprynate
CCL	Cholesteryl caprylate
CHP	1-Cyclohexyl-2-pyrrolidone
COC	Cholesteryl oleyl carbonate
CTAB	Cetyl-trimethyl ammonium bromide, anionic surfactant
DDAM	Dodecyldimethylammonium ethyl methacrylate
DMF	N,N-dimethylformamide

DSCG	Di-sodium cromoglycate
E7	Commercial mixture of cyanobiphenyl and cyanoterphenyl components
E63	Commercial mixture similar to the E7 with added cyclohexanes
EBBA	4-ethoxybenzylidene-4′-butylaniline
EAN	Ethylammonium nitrate (ionic liquid)
FLC-6304	Deformed helix FLC mixture
KCFLC10S	Commercial FLC mixture
LAHS7	Eutectic multicomponent mixture with a phenyl pyrimidine matrix
MBBA	4-methoxybenzylidene-4′-butylaniline
MLC-6290-000	Commercial nematic mixture (Merck)
MR	(2-(N,N-Dimethyl-4-aminophenyl)azobenzenecarboxylic acid), Methyl Red
NMP	N-methyl-2-pyrrolidone
PEG	Polyethylene glycol, hydrophilic polymer
PCPBB	4-pentylphenyl 2-chloro-4-(4-pentylbenzoyloxy) benzoate
PSS	Poly(sodium styrenesulfonate)
PDADMAC	Poly(diallydimethylammonium chloride)
ROTN403/015S	Commercial multi-component mixtures of a nematic LCs with 0.1 wt% of CN as chiral dopant
SDS	Sodium dodecyl sulfate, cationic surfactant
Triton X-100	Polyethylene glycol p-(1,1,3,3-tetramethylbutyl)-phenyl ether, non-ionic surfactant
ZhK-440	Commercial azoxy-component mixture of *p*-*n*-butyl-p-methoxyaz oxybenzene (2/3) and *p*-*n*-butyl-pheptonoiloxyazoxybenzene (1/3)
ZhK-805	Commercial mixture of 4-butyl-(1/2) and 4-hexyl-*trans*-cyclohexanecarboxylic acid (1/2)
ZhK-1282	Commercial mixture of several 4-alkyl- and 4-alkoxy-4′-cyanobiphenyls and nematic cyclohexyl-containing esters
ZLI-811	Benzoic acid, 4-hexyl-,4-[[(1-methylheptyl)oxy]carbonyl]phenyl ester, chiral dopant

10.1 Introduction

In recent years the suspensions on the base of liquid crystals (LCs) doped by carbon nanotubes (CNTs) have been studied intensively. The liquid crystalline self-organization results from anisotropic shape of LC molecules. Supplementary self-organization in LC doped by rod-like CNTs with extremely high aspect ratio, a ($a = L/d \approx 100$–1000 is the length, L, to diameter, d, ratio) is also expected. Many intriguing properties of self-organized CNT systems and LCs doped with CNTs were already reported. Here, we can refer the recent reviews on the problem [1–21].

This chapter reviews behaviour of CNTs in isotropic and LC suspensions. The resent experimental works on different physical properties of CNTs in nematic, smectic, cholesteric and lyotropic LC suspensions are critically analysed.

10.2 A Short Introduction to CNTs

The priority publication related with synthesis of CNTs was issued in 1990 [22] and then many review works related with properties and structure of CNTs and different composites on their base were proposed (see, e.g., [23]).

CNTs can be described as wrapped graphene sheets (sp^2) into the cylinders. There exist single-walled (SWCNTs) and multi-walled (MWCNTs) nanotubes with external diameters d 1−4 nm and 1–100 nm, respectively. The length of CNTs L can vary within 100 nm–50 μm. The way of wrapping of the graphene sheet is important and can result in chirality of SWCNTs. The change of chirality causes significant changes in electrical conductivity (metallic and semiconducting types).

MWCNTs usually display high mechanical strength and a metallic ("ballistic") type of electrical conductivity. Figure 10.1 shows examples of scanning electron microscopy images of MWCNTs. Usually, MWCNTs have wide diversity in their diameters, d, and lengths, L (Fig. 10.1a). The high resolution TEM image of a MWCNT reveals that it consists of multiple concentric tubes of graphene (Fig. 10.1b) [24]. For these MWCNTs the inner and outer diameters were $d_i \approx 5$ nm and $d \approx 10$ nm, respectively and the number of walls n was 7.

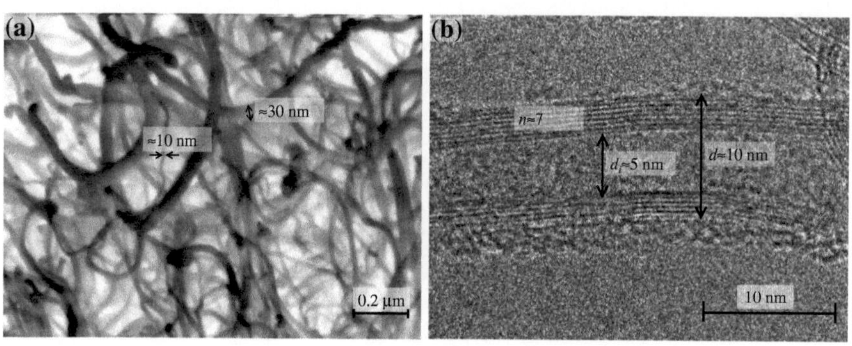

Fig. 10.1 Examples of scanning electron microscopy image of MWCNTs in powder (**a**) and high resolution electron microscopy image of individual MWCNT (**b**) (From [24]. With permission). Here, d_i and d are the inner and outer diameters, respectively and n is the number of walls

10.2.1 Specific Surface Area and Density

The specific surface area S and density ρ of MWCNTs can be theoretically estimated as follows [25]. The weight of one CNT, m, is

$$m = \frac{\pi L}{1315}\left(nd_i + 2d_s\sum_{j=0}^{n-1} j\right) = \frac{\pi Ln}{1315}(d_i + (n-1)d_s) = \frac{\pi Ln}{1315}(d_n - d_s(n+1)). \tag{10.1}$$

Here, $d_s = 0.34$ nm is the inter-shell distance in MWCNTs, $d_i = d - 2nd_s$ is the inner diameter and the specific surface area of graphene sheet (≈1315 m^2/g) [26] was taken into account.

For the high aspect ratio MWCNTs, i.e., when $a = L/d >> 1$, the effects of the end tips can be neglected. The surface ($s = \pi Ld_n$) and volume ($v = \pi Ld_n^2/4$) of one CNT can be treated as depending only on the outer diameter d. So, the specific surface area S and density ρ of MWCNTs can be estimated as:

$$S = \frac{1315}{n(1-(n+1)d_s/d)}, \tag{10.2}$$

$$\rho = \frac{4n(1-(n+1)d_s/d)}{1315d}. \tag{10.3}$$

Figure 10.2 presents the calculated specific surface area S and density ρ as functions of the CNT outer diameter d at different values of n [24]. The values of S and ρ

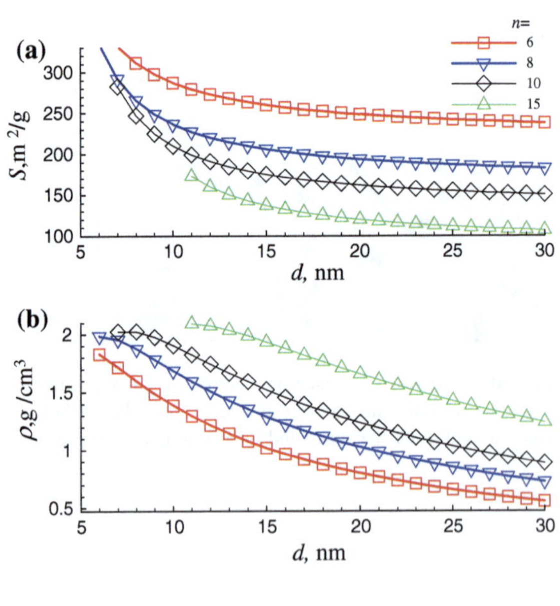

Fig. 10.2 Specific surface area S, (10.2) (**a**) and density ρ, (10.3) (**b**) versus the outer diameter of MWCNTs, d, at different numbers of walls, n

decrease as d increases and these values are also dependent on the number of walls, n.

The density of CNTs is a useful property to find relation between the volume, φ, and mass, C, concentrations of CNT in suspensions. These values are related as

$$\varphi = [1 + (1/C - 1)\rho/\rho_m]^{-1} = C\rho_m/\rho, \tag{10.4}$$

where ρ and ρ_m are the densities of CNTs and host medium. In the limit $C \rightarrow 0$, we get $\varphi \approx C\rho_m/\rho$.

In many practical cases, for composites doped by MWCNTs, the approximation of $\varphi \approx 0.5C$ may be used [27].

10.2.2 Do Carbon Nanotubes Behave Like Rods or Like Polymers?

A very important characteristic of CNTs is their flexibility or waviness. In many cases, the shape of CNTs deviates significantly from rod-like geometry and is rather tortuous when grown randomly by chemical vapor deposition method [28].

For random coil-like CNTs the dependence of the effective size, d_c, of the coil upon the length, L, of CNTs can be approximated as [29]

$$d_c \approx \sqrt{2L_pL + 2L_p^2(\exp(-L/L_p) - 1)}, \tag{10.5}$$

where L_p is the static bending persistence length (i.e., the maximum straight length that is not bent by a permanent structural deformation).

In the rigid random-coil limit (i.e., when $L >> L_p$):

$$d_c \approx \sqrt{2L_pL}. \tag{10.6}$$

In the rigid-rod limit (i.e., when $L \approx L_p$) the effective size of coil, d_c, is comparable with the length of CNTs, L:

$$d_c \approx L. \tag{10.7}$$

Although various theoretical and experimental methods have been used to estimate the persistence lengths of SWCNTs and MWCNTs, the existing data are still controversial. The theory predicts that the persistence length of CNTs can be evaluated as [30]

$$L_p = K/k_BT, \tag{10.8}$$

where $k_BT \approx 4.1 \cdot 10^{-21}$ J is the thermal energy at T = 273 K and K is a bending stiffness defined as

$$K = \frac{\pi}{8}\sum_{j=0}^{n-1} C_j d_j^3. \tag{10.9}$$

Here $d_j = d - 2jd_s$ is the diameter of jth shell (See, (10.1)).

According to ab initio calculations [31], $C = 345$ J/m^2 for a plane sheet, or a little less for CNTs. Finally, we get for the persistence length

$$L_p = \pi Cn(d - d_s(n-1))(d^2 - 2dd_s(n-1) + 2d_s^2 n(n-1))/8k_BT. \tag{10.10}$$

For SWCNTs with diameter d = 1.4 nm this equation gives $L_p \approx 9.1$ μm.

Figure 10.3 presents dependences of the persistence length L_p versus outer diameter of CNTs d at different values of n. Remarkably, the theoretically estimated persistence lengths are of macroscopic magnitude, and may exceed 1 m for the outer diameter of MWCNTs of the order of $d \approx 20$ nm. The data of fluorescence microscopy for SWCNTs in water confirmed that the bending stiffness, K, scales as the cube of the CNT diameter, d, and the values of L_p range from 26 to 138 μm [32].

However, the noticeably smaller values of the persistence length were reported in the literature [33–35]. Neutron scattering data suggested that SWCNTs have a persistent length of about 160 nm [34]. X-ray scattering data evidenced the absence of the rod-like morphology over length scales from 1 nm to 50 μm [35]. Analyses of SEM images of MWCNTs with different diameters 10–50 nm revealed the presence of a strong $L_p(d)$ dependence [33] (Fig. 10.4a).

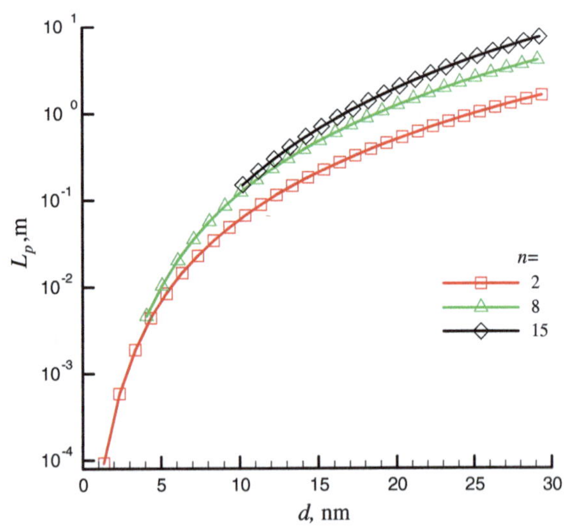

Fig. 10.3 Persistence length L_p versus outer diameter of CNTs d, at different numbers of walls, n. The data were calculated using (10.7)

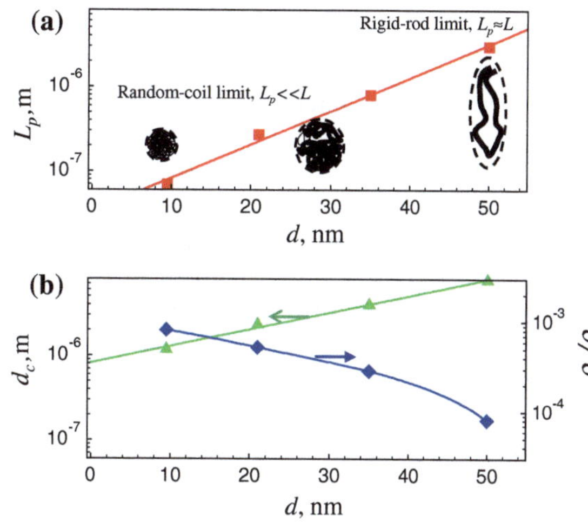

Fig. 10.4 The persistence length, L_p, versus the outer diameter of MWCNTs, d (**a**) [33] and the effective diameter of the coil d_c, (10.6) and its relative density ρ_c/ρ, (10.12), versus d (**b**). The experimental data on $L_p(d)$ were taken from [33]

In the random coil limit the CNTs behave like polymers and may be presented as the porous spheres with diameter d_c. The relative density of CNT rolled up into the coil can be estimated as

$$\rho_c/\rho \approx 3d^2L/(2d_c^3) \approx 3d^2L/(2(2L_pL)^{3/2}). \tag{10.11}$$

Figure 10.4b presents the effective diameter of the coil d_c ($\approx(2LL_p)^{0.5}$) and its relative density ρ_c/ρ versus d for CNT with the length $L = 10$ μm. The increase in CNT diameter d resulted in an increase in d_c and decrease in ρ_c/ρ. The data evidence that the packing densities of CNTs in these coils may be rather low.

In the rigid-rod limit (i.e., when $L \approx L_p$) the CNTs behave as rigid rods, and their shape anisotropy becomes essential.

10.3 CNTs in Isotropic Suspensions

Nowadays, it is generally recognized that composites based on carbon nanotubes (CNTs) dispersed in isotropic media demonstrate many attractive electrical, mechanical and thermal properties with potential of their applications in engineering, electronics and medicine. On the base of these composites the different types of sensors, switches, screens, electrical energy and memory storage devices, supercapacitors, sorbents and filters were already proposed [36–40].

10.3.1 Dispersing of CNTs

Good functionality of different solvents doped by CNTs are mainly determined by their temporal stability and reproducibility of their properties [41–43]. Usually, dispersability of CNTs is rather bad, and they show high tendency to aggregation or formation of bundles [44, 45]. It presents a serious obstacle to good functionality of these materials. The quality of CNT dispersion in suspensions and functional composites based thereon may be particularly important in determination of electrical, mechanical and thermal properties of these systems. Many studies have shown that shearing processing strongly affected the percolation behavior and clustering of CNTs in different composites [46–48].

Colloidal processing is a recognized tool for preparation of CNT-based materials [49, 50] and regulation of their stability [51–54]. In some cases, colloidal suspension of CNTs may serve as a useful intermediate for further processing and preparation of more complex nanotube-based composites. Because of high hydrophobicity of CNT surface, water is a bad solvent for CNTs. Colloid stability of aqueous suspensions may be improved by oxidation or functionalization of CNTs [55–57] and introduction of different surfactants [51–53, 58–61] or polymers [62].

The correlations between electrokinetic potential of MWCNTs, dispersability in different solvents (water, ethanol and hexane), stability of CNT suspensions, surface energy, and oxygen content of MWCNTs modified by functionalization were discussed in details [63]. It was speculated that CNT dispersability in a liquid is affected by the hydrophilicity of CNT surface, which is reflected by the electrokinetic potential value.

The dispersability of CNTs may be noticeably improved in some non-aqueous solvents or their mixtures [44, 45, 64–68]. CNTs can be debundled to a significant degree in such good solvents as DMF, NMP, CHP [45, 64]. e.g., the aprotic CHP can disperse CNTs up to 3.5 mg/mL with very large populations of individual MWCNTs [44]. It was suggested [66, 69, 70] that good dispersion ability of solvent is determined by the values of their Hansen solubility parameters [71]. It was also speculated that good CNT-dispersing ability of solvents is due to the low energetic cost of exfoliation of nanotubes, since the surface tensions of solvents successfully dispersing CNTs are close to 40 mJ/m^2 [44]. Application of these arguments for prediction of CNT-dispersing ability of a mixture of solvents is still unclear.

Recently, the microstructure and electrical conductivity of suspensions of MWCNTs in binary mixture of good (CHP) and bad (water) solvents were studied in details [72]. CHP and water have significantly different total Hansen parameters, δ_T = 47.8 and 20.5 MPa$^{1/2}$, respectively [71]. This difference mainly reflects the differences in dispersion and hydrogen bonding components of Hansen parameters. The experiments have shown that dispersing quality of CNTs in a mixture CHP + water can be finely regulated by adjustment of the composition of the mixture. The electrical conductivity data evidenced the presence of a fuzzy-type percolation with multiple thresholds in the systems under investigation. This behavior was explained by formation of different percolation networks in dependence of MWCNT concentration.

Dispersability of CNTs can be enhanced also by introduction of supplementary colloidal particles into the suspension [73–78]. In particular, it was shown that introduction of charged laponite platelets (Lap) significantly improved the dispersability of MWCNTs in aqueous suspensions [77]. It was shown that Lap platelets prevent the aggregation of MWCNTs and, thus, the Lap could act as a dispersing agent.

Stabilization of MWCNTs in the presence of Lap was explained by strong interactions between MWCNTs and Lap and formation of Lap hydrophilic shells on the surface of MWCNTs (Fig. 10.5). The observed attraction between similarly charged colloidal particles of MWCNTs and Lap was explained by the highly heterogeneous distribution of negatively charged functional groups on the surface of MWCNTs and difference in electrophoretic mobility of MWCNT and Lap particles [24].

10.3.2 Onsager Ordering

Onsager [79] described the appearance of oriented state in the suspension of rods from purely entropic reasons. The free energy of suspensions was assumed to depend only on concentration and shape of the particles, and the eventual phase transition is athermal. The theory predicted the first order nematic/isotropic transition, i.e., with the phase coexistence and jump of the order parameter, s, from 0 to 0.8. In the "Onsager" limit $L >> d$ the concentrations (volume fraction of rods, φ) of the coexisting isotropic and nematic phases are given by $\varphi_i = 3.29/a$ and $\varphi_n = 4.19/a$, where a is the aspect ratio [80].

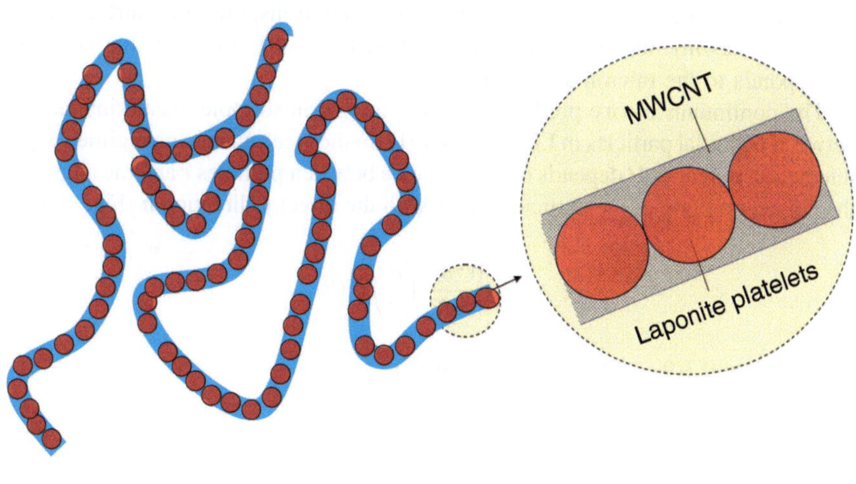

Fig. 10.5 Impact of Lap on stabilization of MWCNTs in aqueous suspensions

The Onsager liquid crystalline ordering in suspensions of CNTs can be expected for rod-like particles, i.e. CNTs with large persistence length, $L_p \geq L$. Several examples of liquid-crystalline phases of CNTs have been reported [2, 15, 81–87]. Observation of liquid-crystalline ordering requires an effective suppression of strong van der Waals attractive interactions between the individual CNTs and a solvent with good ability to disperse CNTs. It can be achieved by using of chemical functionalization of CNT surface or the use of surfactant-covered or polymer-wrapped CNTs [4]. Formation of a lyotropic nematic phase by MWCNTs dispersed in water was reported [81, 82]. The coupling between the degrees of alignment of CNTs with other type of elongated particles (e.g. for rod-like fd virus particles) was noted in binary nematic dispersions [88, 89]. The guest particles (CNTs) were shorter and thinner than the fd virus particles. It was shown that order parameter of the CNTs was systematically lower than that of the fd virus particles for the whole nematic range.

10.4 LC-Mediated Interactions Between Embedded Particles

The extent of the perturbation of the LC structure related to the presence of colloidal particle can be estimated accounting for the elastic penetration length $\xi = K_e/W$, where K_e is a certain average of the Frank elastic constants and W is the surface anchoring energy. In typical cases for the nematic phase $K_e \approx 10^{-11}$ N, $W \approx 10^{-5}$–10^{-10} N/m [90], and the penetration length is of order of $\xi \sim 1$ μm–0.1 m.

For sufficiently small spherical particles with diameter $d << 1$ μm the condition of weak anchoring limit $d/\xi << 1$ is fulfilled. In this limit the nematic LC is practically unperturbed. In the strong anchoring limit $d/\xi >> 1$ the nematic ordering around particles is strongly perturbed. The condition $d \approx \xi$ corresponds to the "transition" between the weak and strong anchoring limits. For large surface energy ($W \approx 10^{-5}$ N/m), the elastic penetration length $\xi \sim 1$ μm, and this condition corresponds to the micron-sized particles.

The continuum theory predicted the long-range anisotropic elastic interactions between colloidal particles in LC medium [91]. In the weak anchoring regime the pair interaction potential U depends on the distance between particles r and the angle θ of the connecting line of the particle centres with the director direction $\boldsymbol{n}$ (Fig. 10.6):

$$U = \frac{W^2 d^3}{256 K_e} \left(\frac{d}{r}\right)^5 f(\theta), \tag{10.12}$$

where $f(\theta) = 0.3(9 - 90\cos^2\theta + (105)\cos^2\theta)$.

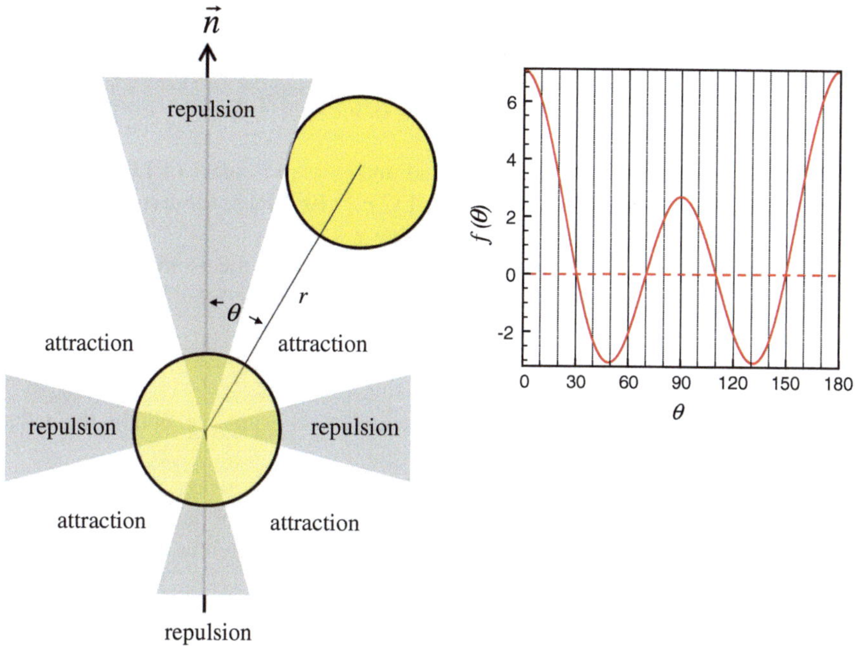

Fig. 10.6 Interaction of two particles in LC medium. The attraction and repulsion may be realised in dependence upon the angle θ of the connecting line of the particle centres with the director direction $\boldsymbol{n}$. $f(\theta)$ is the angular function in the interaction potential $U(r,\theta)$

The angular function $f(\theta)$ follows from an electrostatic analogy for quadrupole–quadrupole (qq) interactions [92]

$$U_{qq} \propto f(q)/r^5. \qquad (10.13)$$

The angular function $f(\theta)$ is presented in Fig. 10.6. The particles are repelled when they are aligned along $\boldsymbol{n}$ or in perpendicular direction and attract each other when placed at oblique angles.

At the distance of the direct contact $d \sim r$ the interaction energy is of order $U \approx W^2d^3/256$ K. The aggregation strength may be characterized by the dimensionless ratio

$$\omega_a = \frac{W^2d^3}{256Kk_BT}, \qquad (10.14)$$

where $k_BT \approx 4.1 \cdot 10^{-2}$ J is the thermal energy at $T = 273$ K.

For the typical values $K_e \approx 10^{-11}$ N and $W \approx 10^{-5}$ N/m ($\xi = 1$ μm) we get $\omega_a = 10$ for suspension with micron sized particles (strong aggregation) and $\omega_a = 1$ for suspension with smaller particles of size $d \approx 100$ nm (weak aggregation).

For the dilute suspensions the time of aggregation (the time required for two particles to move towards each other) was estimated as [91]

$$\tau \approx 0.5\eta K\varphi^{-7/3}/W^2 = 0.5\eta\xi^2\varphi^{-7/3}/K, \tag{10.15}$$

where φ is the volume fraction of a colloid and η is a viscosity of LC.

For the typical viscosity of the nematic LC $\eta \approx 10^{-1}$ Pa s, for a volume fraction of $\varphi = 0.1$ and $\xi \approx 1$ μm, we obtain the value $\tau \approx 10$ s.

The mean field Smoluchowski theory estimates τ in the isotropic solvent as follows [93]

$$\tau \approx (\pi/8)\eta d^3\varphi^{-1}/k_B T. \tag{10.16}$$

For the same values of the parameters and $d \approx 1$ μm we obtain the value $\tau \approx 100$ s.

The pair interaction potential was tested experimentally using a standard video microscopy technique [94] and the estimated experimental value of U agreed qualitatively well with that given in [91]. The proposed potential has the form

$$U = \frac{W^2\pi d^3}{280K}\left(\frac{d}{r}\right)^5 f(\theta), \tag{10.17}$$

where $f(\theta) = 5(7\cos^2\theta - 4)^2/24 - 1$. The value of U is minimized for $\cos^2\theta = 4/7$ and the aggregation strength may be characterized by the ratio $\omega_a = \pi W^2 d/(280Kk_B T)$.

For elongated particles, e.g. rods or ellipsoids, two different effects can be essential:

(i) the elongated particles embedded in the LC solvent may produce the defect structures around them [95] and
(ii) the nematic phase may act as a temperature dependent external field on elongated particles [96, 97].

The extent of these effects is controlled by the ratio $d/\xi \approx 1$, where d is the transverse size of the particle (e.g., diameter of the rod) and ξ is the elastic penetration length of the nematic material.

There is no explicit information about colloidal forces between elongated particles in LC. It can be expected that these forces are long-ranged and anisotropic. The generalizations of the quadrupole–quadrupole potential for particles of arbitrary shapes were derived [98]. The application of mesoscale theory evidenced that the LC-mediated interactions between the nm-sized spherocylindrical particles are rather strong and they could be used to assemble the particles into ordered structures with different morphologies [99]. The numerical analysis of the energetic of elongated objects immersed in a nematic medium revealed complex topological structures around particles, depending on their aspect ratio, a, and the boundary conditions [100].

In the strong anchoring limit $d/\xi >> 1$ the large elongated colloidal particles disturb the alignment of the LC molecules around them. The particles also experience ordering forces that depend on their orientation. So, in this limit the ordering is driven by a director field deformation. The defect structures around a elongated colloidal particle embedded in a nematic LC was studied combining molecular dynamics and Monte Carlo simulation [101]. The torque on a particle tilted with respect to the director was determined [95]. The presence of a coil-rod transition at the isotropic-nematic transition for polymer indebted in the background solvent (nematic phase of rodlike fd virus) was experimentally demonstrated [102]. The topological properties of a nematic 5CB in the vicinity of micro-rods was experimentally studied [103]. The strength and separation dependencies of various pair interaction potentials have been also determined.

The continuum theories for different types of boundary conditions predicted alignment of the elongated particles along the director in the strong anchoring limit [104, 105]. It was stated that the distortion energy of a nematic LC is usually minimal if the long axis of the elongated particle coincides with the director [104]. To put the particle in a perpendicular direction requires an energy of the order of $U = K_e L$, where $K_e \approx 10^{-11}$ N. This energy is rather large ($>>k_B T$) even for nano-sized particles. The following dependence for the energy of the alignment of rods in the nematic medium was obtained later on [105]

$$U \propto \cos^2\theta \tag{10.18}$$

In the weak anchoring limit $d/\xi << 1$ the perturbation of LC around particles is unessential. This limit is valid for the particles of submicron size. However, the alignment of elongated particles is also possible in this limit due to coupling of the nematic LC with the anisotropic interfacial tension of the particles [106].

The diameter of CNTs seems to be rather small, $d << \xi$, and this situation corresponds to the weak anchoring regime. Typically the individual CNTs are not able to produce large distortion inside nematic LC matrix. However, it was noted that the weak anchoring regime may be violated for the chemically treated surfaces of CNTs that resulted in increasing of the anchoring energy, W, or in the vicinity of the nematic-isotropic transition [107]. Near the nematic-isotropic transition the elastic constant decreases as $K_e \approx k s_{LC}^2$ and the anchoring energy decreases as $W \approx w s_{LC}$, where s_{LC} is the order parameter in the nematic phase. As the result the ratio $dW/K_e \sim 1/s_{LC}$ may be very large in the vicinity of the nematic-isotropic transition.

The dependence similar to presented by (10.18) was obtained also in the weak anchoring limit [90]. However, in this limit the preferred orientation of particles with respect to the nematic director may depend on the boundary conditions [105].

The induced orientational ordering in the diluted suspension of rods is controlled by the dimensionless ratio of the surface and thermal energy [90]:

$$\omega_s = LdW/k_BT. \quad (10.19)$$

In the limit of weak surface anisotropy $\omega_s << 1$ the induced order parameter of rods was estimated as

$$s_n = \omega_s\pi/15. \quad (10.20)$$

Accounting for the proportionality of anchoring energy to order parameter inside the nematic LC, $W \approx wS_{LC}$ we get a relation between the order parameters of the two components [90]

$$s_n = (\pi/15)LdWs_{LC}/k_BT. \quad (10.21)$$

It was demonstrated that orientational order parameter, s_n, may be tuned by varying the particle volume fraction, φ, and the temperature, T [106].

10.5 CNTs in Nematic LCs

10.5.1 Phase Transitions

The effect of presence of CNTs on the thermal stability of the mesophase (i.e., on temperatures of the nematic-isotropic and other mesomorphic transitions) was discussed in several works. Typically, the non-mesogenic dopants decrease thermal stability of mesophases. On the other hand, CNTs, as large anisometric quasi-macroscopic objects, could play the role of a sort of orienting substrates favoring the liquid crystal ordering. One can recall the so-called "epitropic mesomorphism" or ordering effects of annular proteins on the LC phase of phospholipid membranes [108].

The noticeable increase in the nematic-isotropic transition temperature, T_{ni}, of 5CB (by ≈10 K) upon introduction of MWCNTs in a narrow concentration range (≈0.1–0.2 wt%) was observed [109]. However, numerous attempts to reproduce these results in many laboratories did not succeed. No statistically reliable changes in the nematic-isotropic transition temperature T_{ni} for 5CB + MWCNTs suspensions at concentrations up to 0.15 % were observed [110].

At the same time, for nematic matrices of azomethines (MBBA/EBBA) lowering of T_{ni} by ≈0.5–1.5 K was noted both in spectrophotometric experiments and DSC measurements. However, this decrease could also be explained by partial decomposition of the LC molecules due to known chemical instability of azomethines. Introduction of 0.05 wt% MWCNTs to EBBA caused an increase in T_{ni} by ≈1 K, though subsequent increase in the concentration (up to 0.5 wt%) made this effect about two times weaker [111].

Introduction of 1.5–4 wt% MWCNTs to the nematic matrix 4-methoxyphenyl-4′-propylbenzoate decreased T_{ni} by several degrees, which was recorded by thermogravimetry [112]. The authors also did not exclude partial thermal decomposition of the LC matrix. Impact of CNTs on the thermal stability of the LCs was discussed in detail in literature [113–116]. The two tendencies were considered:

(i) the CNTs additionally align the adjacent layers of nematic molecules, forming a region with locally enhanced orientational order, and
(ii) the thermal stability may be affected by stronger decomposition of LC and introduction of ionic impurities in presence of CNTs.

In general, introduction of the CNTs usually resulted in insignificant decrease of the measured T_{ni}. We can also recall the paper [117], where MWCNT introduction to a cholesteric (helically twisted nematic) LC led to an insignificant decrease in the transition temperature from cholesteric to smectic-A phases (a few tenths of K).

10.5.2 Surface Anchoring and Alignment of CNTs

The theoretical calculations predicted the strong surface anchoring due to π-π interactions between CNTs and LC molecules with a binding energy of about 2 eV [118, 119]. The density functional calculations evidenced also the formation of helical wrapping enhancing the hexagon–hexagon π-overlapping and charge transfer from LC molecule to CNT. Molecular dynamics simulations have shown that the mobility 5CB molecules becomes more restricted due to interaction with the CNT surfacc [120]. The strong interaction between carbon surface and LC molecules was also experimentally confirmed by the spontaneous alignment of nematic liquid crystals on oriented carbon fibers [121] or SWCNTs [122, 123].

It was demonstrated that the strong interaction of 5CB molecules with the surface of NWCNT aggregates is responsible for the formation of micron surface liquid crystal layers with an irregular field of elastic stresses and a complex structure of birefringence [124].

Figure 10.7 shows examples of the microphotographs of MWCNT clusters formed in the nematic (a) and isotropic (b) states. As it can be seen, the surface liquid crystal layers with a changed structure exist in the nematic phase (Fig. 10.7a).

However, they disappear in the isotropic phase as shown in Fig. 10.7b, where elastic boundary stresses are absent. It is remarkable that small micron-sized aggregates of MWCNTs become invisible in this case. Thus, the boundary layer of the 5CB nematic phase visualizes submicron MWCNT clusters, which are below the resolution threshold of the usual optical microscope. The effect of perturbation of the LC structure in the interfacial shells surrounding the MWCNT aggregates

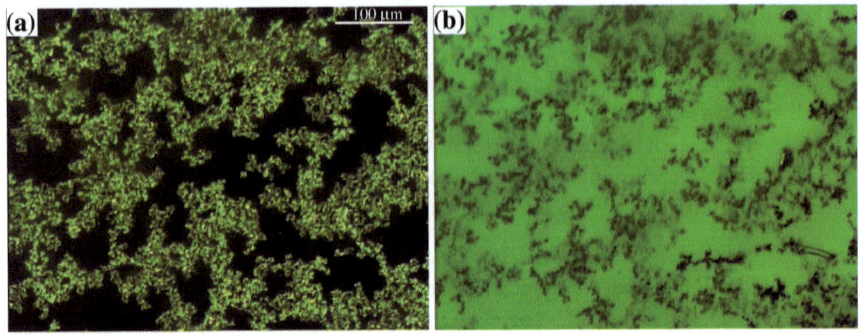

Fig. 10.7 Microphotographs of 5CB + MWCNT suspensions (C = 0.025 wt%) in the nematic state at T = 304 K (**a**), and liquid state at T = 308 K (**b**). The data were obtained for cell with thickness of h = 20 μm with planar orientation of 5CB. The polarizer was oriented along direction of the director and the analyzer was crossed

was visually demonstrated by analyzing the optical microscopy images of aggregates at different values of applied AC voltage, u [125].

The examples of such images for 5CB + MWCNT suspension (C = 0.05 wt%) at different u are presented in Fig. 10.8. The presence of enlightened shells near the surface of MWCNT aggregates clearly reflected the strong anchoring of 5CB molecules to the surface of MWCNTs. The strong anchoring between 5CB molecules and lateral surface of MWCNTs reflect the similarity of carbon hexagons in 5CB and MWCNTs. The observed enlightened shells can be explained by the presence of complicated three-dimensional 3d fields of elasticity strength inside the layers of anchored molecules of 5CB. It results, in turn, in perturbation of refractive index distribution near the surface of MWCNTs.

The thickness of interfacial shells surrounding the MWCNT aggregates, Λ, as function of the applied AC voltage, u, was estimated (Fig. 10.9) [124].

The value of Λ reached the maximum near some threshold value of $u \approx 3.5$ V, and at higher voltages it decreased. The observed effect of the applied voltage on the thickness of interfacial LC shells surrounding the MWCNT aggregates, possibly, reflects the electric field-driven enhancement of LC structure perturbation in the LC interfacial shells near the Freedericks transition. At high voltage ($u > 3.5$ V) application of the crossed electric field causes the Freedericks transition of nematic molecules and they can rotate along field direction even in the interfacial LC shells, and the value of Λ decreases.

It was also demonstrated that the CNT surface chirality can be transmitted into the adjacent achiral LCs [126–129]. In the nematic 5CB the chiral pitch length λ was estimated by measuring the curvature of reverse twist disclination lines in a 90° twist cell [127]. At small concentration of CNTs ($C < 0.15$ wt%) the inverse pitch λ^{-1} appeared to be approximately linear in C and above 0.2 wt% the apparent saturation of λ^{-1} was observed.

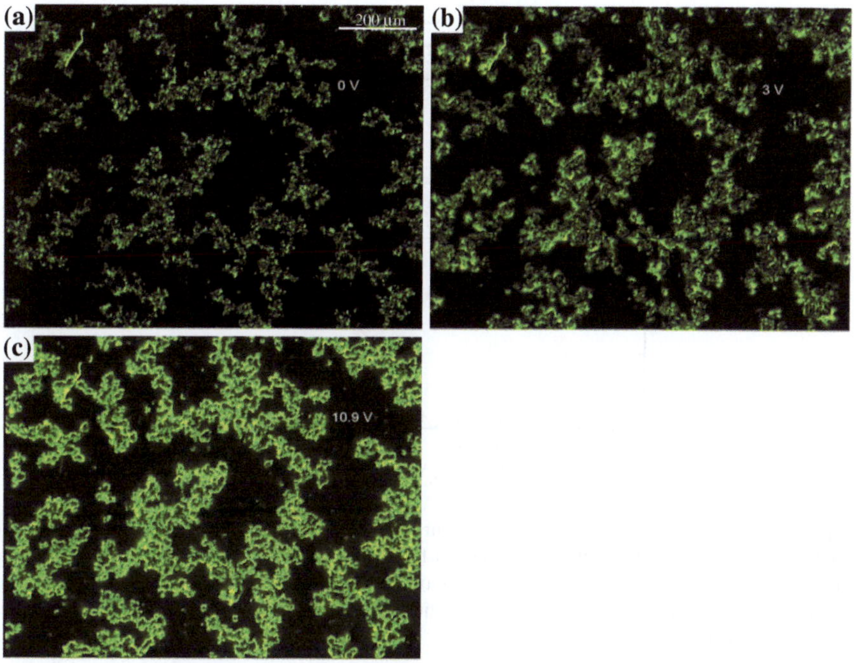

Fig. 10.8 Microphotographs of 5CB + MWCNT suspensions (0.05 wt%) at different applied voltages: u = 0 V (**a**), 5 V (**b**) and 11 V (**c**). The data were obtained for cell with thickness of h = 20 μm. The AC voltage, u, at frequency of f = 10 kHz was applied. The polarizer was oriented along direction of director for planar oriented 5CB molecules and the analyzer was crossed

This saturation was explained by aggregation of CNTs. The induced chirality was compared in LCs having either a thioester or oxoester linkage group in the core [129]. The more pronounced chiral properties (an electroclinic effect and a macroscopic helical twist of the director) were observed for the thioester linkage as compared to the oxoester linkage. This effect was explained by the differences in noncovalent interactions of LCs with the chiral dopant. The induced chirality in the smectic-A phase of 8*S*5 was also probed by means of the electroclinic effect (a rotation of the liquid crystal director perpendicular to an applied electric field) [126]. A bulk-like electroclinic effect in the nematic phase of 8*S*5 was also observed [128]. The magnitude of the effect showed shown significant pre-transitional behavior near the nematic-smectic-A transition.

Effects of aligning and reorientation of CNTs in LC medium have been also intensively studied experimentally during the last decades. These processes can be noticeable for the rod-like particles, i.e. CNTs with large persistence length, $L_p \geq L$. This assumption was used in a majority of works devoted to the problem. The presence of the orientational ordering of CNTs dispersed in 5CB (for small concentration of CNTs, $\approx 10^{-4}$ %, i.e., 1 nanotube particle per 10^9 molecules of 5CB) was demonstrated for the first time by Harte in 2001 [130]. He used the

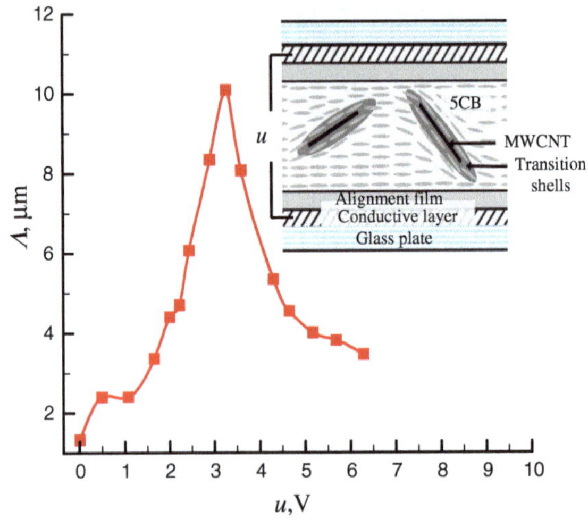

Fig. 10.9 The thickness of interfacial shells surrounding the MWCNT aggregates, Λ, versus the applied AC voltage, u. The data were obtained for 5CB + MWCNT suspensions (0.025 wt%) in the electro-optical cell with thickness of $h = 20$ μm, the AC voltage at frequency of $f = 10$ kHz was applied. The polarizer was oriented along direction of director for planar oriented 5CB molecules and the analyzer was crossed

external magnetic field to align molecules of 5CB, and the effects of alignment of nanotubes in the same sample on birefringence and the threshold voltage for the Freedericks transition were observed. The presence of CNTs produced a consistent 10–30 % increase in the Freedericks transition field.

The orientational ordering of SWCNTs and MWCNTs dispersed in nematic liquid crystal (LC) was also experimentally observed [131]. CNTs were dispersed in a LC solvent (5CB and E7) using ultrasonication, deposited onto a porous polycarbonate membrane substrate. The bulk LC was aligned using a grooved surface or external (electric or magnetic) fields and then the LC was drained through the porous membrane. The AFM images showed that in the resulting films the CNTs were strongly oriented.

Impact of external magnetic fiend on the structure of self-organized LC + CNT suspension may be rather ambiguous. e.g., the effects of magnetic field on 5CB + MWCNT suspension were studied by X-ray scattering [132]. It was shown that in the presence of CNTs the nematic director was aligned perpendicular to the magnetic field, whereas it was parallel to it for pure 5CB.

In principle, it is possible to *imprint* nematic (uniaxial) order on thin films of elongated particles grown at the LC-solid interface, and this technique of alignment was noted as "liquid crystal imprinting" [133].

Experimental data evidence that CNTs always try to be aligned parallel to the local director orientation [134–136]. Orientation of SWCNTs and MWCNTs in E7 was analyzed using polarized light microscopy. The estimated order parameter was rather high, $s_n \approx 0.9$. For comparison, similar dispersions of CNT in glycerol (an isotropic viscous liquid) were used. It allowed accounting for effects related to orientation of the nanotubes when the CNT suspensions were introduced between planar walls by capillary forces. Such qualitative evaluation gives the values of the CNT orientational order parameter at the level of the corresponding values for the undoped nematic.

For SWCNTs, more definite data on their orientation in the nematic suspension were obtained by polarization spectroscopy of Raman scattering [137]. The intensities of the corresponding bands were noticeably different for light polarized in directions parallel and perpendicular to the director. This method allows "monitoring" of molecular reorientation during the Freedericks transition under continuously increased voltage applied to the cell. In the presence of CNTs, the transition from planar to homeotropic texture became less pronounced and required higher threshold electric fields.

A non-trivial case of electrically induced supramolecular ordering in LC + MWCNT suspensions was described in [138]. Under electric field, elongation of aggregates was observed (by about four times), and some nanotubes could slide out of the bundles. More complex processes of electrically induced structuring were also observed.

Reorientation of CNTs in 5CB suspensions in 50–70 μm thick cells under applied DC electric field (up to 25 V) was studied [139]. Capacitance and electrical conductivity were measured at 1 MHz with AC voltage of ~0.25 V. On reaching the threshold voltage ensuring Freedericks transition from planar to homeotropic orientation, increase in capacitance and significantly higher conductivity were observed. As distinct from the earlier cited papers [135, 136], here the reorientation of the LC + CNT system was achieved by DC (and not AC) electric field. Further studies showed [110] that under these conditions the sharp rise of electrical conductivity above a certain voltage value was observed in different nematic matrices with $\Delta\varepsilon > 0$ (planar to homeotropic reorientation of the nematic molecules) and $\Delta\varepsilon < 0$ (when either LC planar structure is stabilized by the electric field, or electrohydrodynamic instabilities appear).

In the latter case, the required "threshold" voltage was notably higher and CNTs were oriented along the electric field themselves (as particles with large positive dielectric anisotropy). This phenomenon was studied in detail [113, 114]. Suspensions of MWCNTs in nematic 5CB (0.005 wt%) in a standard LC cell of 20 μm thickness were subjected to AC electric field sufficient for realization of the Freedericks transition. The measured dielectric permittivity ε significantly increased. When the field was removed, the dielectric permittivity returned (with a certain hysteresis) to its initial values. The same experiment was carried out at higher temperatures corresponding to the isotropic phase and there was no restoration of the initial ε values after the voltage was switched off. In the nematic phase orientational order of the nematic molecules controls the orientation of CNTs below and above the Freedericks transition. In the isotropic phase the electric field oriented the nanotubes themselves. Because of the absence of orientational order, there was no stimulus for the return of the dispersed nanotubes to their initial state. An interesting model of the ordering of several layers of nematic molecules adjacent to the CNTs was proposed [113, 114].

The two relaxation times were observed (by the measured fall of conductivity) after Freedericks transition and subsequent removal of the electric field in suspensions of SWCNTs in nematics with $\Delta\varepsilon > 0$ [115]. The faster and slower relaxation processes were associated with the matrix and the nanotubes (that return to the initial state following the nematic director), respectively.

10.5.3 Aggregation and Incubation

The LC ordering may strongly affect the aggregation of colloidal particles indebted into LC medium. The experimental and numerical simulation data for suspension of spherical particles in a micellar nematic revealed rather complex regimes of aggregation in dependence on the particle size [94]. Evidently the aggregation of elongated particles, such as CNTs, may be even more complex.

The aggregation of CNTs in LC suspensions is a well-known phenomenon. The structure of MWCNT aggregates in LC suspensions composites is not stable in time and change continuously after the sample preparation [140, 141]. Typically, at small concentration (<0.001 wt%), the individual CNTs are rather uniformly distributed inside nematic matrix. However, at higher concentration CNTs tend to be associated into more and more large aggregates. The high quality dispersions of CNTs in LCs and stabilization of properties of these systems in the equilibrium state is rather important task [142]. Figure 10.10 presents the microphotographs of the 5CB + MWCNT (0.1 wt%) suspensions that were "fresh" (a) and "incubated" for one week at $T = 298$ K (b) [141].

After long term incubation the obvious changes in the structure of aggregates were observed. In "fresh" suspension a large quantity of nearly homogeneously distributed aggregates were formed at $C = 0.1$ wt% (Fig. 10.10a).

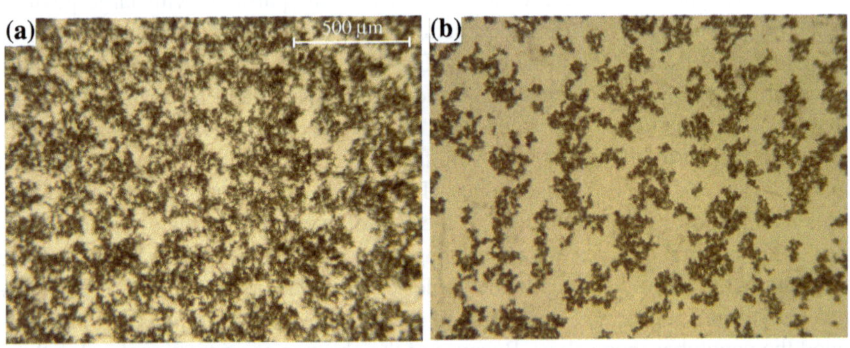

Fig. 10.10 Examples of microphotographs of "fresh" and "incubated" (for one week at $T = 298$ K) 5CB + MWCNT suspensions ($C = 0.1$ wt%) in the cell with thickness of $h = 50$ μm (From [141]. With permission)

During the incubation the most important was transformation to the segregated state with separated regions filled with compact aggregates and "lakes" of 5CB depleted of MWCNTs (Fig. 10.10b). This anomaly can be explained by existence of the mechanism causing compacting of the ramified aggregates formed in the "fresh" suspension after intensive sonication.

The presence of incubation processes was also quantitatively supported by data on differential scanning calorimetry (DSC) and changes in optical transmission near the nematic-isotropic transition [141]. Figure 10.11 presents the DSC traces in the heating mode with pure 5CB and with 5CB filled by MWCNTs (0.1 wt%). For pure 5CB, the temperature of nematic-isotropic transition T_{ni} (estimated from the maximum of heat flow curve $H(T)$) was ≈308.85 K.

Filling of 5CB with MWCNTs (0.1 wt%) resulted in changes in T_{ni} and broadening of $H(T)$ peaks. For "fresh" suspension the value of T_{ni} was slightly (by 0.15 K) lower compared to that of pure 5CB. The observed behaviour can be explained by formation of lyotropic pseudo-nematic phase in the vicinity of MWCNT surface [113, 114]. However, for "incubated" suspension the value of T_{ni} was approximately the same as that for pure 5CB. It can be concluded that impact of MWCNTs on the structure of LC medium become less essential after the incubation.

The similar incubation processes were also observed for EBBA filled by MWCNTs [140]. These effects reflected strong agglomeration and rearrangement of nanotubes during the thermal incubation. The estimates have shown that that the time required to translate over the distance, corresponding to the MWCNT length, may be rather large, of order of 250 s at $T_{ni} \approx 352$ K for EBBA. Hence, the nature of incubation can reflect the aggregation caused by Brownian motion of MWCNTs.

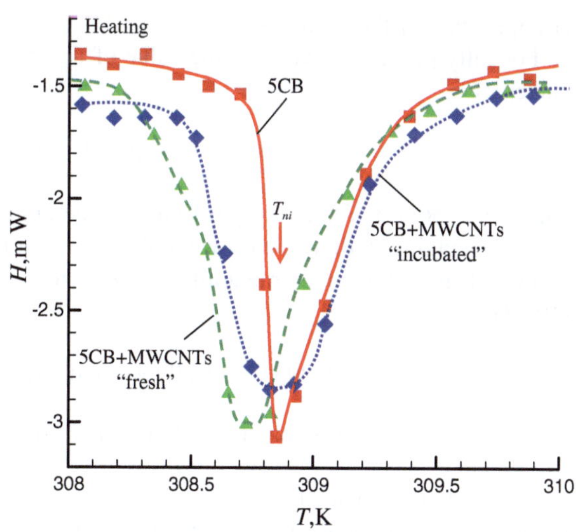

Fig. 10.11 Heat flow H versus temperature T in differential scanning calorimetry experiments (heating mode) with pure 5CB and 5CB + MWCNT suspensions (0.1 wt%), "fresh" and "incubated" 8 h at $T = 298$ K (From [141]. With permission)

10.5.4 Percolation Transition and Electrical Conductivity

10.5.4.1 Percolation for Partially Oriented Elongated Particles

Many experimental data for nematics filled by CNTs evidenced in favor of the percolation mechanism of electrical conductivity [13, 111, 124, 125, 140, 143]. The percolation is accompanied by the abrupt changes in different physical properties of LC + CNT composite, and this transition may be identified as a geometrical phase transition related with formation of the spanning network of CNTs [144–147]. The percolation phenomena were also observed in investigations of magnetic, diffusion, thermal, optical and many other properties of different composite systems filled by CNTs. The increase of the CNT content usually results in formation of clusters of particles, and the percolation threshold corresponds to the concentration (weight C_c or volume $\varphi_c \approx 0.5C_c$) when the largest cluster reaches the size of the whole system.

The percolation transition is accompanied by the scaling laws that are typical for phase transitions. The percolation theory predicts the following power behaviour of the electrical conductivity σ near the percolation threshold C_p:

$$\sigma \propto (C_c - C)^{-s}, \text{ at } C < C_c, \tag{10.22a}$$

$$\sigma \propto (C_c - C)^{t}, \text{ at } C > C_c, \tag{10.22b}$$

where s and t are the exponents of the electrical conductivity. For the random percolation (when the particles fill the space at random) $s = t \approx 4/3$ for two-dimensional (2d) and $s \approx 0.73$, $t \approx 2$ three-dimensional (3d) systems [146].

The percolation exponents are universal and their values depend only on the dimensionality of the system and presence of correlations between the spatial arrangements of the particles.

For fully penetrable particles the value of C_c or φ_c can be estimated using c the values of φ_c:

$$1 - \exp(-F) < \varphi_c < 1 - \exp(-2F), \tag{10.23}$$

where $F = 56(1 + a\langle \sin\gamma \rangle/\pi)/5$. Here γ is the angle between the rods, and <…> means averaging with account for the distribution of orientations.

For ideally aligned or randomly oriented rods $\langle \sin\gamma \rangle = 0, \langle \sin\gamma \rangle = \pi/4$, and (10.23) gives:

$$0.161 < \varphi_c < 0.295 \text{ for ideallyoriented rods, } \langle \sin\gamma \rangle = 0, \tag{10.24a}$$

$$0.7/r < \varphi_c < 1.4/r \text{ for randomly oriented rods, } \langle \sin\gamma \rangle = \pi/4. \tag{10.24b}$$

For partially oriented rods the $\langle \sin\gamma \rangle$ versus the order parameter $s_n = 0.5$ $(3\cos^2\theta - 1)$ dependence was numerically estimated (Fig. 10.12) [148]. The case of $s_n = -0.5(\langle \sin\gamma \rangle = 2/\pi)$ corresponds the situation when all the rods have random

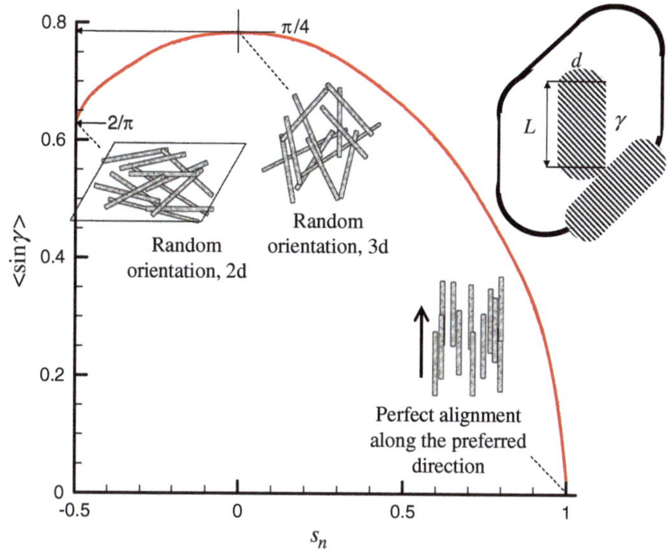

Fig. 10.12 The averaged value of $\langle \sin \gamma \rangle$ versus order parameter s_n for partially oriented rods. Here, <…> corresponds to an averaging, accounting for the distribution of orientations. *Inset* shows the excluded volume for two rods with the angle γ between them

orientation in the plane. In the case of $s_n = 0(\langle \sin \gamma \rangle = \pi/4)$ all rods are randomly oriented in the volume and the case of $s_n = 1(\langle \sin \gamma \rangle = 0)$ corresponds to the perfect orientation along the preferred direction. The value $\langle \sin \gamma \rangle$ goes through the maximum with increasing s_n and approaches zero at $s_n \rightarrow 1$.

Figure 10.13 presents the intervals of the volume percolation concentrations φ_c as function of the aspect ratio a for partially oriented rods [148]. For the perfectly oriented rods the percolation concentration does not depends on the aspect ratio, a. However, even a weak disorder (e.g., $s_n = 0.95$ in Fig. 10.13) may result in strong $\varphi_c(a)$ dependence and significant decrease of φ_c for the particles with large aspect ratio. The theory predicts that the values of order $\varphi_c \approx 0.1$ % may be easily realized for $a = 1000$ and $s_n = 0$.

Monte Carlo simulations were applied for estimation of the value of percolation threshold for partially penetrable rods with the core-shell structure [149]. The degree of impenetrability k ($k = 0$–1) was defined as the ratio of the radius of the core to the outer radius of the soft shell with $k = 0$ for the soft-core limit and $k = 1$ for the hard-core limit.

The results evidenced that for the long particles ($a = L/d > 500$) the dependence of φ_c upon a and k become unessential. The conductive soft shells around CNT particles allow accounting for the tunnelling distance of order 5 nm as suggested in [150]. The effect of the waviness of rods on the percolation onset φ_c was also studied [151]. Enhancement of waviness resulted in increased values of φ_c.

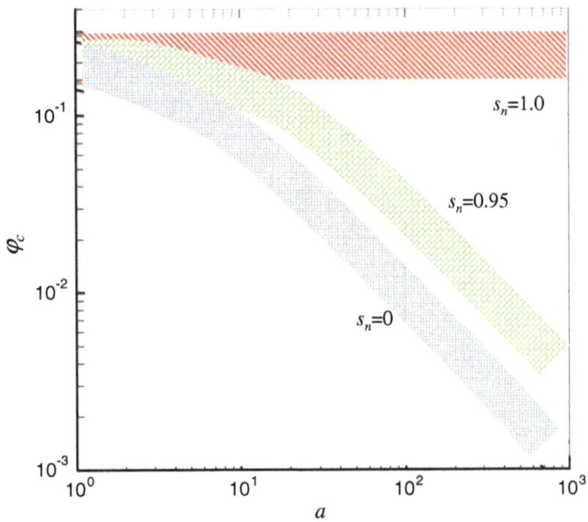

Fig. 10.13 Intervals of the volume percolation concentrations φ_c versus the aspect ratio a for partially oriented rods

10.5.4.2 Electrical Conductivity

Commonly, the electrical conductivity of LC doped by CNTs displays a rather complex behaviour versus concentration of nanotubes, phase state of the LC matrix and temperature. Experimental studies of these effects were discussed in details [13, 111, 124, 125, 140, 143].

Figure 10.14 shows the typical curves of electrical conductivity σ as a function of concentration of MWCNTs, C, in EBBA [111] and 5CB [124] obtained at the fixed temperatures that were just above the solid-to-nematic phase transition temperatures. At concentrations exceeding 0.05–0.1 wt% the percolation transition to a high conductivity state was observed. Such behaviour reflects formation of a continuous conductive path from conductive nanoparticles inside the isolative LC matrix. When the content of MWCNTs outreached the percolation threshold concentration, the electrical conductivity jumped up by many orders of magnitude.

Detailed analysis has shown that concentration dependencies of electrical conductivity of LC + MWCNT suspension above percolation thresholds can be well fitted by the power law, equation (10.22b) [13]. It should be noted that, due to technical difficulties of preparing LC + CNT suspensions with exact values of low CNT concentrations, C, it is rather difficult to determine the precise values of C_c. For 5CB + MWCNT suspensions an increase of the percolation threshold, C_c, with temperature, T, was observed, especially in the vicinity of the nematic-isotropic transition ($T_{ni} \approx 308$ K).

Such character of $C_c(T)$ dependence can be explained by changes in intensity of Brownian motions of MWCNTs, changes in the thickness and electrical conductivity of interfacial shells surrounding the MWCNTs, and impact of LC on orientational ordering of MWCNTs. These factors can affect the connectivity between the MWCNTs and, as a result, the value of the percolation threshold C_c [152].

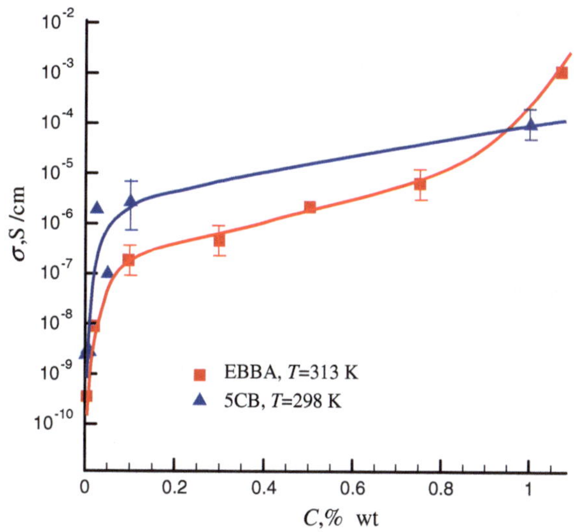

Fig. 10.14 Electrical conductivity, σ, versus a concentration of MWCNTs, C, in EBBA and 5CB (compiled from the data presented in [111, 124])

Temperature changes in the conductivity exponent t (Equation (10.22b)) for 5CB + MWCNT suspensions have been also observed [13]. The value of t noticeably exceeded its classical value ($t = 2$) in the nematic phase and approached $t \approx 1$ for the temperatures above nematic-isotropic transition.

Impact of the alignment of CNTs on percolation behavior can be expected for the rod-like particles, i.e. CNTs with large persistence length, $L_p \geq L$. It was experimentally demonstrated that the electrical conductivity parallel to the alignment direction displays a non-monotonic dependence on the degree of CNT alignment, with the highest conductivity observed for slightly aligned nanotubes rather than for isotropic case [153]. Similar results were also obtained in the Monte Carlo simulations of the effect of the alignment of CNTs on percolation and electrical conductivity [154–157]. These models also predicted that the peak value of the conductivity should occur for partially aligned rather than perfectly aligned CNTs. The peak value of electrical conductivity was significantly greater than that of perfectly aligned CNTs and was also larger than that of randomly distributed CNTs.

At volume fractions φ above percolation threshold the simulations exhibited an optimum order parameter s_n^o above which the electrical conductivity decreases dramatically. It can be expected that for perfectly or highly aligned CNTs the number of intersections between them is small and the connective paths are not formed.

The temperature dependences of σ for EBBA doped with MWCNTs are presented in Fig. 10.15 [140]. They appeared to be rather anomalous in the vicinity of the percolation threshold. The noticeable heating–cooling hysteresis and the effect of negative temperature coefficient (NTC) of conductivity were observed near the temperature of the crystal-nematic transition, $T_{cn} \approx 309$ K.

The NTC effect was accompanied by a drastic decrease in electrical conductivity up to the melting point. The nature of the NTC effect was explained accounting for

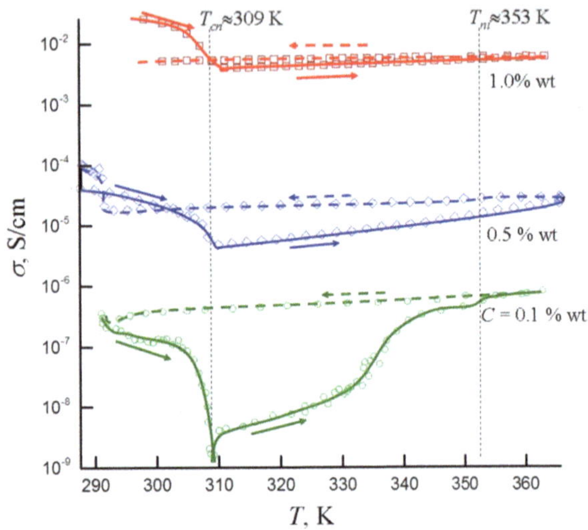

Fig. 10.15 Electrical conductivity σ_ψ versus temperature T_ψ in the heating (→) and cooling (←) cycles for EBBA doped by MWCNTs at different concentrations. Here, $T_{cn} \approx 309$ K, $T_{ni} \approx 353$ K are the temperatures of crystal-nematic and nematic-isotropic phase transitions, respectively

thermal expansion of the LC matrix [111]. The thermal expansion causes damage of percolation networks, which results in a decrease in electrical conductivity. The maximum of the NTC effect was observed near the percolation threshold ($C \approx \leftarrow 0.1$ wt%), but at higher concentrations ($C = 0.5$–1.0 wt%) the multiply connected networks start to form, and the NTC effect becomes less significant [140].

Important information about electrical conductivity mechanism in LC doped by CNTs can be obtained from the study of voltage and frequency dependencies of electrical conductivity.

Figure 10.16 presents voltage dependencies of electrical conductivity of EBBA doped by MWCNTs at different concentrations, C, and temperatures, T [111]. The pronounced growth of σ with increase of the measuring voltage u was observed. The similar effects were also observed for E7 doped by SWCNTs and MWCNTs [135] and 5CB doped by MWCNTs [125]. A possible mechanism of non-linear behaviour of $\sigma(u)$ can involve effects of transport through hopping junctions between different CNTs, induced by electric field [111]. The electric field can localise on the non-conductive gaps between different MWCNTs and it can enhance the hopping junction transport through the LC gaps. The most pronounced non-linear effects were observed in the crystalline phase (see, Fig. 10.16 for EBBA at $T = 298$ K). It can be explained by segregated distribution of MWCNTs on the surface of microcrystals with small gaps between different MWCNTs. In the nematic and isotropic phase the average distance between MWCNTs increases and the non-linear effects can diminish, which was in accordance with the experimental observations (Fig. 10.16).

Hopping/tunneling mechanism of charge transfer through LC interfacial shells between different CNTs is rather important for explanation of anomalies in electrical conductivity of LC + CNT suspensions. The electrical conductivity of composites doped with CNTs can be limited by tube-tube junction conductivity or own

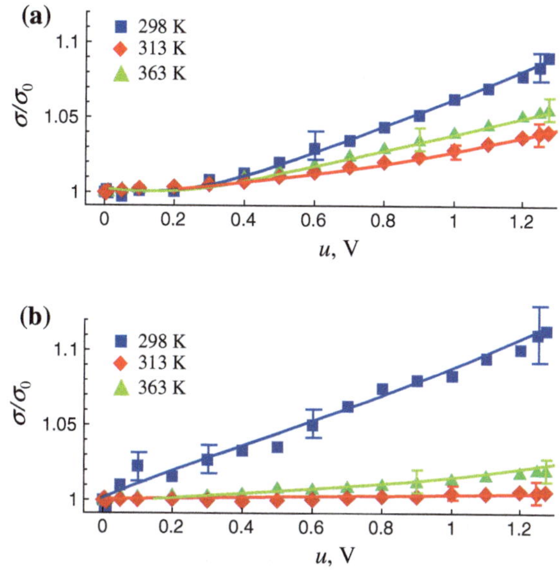

Fig. 10.16 The voltage dependencies of the relative electrical conductivity σ/σ_0 of EBBA doped by MWCNTs at different concentrations, $C = 1$ wt% (**a**) and $C = 0.1$ wt % (**b**). Here, σ_0 is electrical conductivity in the limit of $u \rightarrow 0$ V. The measurements were done at different temperatures, the frequency was fixed at $f = 1$ kHz (From [111]. With permission)

conductivity of CNTs [156, 157]. The experiments evidenced that the CNT-CNT junction conductivity is much smaller than the conductivity of the CNT themselves. In these systems the electrical transport can be classified as a CNT-CNT junction limited transport. The anomalous behaviour of electrical conductivity can reflect impact of CNTs spatial distribution and alignment on the density of low-conductive contacts and the total length of the conduction paths [157].

The detailed studies of electrical conductivity 5CB + MWCNT [13] and PCPBB + SWCNT [158] suspensions as a function of frequency, f, have shown that up to a certain frequency f_c, the value of σ is essentially independent of f and starts to increase at higher frequencies. The most pronounced frequency dependence of σ were observed near the percolation threshold [13]. The effect of dual frequency conductivity switching was observed when the current through the suspension can be field-driven by changing the frequency of the applied voltage [158].

The electrical conductivity of cells filled with LC (nematic mixture MLC-6290-000 with positive dielectric anisotropy, $\Delta\varepsilon > 0$) + MWCNT (0.0001–0.01 wt%) suspensions was studied [159]. The changes in electrical conductivity as a function of the time elapsed from field switching were observed.

Here, we would note some aspects that, in our opinion, were largely neglected or rarely discussed in most studies of electrical conductivity behavior for LC + CNT suspensions. These aspects are related with impact of length of CNTs, L, the measurement direction of electrical conductivity and the thickness of the measurement cell, h. In a number of papers, the CNT length varied from ~0.5 μm to 10–20 μm and more, but seemingly no need to account for this factor appeared in the given interpretation of the results. For 5CB + MWCNTs suspensions the differences in percolation thresholds, C_c, for long and short CNTs were explained by the impact of aspect ratio on the value of C_c [141]. The noticeable effects were found for "macroscopic" length of CNTs, $L \approx 60$ μm, that exceeded the thickness of the electrooptical LC cell [160]. For nanotubes oriented along the electric field, "shunting" of the cell was observed and Joule heating resulted in nematic-isotropic transition in the local heating areas.

For LC + CNT suspensions the electrical conductivity σ dependence upon the measurement direction was rarely discussed. The temperature variations of the low frequency (100 Hz) electrical conductivities, parallel ($\sigma_{||}$) and perpendicular ($\sigma_{\perp}$) to the nematic director were measured for LC (PCPBB) + SWCNT (0.036 vol%) suspensions [158]. The value $\sigma_{||}$ was slightly higher than the value $\sigma_{\perp}$ for both pure and doped LC suspensions. The same tendency was observed for the the dielectric constant parallel ($\varepsilon_{||}$) and perpendicular ($\varepsilon_{\perp}$) to the director. Monte Carlo simulations predicted the presence of a strong dependence of electrical conductivity on the measurement direction for aligned CNTs [155]. Experimental data for epoxy resin + MWCNT composite evidenced that the electrical conductivity in the aligning direction was much higher than perpendicularly to the CNTs orientation [161].

The dependence of electrical conductivity σ on the distance between electrodes h for the random carbon nanotube networks was analyzed [162]. It was noted that for the short channels, when $h \approx L$, the individual CNTs can directly bridge the electrodes and thus increase the electrical conductivity.

10.5.5 Dielectric Properties

Dielectric properties of 5CB + MWCNT and EBBA + MWCNT suspensions were studied in detail [13]. The measurements were done at the temperatures within nematic phase of LC medium, $T = 297$ K (5CB) and $T = 313$ K (EBBA). Three different frequency ranges, namely, $f < 10^2$ Hz (A), 10^2 Hz $< f < 10^5$ Hz (B) and $f > 10^5$ Hz (C) have been distinguished. The frequency range (A) reflected the near-electrode processes and electron exchange between electrodes and ions. The frequency range (B) corresponded to the bulk polarization and charge transfer. The frequency range (C) corresponded to the relaxation process caused by transition from the electronic and dipole polarization to only electronic polarization of LC phase. At low frequencies ($f < 0.5$ Hz), increase of the imaginary (AC conductance) component ε'' was observed by doping of LC (both 5CB and EBBA) with MWCNTs. It was argued that MWCNTs served as shunts of the double electrical layers providing paths for the electron exchange between electrodes and impurity ions inside LC medium.

Electrical impedance model of MWCNT-doped LC cells was recently proposed and Cole-Cole plots (100 Hz–10 MHz) of LC (MLC-6290-000) + MWCNT (0.01 wt%) suspension at different DC bias voltages (0.5 Hz, 0–8 V) were analysed [163]. The initial LC alignment was planar (with the LC director oriented parallel to the plane of the electrodes). It was demonstrated that the doped LC cell behaves as a capacitor in parallel with a resistor. The electrical conductivity increased dramatically for voltages above Freedericks threshold. It was explained by a reorientation of the MWCNTs, creating multiple electrical paths between the electrodes. The data also evidence that a small fraction of MWCNTs is not oriented as the liquid crystal, but in tilted positions.

The dielectric behaviour of LC filled by SWCNTs and MWCNTs was compared [164]. Figure 10.17 presents frequency dependences of real $\varepsilon'_{\perp}$ (a) and imaginary $\varepsilon''_{\perp}$ (b) perpendicular components of dielectric permittivity for pure nematic 5CB and 5CB doped by 0.01 % SWCNTs and 0.01 % MWCNTs. Large differences in the obtained values of $\varepsilon'_{\perp}$ and $\varepsilon''_{\perp}$ for the pure 5CB and 5CB doped by CNTs were observed in the low-frequency part of the dielectric spectrum. These differences were more pronounced for SWCNTs than for MWCNTs, which can be easily explained by large difference in the specific surface areas for SWCNTs and MWCNTs. The relaxation effects were also clearly observed in $\varepsilon''_{\perp}$ (f) plots at higher frequencies above 1–10 Hz. Characteristic relaxation frequencies were of ≈8 MHz, ≈15 kHz and ≈3 kHz for pure 5CB, 5CB + SWCNT, 5CB + MWCNT samples, respectively (Fig. 10.17). The observed characteristic frequency of pure 5CB was in full correspondence with the data reported in [165]. The smaller relaxation frequencies of 5CB + SWCNT and 5CB + MWCNT samples, possibly, reflected the effect of CNTs on restriction of director dynamics inside nematic LC.

The dynamic response of the average dielectric constant for MWCNTs (0.005 wt %) dispersed in 5CB as a function of applied AC field (E = 0–250 kV/m, 1 MHz, 30 s duration) in both the nematic and isotropic phases was studied [114]. The observed dielectric relaxation was explained by a mechanical relaxation mechanism of the director on turning off the electric field. The measured relaxation time decreased as E increased and the doped LC system relaxed back faster (for $E > 80$ kV/m) than pure 5CB. It was explained by trapping of ions present in the LC by CNTs. The presence of dielectric hysteresis by cycling the ac field magnitude

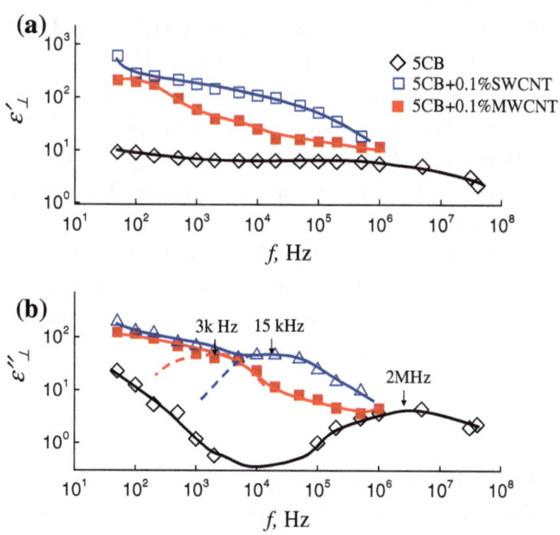

Fig. 10.17 Frequency dependences of real $\varepsilon'_{\perp}$ (**a**) and imaginary $\varepsilon''_{\perp}$ (**b**) perpendicular components of dielectric permittivity. Data are presented for pure nematic 5CB, and 5CB doped by 0.01 wt% SWCNTs and MWCNTs, T = 301 K. The cell thickness was h = 50 μm and the texture of LC was planar

(100 kHz) ranging inside 25 ↔ 250 kV/m was also observed. The field-induced dielectric response as a function of applied AC voltage (0–7 V) and probing frequency (1–100 kHz) for MWCNTs (0.007 wt%) dispersed in 5CB was also studied [166]. It was shown that doping of 5CB by MWCNTs may result in dramatic increase in the dielectric anisotropy of the system. 5CB has a positive dielectric anisotropy, ($\Delta\varepsilon = \varepsilon_{\|} - \varepsilon_{\perp} \approx 10$) and so, the applied electric field reorients the director parallel to it. For the 0.007 wt% suspension of CNTs in 5CB the dielectric anisotropy at 1 kHz was $\Delta\varepsilon = +13.8$, indicating an improvement in the orientational order s_{LC} in 5CB as $\Delta\varepsilon \propto s_{LC}$ [166].

The anomalous behavior of dielectric properties of 5CB + CNT suspensions was also observed in isotropic phase above the temperature of nematic-isotropic transition. The data on the dynamic response of the average dielectric constant evidence that in the isotropic phase the LC molecules and CNTs cooperatively form local pseudonematic domains due to strong LC-CNT interactions [114, 167]. Moreover, for 5CB + CNT suspensions non-zero value of dielectric anisotropy $\Delta\varepsilon$ was observed even in isotropic phase that also evidence the presence of local anisotropic pseudo-nematic domains [113]. These anisotropic domains caused a large dielectric hysteresis effect in isotropic phase. The observed effects may be attractive for applications in memory devices.

10.5.6 *Optical Properties*

10.5.6.1 Optical Transmission

UV-Vis absorption spectra of carbon nanotubes in different solvents were intensively studied (see, e.g. [168]). The practical importance of LC-based optoelectronic devices also stimulated the detailed studies of optical transmission in different LC systems doped with SWCNTs and MWCNTs.

Figure 10.18 presents examples of UV-Vis absorption spectra (200–800 nm) of pure 5CB and 5CB + SWCNT suspension (0.02 wt%) measured in the isotropic ($T = 310$ K) and nematic ($T = 301$ K) phases [164]. The cut-off region of the pure 5CB is located below 350 nm (see, e.g., [169–171]. This very broad absorption band of the LC matrix can be related to the aromatic nature of the substance with π-electrons.

For LC doped with CNTs another contribution in this range comes from absorption of CNTs themselves that could be due to π-plasmon absorbance. The absolute wavelength of the π-plasmon absorbance has been shown to vary with nanotube diameter d (nm) as [168]

$$\lambda_p(\text{nm}) \approx 1240/(4.8 + 0.7/d^2). \tag{10.25}$$

This empirical relationship gives ≈255 nm for a single sheet of graphene and ≈243 nm for nanotubes with diameter of 15 nm.

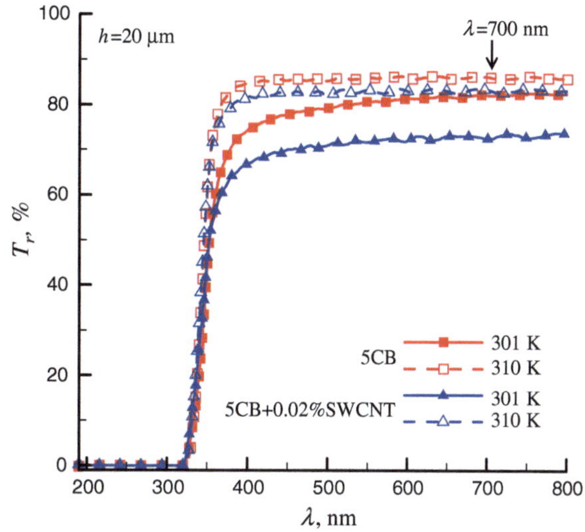

Fig. 10.18 UV-Vis absorption spectra (i.e., optical transmission T_r versus wavelength λ, 200–800 nm) of a 20 μm thin film for pure 5CB and 0.02 % single-walled carbon nanotubes (SWCNTs) dispersed in 5CB in isotropic ($T = 310$ K) and nematic ($T = 301$ K) phases. The cell thickness was $h = 20$ μm and the texture of LC was planar

The introduction of MWCNTs into the 5CB results in a noticeable decrease of the optical transmission in all the spectral range. Such behaviour was typical for LC matrices doped with CNTs [143, 172]. The cut-off wavelength can slightly change if we pass from cyanobiphenyl 5CB to nematics of different chemical nature. However, the general picture of Fig. 10.18 remains the same with different nematics and different types of CNTs (SWCNTs and MWCNTs).

It is generally known that in conventional nematics the transmission, T_r, in the nematic phase is always substantially lower than in the isotropic phase. This can be explained by strong light scattering on spontaneous fluctuations of the LC director [173]. Upon introduction of CNTs, the measured transmission in the isotropic phase was only slightly lowered. However, in the nematic phase the transmission decrease was much stronger. The contribution of CNTs to the total value of transmission (i.e., absorption+reflectance/scattering) of the LC doped by CNTs may be estimated as the difference $T_r^0 - T_r$ where T_r^0 and T_r are the measured optical transmissions of the pure LC and LC + CNT systems [110, 140, 164, 172]. In order to exclude possible impact of the absorption bands of LC matrix and CNTs, the optical transmission T_r was analyzed at 700–800 nm wavelength, i.e., far away from the mentioned cut-off regions. The difference between $T_r^0 - T_r$ values below and above the nematic-isotropic phase transition temperature T_{ni} can be considered as a measure of the degree of incorporation of the CNTs into the joint ordered LC structure.

Figure 10.19 presents examples of stepwise changes in $T_r^0 - T_r$ at the nematic-isotropic transition (the "transmission jump") [172]. The supramolecular

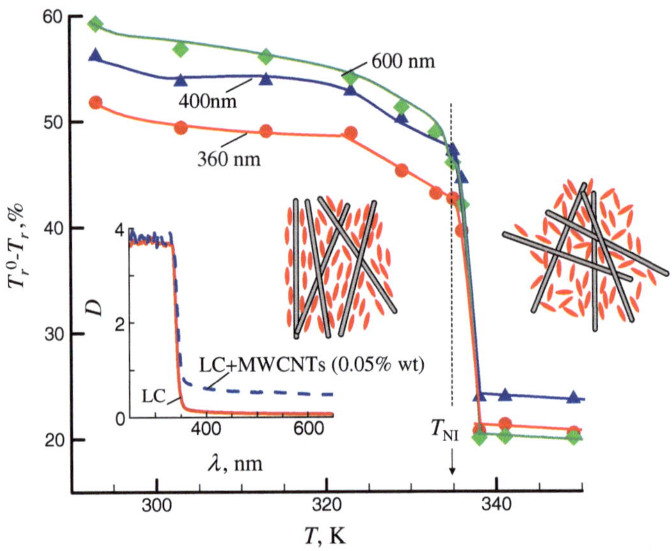

Fig. 10.19 Difference $T_r^0 - T_r$ in the optical transmittance of the nematic LC (ZhK-1282) and the LC + CNT system (0.05 wt%) as function of temperature T in the nematic and isotropic phases; T_{ni} is the temperature of nematic-isotropic transition. *Inset* shows transmission spectra of pure LC and LC + CNT suspension at T = 333 K

ordering in LC + CNT suspensions in the nematic and isotropic phases is also schematically shown in Fig. 10.19.

Figure 10.20 presents $T_r^0 - T_r$ differences for the temperatures in the isotropic phase (T = 310 K) and in the nematic phase (T = 301 K) of 5CB [164]. The values of $T_r^0 - T_r$ values were much smaller in the isotropic phase than in the nematic phase. This difference may be a clear indication of strong perturbations of 5CB medium in the vicinity of carbon nanotube aggregates. The presence of such perturbations was demonstrated by recent electrooptical investigations of interfacial 5CB layers trapped by CNTs [124, 125, 174, 175]. The difference in transmission between undoped and CNT-doped samples became somewhat larger when the measurement wavelength is increased. i.e., when we move farther away from the absorption range of the LC matrix, and the contribution from CNT is felt in a more "pure" form. In the isotropic phase the effects of structural ordering related to the LC state became irrelevant. The values of $T_r^0 - T_r$ were much lower than in the nematic phase, reflecting the effects of spatial organization of CNTs in the ordered media. We could hardly expect such sensitivity of optical transmission to the changes in phase state in the situation of mechanical mixing CNTs with LC (i.e., without interaction between CNTs and LC). It seems natural to assume that CNTs interact in some way with the LC matrix on the molecular level and are integrated/incorporated into a joint ordered structure formed in the LC + CNT suspension. Possible effects of the LC matrix chemical nature on the ordering of CNT in suspensions was analyzed [110].

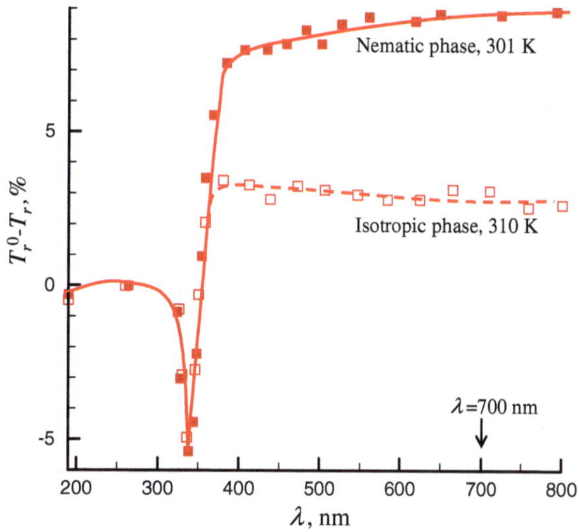

Fig. 10.20 The difference in optical transmissions $T_r^0 - T_r$ of pure 5CB and 5CB + SWCNTs (0.02 wt%) versus wavelength λ, 200–800 nm at temperatures corresponding to the isotropic (T = 310 K) and nematic (T = 301 K) phases

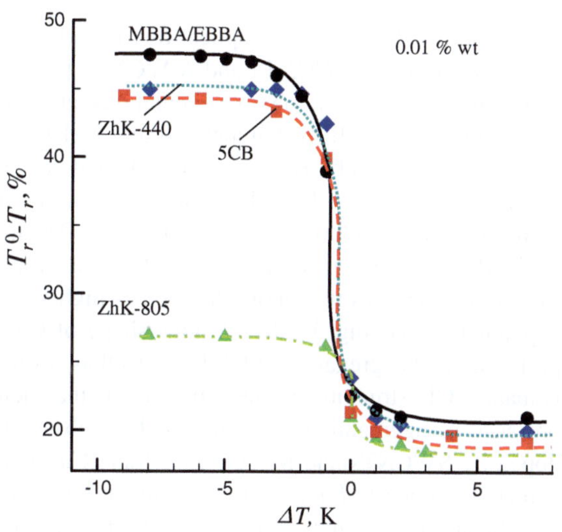

Fig. 10.21 Difference between optical transmissions of pure and CNT-doped nematics $T_r^0 - T_r$ as function of temperature differences $\Delta T = T - T_{ni}$ close to the nematic-isotropic transition temperatures for different LC matrices. The concentration of MWCNTs was C = 0.01 wt% (From [110]. With permission)

Figure 10.21 shows $T_r^0 - T_r$ as function of $\Delta T = T - T_{ni}$ for different nematic matrices doped with 0.1 wt% MWCNTs. The changes in transmission at the nematic-isotropic transition were practically fully reversible in subsequent heating and cooling cycles. For aromatic matrices the measured "transmission jump" was rather large, which could be due to strong interaction of π-π electron systems of LC molecules and nanotubes [138]. For a non-aromatic nematic (mixture of alkyl-cyclohexanecarboxylic acids) the changes in $T_r^0 - T_r$ were substantially smaller, probably reflecting weaker anisotropic interaction in the system.

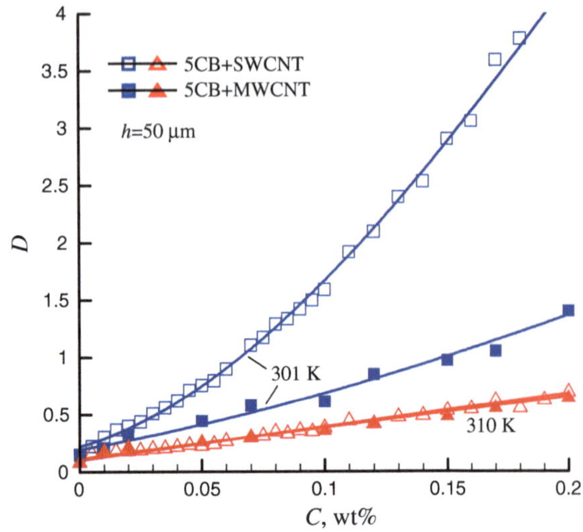

Fig. 10.22 Optical density D versus concentration, C, of SWCNTs and MWCNTs in 5CB at two different temperatures $T = 301$ K (nematic phase) and $T = 310$ K (isotropic phase). The thickness of cell, h, was 50 μm (From [164]. With permission)

Figure 10.22 presents the optical density, D, versus concentration of CNTs, C, for 5CB doped by SWCNTs and MWCNTs [164]. The $D(c)$ plots were close to linear at small C with obvious deviations at higher concentrations. To a first approximation, the Beer–Lambert–Bouguer (BLB) law can be applied at small concentration of CNTs. However, the obtained data evidenced significant violations of BLB law both in cell thickness and concentration dependences [164].

In the isotropic phase the effects of SWCNTs and MWCNTs were nearly identical, while in the nematic phase the effective extinction coefficient (the $D(c)$ plot slope) for SWCNTs was about three times higher. These experimental results were explained accounting for the tortuous shape of CNTs (CNT coils), their physical properties and aggregation of CNTs, as well as strong impact of perturbations of the nematic 5CB structure inside coils and in the vicinity of CNT aggregates. The presence of aggregation was supported by observation of micro-patterns of 5CB doped by SWCNTs and MWCNTs. Monte Carlo simulation data were applied to explain the impact of aggregation on direct transmission and violation of BLB law.

Similar changes in $T_r^0 - T_r$ values at the isotropic transition were also observed LC suspension loaded be nanoplatelets [176]. Data for nematic suspensions of MWCNTs and platelets of organomodified montmorillonite (MMTO) are compared in Fig. 10.23. For hybrid 5CB suspensions filled by mixture of particles with different shapes (rod-like MWCNTs and plate-like MMTO), changes in optical transmission were approximately additive over concentrations of the dispersed components [176].

On the contrary, the effects of MWCNT and MMTO nanoparticles on reorientation of 5CB ($\Delta\varepsilon > 0$) were essentially different. It was demonstrated that doping by MWCNTs facilitated and doping by MMTO hindered the electric field induced Freedericks transition from planar to homeotropic state in the LC cell.

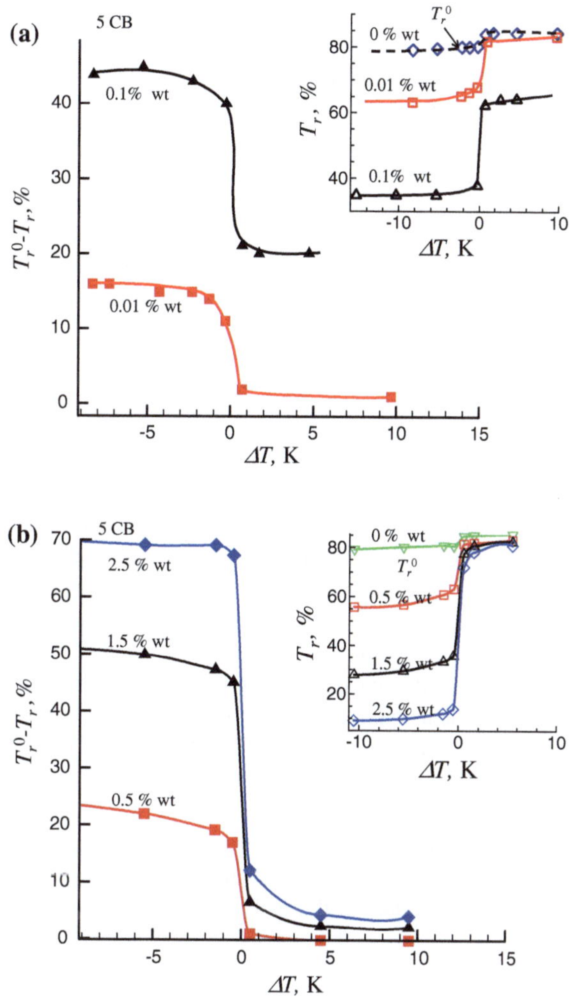

Fig. 10.23 Difference between optical transmissions of pure and CNT-doped nematics $T_r^0 - T_r$ as function of temperature differences $\Delta T = T - T_{ni}$ for different concentrations of MWCNTs (**a**) (From [110]. With permission) and nanoplatelets of MMTO (**b**) [176] dispersed in 5CB. *Inserts* show the measured values of optical transmission T_r

10.5.6.2 Electro-optical Applications and Memory Effects

Doping of LC by CNTs was used for testing of different electro-optical applications of LCs. The presence of undesired impurity ions in LCs can negatively affect the performance characteristics of LC electro-optical cells. CNTs were shown to be the good candidates as agents for efficient ion trapping in LC materials [177]. It was demonstrated that the MWCNTs have good ion trapping characteristics. The CNTs can serve as traps for charge carriers in LC, which substantially decreases electric current due to charge transfer (e.g., movement of ions) through the LC cell, thus eliminating undesirable stray currents in AC field effects [118, 178–181]. These improved characteristics of electro-optical LC cells and LC displays, in particular, lowered the working voltages and allowed achieving the better image clarity. Also, introduction of CNT can decrease rotational viscosity and achieve shorter response times of electro-optical effects [182, 183]. These effects are not directly related to orientation of CNTs in LC matrix and they can be achieved of other types of nanofillers, e.g. fullerenes, grapheme [3] of nanoplatelets of MMTO [184].

The electro-optical effect in 90° twisted nematic cells of MWCNTs doped liquid crystals (≈0.02 wt% in 5CB or E7) under an applied dc voltage was studied [185]. It was shown that doping with CNTs reduced the dc driving voltage and improved the switching behavior. The field-induced switching time for the hybrid CNT + LC system was tested by means of the electro-optic response. The data evidence that doping of LC by CNTs results in a reduced rotational viscosity, in enhanced dielectric anisotropy as well as in acceleration the field-induced switching [186].

Doping of LC by MWCNTs increased the threshold voltage of electrohydrodynamic instability in nematics with $\Delta\varepsilon < 0$ (MBBA + EBBA) and initial planar orientation [110]. Electro-optical memory effect was observed in EBBA + MWCNT suspensions [13, 187, 188]. The initial orientation of LC with $\Delta\varepsilon < 0$ was homeotropic before the action of the electric field. Upon the application of electric field, the CNT-containing nematic was reoriented to the planar texture with a "memory" effect. Irreversible response on the applied electric field (electro-optical memory) was revealed. After the switch-on and subsequent switch-off of the field, the optical transmittance of suspension substantially increased compared to the initial transmittance typical for homeotropic orientation of LC. The efficiency of electro-optical memory was depended on the concentration of MWCNTs, C, and reached its maximum at $C = 0.02$–0.05 wt%. This effect was even stronger when the nematic matrix was additionally doped with a small quantity of chiral dopant, which stabilized the planar texture [189]. The observed data were explained accounting for possible stabilization of the planar state of LC by the network of CNTs formed upon the disintegration of aggregates under the action of electro-hydrodynamic fluxes [188].

10.5.6.3 Related Effects

The diffraction gratings and the orientational photorefractive effect in cells of the homogeneously aligned liquid-crystal E7 doped with MWCNTs were studied [190–192]. It was demonstrated that the permanent gratings was associated with periodically distributed CNTs adsorbed on the inner surfaces of the cells under prolonged illumination to the 514.5 nm beams [193].

Recently great attention was paid to the dye-doped LC systems. These compositions are attractive for applications in guest-host liquid crystal displays with high quality color and brightness. For these purposes, a high ability of dye to orientation in the LC matrix is required. In many works, it was demonstrated that addition of CNTs has positive effect on functionality of dye-doped LC composite.

The effects of SWCNTs (0.002 wt%) on diffraction efficiency in the azo dye MR–doped LC (E7) were studied [194–196]. An order parameter of $s_{LC} = 0.605$

was estimated using polarized absorption measurements. It was shown that the presence of CNTs enhanced the diffraction efficiency by a factor of 1.8. Permanent gratings were written in planar nematic LC cells containing hybrid mixture MR + E7 + SWCNTs [195]. The grating formation was caused by *trans-cis* photo-isomerisation of the dye molecules, followed by surface adsorption of the *cis*-isomer [196]. It is important that these photonic devices are functional without the application of external electric fields. Maximum relative diffraction efficiency was noticeably higher for cells doped with MR and SWCNTs (67 %) than for cells doped only with MR (28 %). The permanent holographic images in a hybrid cells filled with MR (0.6 wt%) + E7 + SWCNTs (0.002 wt%) hydrid material were recorded [197]. The quality of images were higher for the hybrid cells than for cells filled with MR (0.6 wt%) + E7. Cells filled with hydrid material showed increased conductivity and capacitance, slower ac-rise switching time and dc-backflow on turn-off [198]. The correlations between ac-field-enhanced diffraction efficiency and Freedericks threshold voltage were observed.

The effects of addition of SWCNTs on guest-host interaction of some dye-doped LC was studied [199–205]. It was demonstrated that addition of SWCNTs resulted in an increase in order parameter s_{LC}. e.g., it was demonstrated that in hybrid system dye (Red 60) + LC (E63) + SWCNTs the order parameter of $s_{LC} \approx 0.77$ can be attained [202].

The effect of supra-optical nonlinearity and a large electro-optically induced photorefractive effect in E7 doped by SWCNTs was reported [206, 207]. It was stated that such supra-nonlinearities are promising for various holographic and image processing applications and as low-cost alternatives to conventional liquid-crystal spatial light modulators.

10.6 CNTs in Other Types of LCs

10.6.1 Hydrogen Bonded LCs

Several studies were devoted to the hydrogen bonded LCs doped by MWCNTs. Dielectric relaxations studies of pure and doped hydrogen bonded ferroelectric LCs revealed that doping elevated the activation energies considerably [208]. The bi-stable electrical states were observed in doped LC systems (mixtures of levo tartaric acid and undecyloxy benzoic acid) and was explained by alignment of CNTs in the nematic phase [209]. Optical textural observations of these two states were done and the application of external field allowed realizing an optical shutter. Considerable hysteresis in dielectric permittivity has been observed in doped hydrogen bonded LCs (mixture of succenic and pentyloxy benzoic acids) [210]. Moreover, the electric field induced transition in a nematic phase was reveled by behaviors of electrical conductance, permittivity, and helicoidal structure deformations. These hybrid systems are promising for light modulation applications.

The hydrogen-bonded LC (7OBA) + SWCNT (0.01–0.007 wt%) suspensions were studied using optical microtexture analysis, Raman spectroscopy and differential scanning calorimetry [211]. A cascade of phase transitions and phases not typical for pure 7OBA were observed. The induction of chirality in all LC phases was observed in doped systems, whereas pure 7OBA was typically achiral.

10.6.2 Ferroelectric LCs

Ferroelectric liquid crystals (FLCs) possess promising characteristics such as low driving voltage, better optical contrast and faster response. Several investigations were devoted to the effects of CNT doping on different properties of FLCs. In general, the hybrid systems on the of FLCs doped by CNTs have enhanced parameters for use in practical application, have large electro-optic coefficient, high dielectric constant, fast switching response [212]. These materials may be promising for using in memories, capacitors and display devices. In general the enhanced electro-optic and dielectric responses of CNT doped FLCs were explained accounting for the screening the spontaneous polarization of the FLC by the π-π electron system of CNTs and by trapping the ionic impurities by CNTs. The latter factor may result in significant modification of the internal electric field inside the cells.

Effect of SWCNTs on dielectric and electro-optical properties of doped FLCs has been studied [213–215]. A noticeable changes in relaxation behavior as well as increase of spontaneous polarization and relative permittivity with slight slower response has been observed for the doped system [214]. The high loss factor in doped systems was explained by deformation of helix by movement of long CNTs [215]. It was concluded FLCs doped with low concentration of CNTs are suitable for enhancement the performance of LC devices operating at low applied electric fields.

The effect of SWCNTs (0.002 wt%) on phase behavior of antiferroelectric LCs was studied [216]. It was shown that addition of CNTs resulted in radical altering of phase sequences in the studied systems. The effect of alignment of SWCNTs along the smectic layers (smectic A phase) of FLC ([4-(3)-(S)-methyl-2-(S)-chloropentanoyloxy)]-4′-nonyloxy-biphenyl) the was observed [217]. Fourier transform infrared and Raman spectroscopy data evidenced the presence of charge transfer between contacting hexagonal rings of CNTs and the C = O groups of the FLC molecules. The effect of alignment was explained by the π-π stacking and charge transfer effects.

Dielectric and electro-optic properties of planar aligned FLC LAHS7 doped by MWCNTs revealed that even a small concentration of CNTs greatly affects the performance of the LC cells [218]. Effect of MWCNTs (0.1 wt%) on the dielectric spectra (50 Hz–1 MHz) of FLC KCFLC10S was studied at different bias voltages [219]. The increased effect of the bias voltage on the FLC with the addition of CNTs was observed and the theory for its explanation was proposed. The DC and AC electrical properties of FLC + MWCNT suspensions were investigated [220].

In the doped suspensions AC relaxation was observed and this relaxation was absent in pure FLC.

The dielectric studies of MWCNT doped FLCs have been done [221]. A decrease in dielectric permittivity (≈40 %) in 0.1 wt% doped FLC sample was found over 0.05 wt% doped sample. Effect of MWCNTs (0.00–0.03 wt%) electro-optical and dielectric responses of FLC (KCFLC10R) have been studied [222]. Doping resulted in fastening of switching time about ≈80 %. It was concluded that MWCNTs doping has good response in terms of increased contrast ratio and low threshold voltage.

The studies of dielectric dynamic responses (for frequencies up to 1 MHZ) of MWCNT doped FLC were done [223]. The increase in permittivity in 50 Hz–1 kHz range was observed. It was attributed to the enhanced orientational order due to strong anisotropic interactions.

Dielectric and rotational viscosity measurements for FLC (smectic cyclohexane derivatives) doped (0.01 wt%) by silica nanoparticles (5–15 nm in diameter) and MWCNTs (110–170 nm in diameter, 5–9 μm in length) were carried out [224]. The lower value of the rotational viscosity and higher values of permittivity and dielectric losses were observed for silica-doped sample than those of MWCNT doped sample. MWCNTs deformed the helix more effectively and exerted stronger anisotropic interactions on FLC molecules as compared to silica nanoparticles.

The relaxation time, relaxation frequency and dielectric strength for FLC + MWCNT (0.03–0.10 wt%) suspensions have been studied [225]. Thermo-optic studies revealed a small shift (≈ 5 K) in the clearing temperature of FLC.

The temperature (323–353 K) and concentration (0–0.04 wt%) dependences of electro-optic and dielectric properties of FLC + MWCNT suspensions were investigated [226]. The data evidenced that despite a decreasing tilt angle, the spontaneous polarization increases and the electro-optic response times slow down as a function of MWCNT concentration. These effects were explained accounting for the possible dipole moment due to the presence of the MWCNTs, behavior of rotational viscosity and other characteristics.

10.6.2.1 Functionalized and Metal Nanoparticle Decorated CNTs

FLC doped with functionalized (–COOH, –OH and –NH2 groups) MWCNTs were studied using dielectric, electro-optical and photoluminescence measurements [227]. It was demonstrated that functionalized CNTs have enhanced the dielectric properties and the photoluminescence intensity remarkably. The order parameter of the FLC material also increased depending on the functionalizing.

The introduction of copper oxide decorated MWCNTs in FLC allowed fastening of the response of display device [228]. The dielectric and electro-optical studies were done and the observed effects have been attributed to decrease in rotational viscosity of the FLC material and to trapping of ions on the surface of copper oxide decorated CNTs. The gold nanoparticle-decorated MWCNT doped FLC LAHS 22 material was studied using dielectric relaxation spectroscopy, differential scanning calorimetry and polarization optical microscopy [229]. The doping resulted in enhancement of the dielectric and electro-optical characteristics. e.g., the increments in dielectric dispersion, absorption, spontaneous polarization and rotational viscosity of the FLC material were observed.

10.6.2.2 Deformed Helix FLCs

The fastening of the electro-optic response in MWCNT-doped deformed helix FLC hybrid suspensions has been observed and attributed to the decrease in rotational viscosity and increase in anchoring energy [230]. The non-zero spontaneous polarization in para-electric phase for the same system was also reported [231]. It has been attributed to the possible short range orientational order of LC surrounding the CNTs. Doping of deformed helix FLC FLC-6304 by CNTs significantly affect the dielectric parameters in the chiral smectic C phase [232].

10.6.2.3 Chiral Single Walled CNTs

The FLC + chiral SWCNT suspensions were studied using electrooptic and dielectric measurements [233]. Doping affected greatly the performance of FLC cells. e.g., the spontaneous polarization, the Goldstone mode dielectric strength and the rise time were decreased. A switchable grating based on FLC + chiral SWCNT suspensions was proposed [234]). The presence of CNTs improves the diffraction profile of the pure FLC. The diffraction efficiency was rather high (>100). The effect has been explained by the decrease in ferroelectric domain periodicity and optical activity of the chiral CNTs.

10.6.3 Smectic and Cholesteric LCs

Recently, the phase transitions in smectogenic liquid crystal BBBA doped by MWCNTs were studied by methods of optical transmission, differential scanning calorimetry (DSC), measurement of electrical conductivity and analysis of microscopic images. The concentration of CNTs was varied within 0–1 wt%.

Non-monotonous (extremal) changes in temperature, enthalpies and half-width of the DSC peaks of transitions between different phases (smectic, nematic, isotropic) were observed for CNT concentrations between 0.05 and 0.1 wt%. A noticeable increase of electrical conductivity σ in the same concentration interval evidenced the presence of percolation transition and formation of conductive CNT networks. The detailed analysis of behavior in the whole concentration interval 0–1 wt% revealed the presence of a fuzzy type percolation with multiple thresholds in the studied BBBA + CNT suspensions. The percolation behaviour was strongly dependent on the temperature, and a noticeable step-like drop of σ in the vicinity of isotropic-nematic transition was observed after the multiple heating-cooling cycles.

Figure 10.24 presents optical transmission T_r versus temperature T for BBBA doped with CNTs in the vicinity of the nematic-isotropic transition (a) and smectic transitions (b) [235]. The increase in CNT concentration C leads to non-monotonous changes in nematic-isotropic transition temperatures T_{ni} (Fig. 10.24a). At certain concentration, $C \approx 0.05$–0.1 wt%, T_{ni} decreases to the minimal value of ≈345.5 K and then increases up to ≈347.7 K at $C = 0.2$ wt%. The significant step-wise changes in optical transmission were observed also at temperatures close to smectic phase transitions (Fig. 10.24b). Several extremes were present in the temperature dependence of T_r of pure BBBA inside the temperature range of smectic S_B and S_G phases.

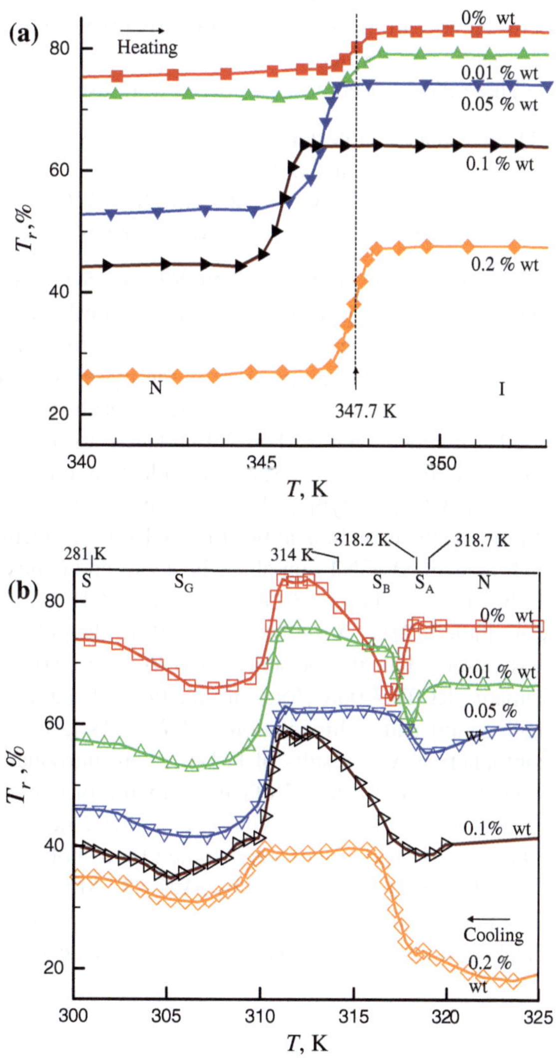

Fig. 10.24 Optical transmission T_r versus temperature T for BBBA + MWCNT suspensions in the vicinity of nematic-isotropic transition (**a**) and smectic transitions (**b**). The phases and temperatures are shown for pure BBBA (C = 0 %). A sandwich-type LC cell (with cell thickness h = 50 μm) was used for optical transmission measurements (From [235]. With permission)

Doping of BBBA by CNTs affected positions of these extremes and resulted in their broadening. The observed changes in the temperature dependence of T_r reflected evidently the perturbations of BBBA structure induced by CNTs. These perturbations can affect the positional order inside the different smectic phases, e.g., partially destroy the network of hexagons within the layers in S_B phase or frustrate three-dimensional order and correlations between the various layers in S_G phase. As a result, the changes in temperature ranges of different smectic phases were observed with introduction of CNTs.

Accounting for the super-cooling effects, the temperature range corresponding to the highest optical transmission T_r (Fig. 10.24b) may be attributed to the S_B phase.

Introduction of MWCNTs into cholesteric LC [110] apparently resulted in orientation of the dispersed MWCNTs in quasi-nematic layers. The general picture of $T_r^0 - T_r$ changes was similar to that observed in nematics, with the same differences between non-aromatic (cholesterol derivatives) and aromatic (5CB + chiral dopant) cholesteric LC matrices.

Three different cholesteric media were used:

(1) an induced cholesteric composed of 70 wt% of nematic ZhK-1282 and 30 wt% of chiral dopant CB-15;
(2) a mixture of 80 wt% COC and 20 wt% CC;
(3) a mixture of 60 wt% CN, 20 wt% CCN and 20 % CCL.

In the first two mixtures, MWCNTs (0.05–0.1 wt%) did not significantly affect the helical twisting properties. In the third case, certain effects on selective reflection spectra were noted, which, in fact, reflected lowering of the cholesteric—smectic-A transition temperature in the presence of MWCNTs. This could suggest that CNTs promoted translational disordering in the smectic-A phase. This is in good agreement with MWCNT-induced lowering of the nematic—smectic-A transition temperature [235]. The selective reflection spectra were just slightly smeared (which is typical for introduction of non-mesogenic dopants), and practically no effects on helical twisting (selective reflection maximums) were observed.

Effects of SWCNTs on phase behavior and helical pitch of cholesterics (mixture of 46 wt% CN, 44 wt% COC and 10 wt% CB with different concentrations of CN) was studied [236]. A well-defined shift of the helical pitch versus temperature plot was observed in the presence of SWCNTs (0.014 wt%). For the mixtures of a nematic ROTN-403/015S, introduction of larger amounts of CN accelerated aggregation and sedimentation of SWCNTs [236]. However, these results, are in contradiction with results of [237], where introduction of chiral components substantially stabilized the CNTs in a mixture of nematics and cholesterol esters.

The electrical conductivity σ and dielectric anisotropy, $\Delta\varepsilon$, changes (in the 100 kHz–10 MHz range) were studied for a cholesteric liquid crystal (nematic E7 + chiral dopant ZLI-811) doped by SWCNTs (0.5 wt%) [238]. The pure CLC exhibits high positive dielectric anisotropy ($\Delta\varepsilon > 0$). With increasing frequency, the electrical conductivity first increased and then reached saturation. It was explained by reorientation of SWCNTs along the director field. Also, SWCNTs substantially affected the dielectric anisotropy $\Delta\varepsilon$.

One should also mention the studies of lyotropic cholesteric LC doped with SWCNTs [239]. The lyotropic LC was formed by dispersing double-stranded DNA in water solution. The phase behavior and microstructure depended on the relative concentrations of DNA and SWCNTs. It was shown that such suspensions could be used for preparation of carbon CNT films with controlled morphology.

It was reported that a hybrid material based on a cholesteric LC mixture and nanotubes could be used for detection of acetone vapors [240]. When the concentration of acetone (absorbed from air by the liquid crystal layer) is relatively low, the selective reflection maximum is shifted due to helical pitch variation. At higher acetone concentrations, when cholesteric structure is destroyed, conducting networks of nanotubes are formed, and concentration is measured by increased conductivity. Thus, such device is operational as gas detector for a very wide dynamic range of concentrations.

10.6.4 Lyotropic LCs

10.6.4.1 Typical Aqueous LCs

The surfactant can be effectively used for assisting the dispersing of CNT in water. SWCNTs (0.05–0.1 wt%) were dispersed in aqueous solution of surfactant (SDS, 1 wt%) [241] [242]. By means of resonant Raman spectroscopy it was demonstrated that CNTs can be macroscopically aligned and simultaneously well dispersed inside self-organized aqueous LLC. It was supposed that SWCNTs are covered with adsorbed surfactant molecules and form very long rod-like or disk-like micelles. The estimated order parameter for SWCNTs was $s_n \approx 0.6$. The stability of suspensions was very high over time scales of months or longer.

SWCNTs were integrated into hexagonal phase of lyotropic liquid crystal (LLC) [243]. The hexagonal LLC was prepared by adding of surfactant Triton X-100 (50 wt%) water, the concentration of SWCNTs was changed within the interval 0–0.25 wt%. Light microscopy and small-angle X-ray scattering (SAXS) data indicated that the SWCNTs are well dispersed and aligned along the LC director. Data of SAXS also evidenced that SWCNTs are intercalated within the cylinders of the hexagonal LLC.

Many experiments evidenced that incorporating of CNTs in a LLC allowed obtaining the uniaxial alignment of the CNTs [11, 244, 245]. A new microfluidic-based method for transferring the LC-dispersed CNTs to a substrate high alignment of CNTs was described [11, 246]. The possibility of the extraction of very thin and long filaments with high alignment of CNTs was demonstrated [245]. Suspension of single-walled CNTs (0.01 wt%) dispersed LLC was studied by SAXS [247]. LLC was based on 25 wt% aqueous solution of surfactant CTAB.

This LLC system exhibited nematic (N), hexagonal (H) and isotropic (I) phases upon heating. SAXS data evidenced that in the nematic and hexagonal phases, the CNTs exhibit 2-D hexagonal ordering and in the isotropic phase the domains of CNTs with 1-D ordering are formed.

10.6.4.2 Chromonic LCs

The chromonic LCs are composed of disc-like or plank-like rigid molecules of aromatic nature. The formation of the supramolecular lyotropic structures in chromonic LCs is governed by π-π interactions between flat core areas of chromonic molecules. The different type of chromonic aggregates and LC phases (e.g., hexagonal and nematic phases) can be realized in the chromonic systems [248]. Nematic chromonic LC DSCG + SWCNTs suspensions were studied [249]. It was demonstrated that chromonic LCs can be considered as the excellent host for individual CNTs. In water the molecules of DSCG self assemble into the nematic or hexagonal phase at room temperature and the individual SWCNTs align parallel to the LC director. The order parameter of CNTs extracted from the polarized coupled Raman and photoluminescence spectra was unusually large, $s_n \approx 0.9$. This value was close to the order parameter of the nematic phase, $s_{LC} = 0.97$.

10.6.4.3 Composite Polymers with Well Oriented CNTs

The good integration of nanotubes in LLCs can be used for production of nanocomposites of vertically aligned SWCNTs by magnetic alignment [250]. It was supposed to orient mesophases of hexagonally packed cylindrical micelles by an applied magnetic field and template the alignment of SWCNTs covered by the micellar cores. The mesophase can be then polymerized to form the composite polymer. The similar approach was discussed for the SWCNTs that were dispersed in LLC on the base of a polymerizable surfactant DDAM in water [251]. At the concentration of 70 wt% of DDAM, the LLC showed columnar hexagonal phases. X-ray diffraction analysis evidenced that the CNTs were well incorporated in the hexagonal phase and induced a swelling of the structure. The suspensions were photopolymerized by photo-irradiation and a mechanically stable, polymeric LC + CNT composites were produced. The proposed method is attractive for preparation of polymeric composited with well-dispersed and highly oriented CNTs.

For effective incorporation of SWCNTs into LLCs the method of phase separation in the presence of water soluble polymers was proposed [252–254]. The method consist of the initial incorporation of CNTs into the LLC phase formed by an anionic surfactant SDS or a cationic surfactant CTAB [252]. In both systems, the surfactant phase was condensed into a hexagonal lattice by adding the anionic polyelectrolyte PSS for the case of SDS and in the presence of a cationic polyelectrolyte PDADMAC for the case of CTAB. The obtained LLC + CNT hybrid systems were studied by polarized optical microscopy (POM) and SAXS. The produced hybrids showed considerable stability against temperature rise. This is important for the LLC + CNT hydrids with practical applications at the elevated temperatures. In the similar work the CNTs were initially incorporated in LLC formed by $C_{12}E_6$. Then the spontaneous phase separation was induced by addition of nonionic hydrophilic polymer PEG [253]. The hexagonal and lamellar phases were observed in dependence of the ratio of PEG to $C_{12}E_6$. It was shown that the quality of CNT + LLC hydrid was dependent on the temperature and at higher temperature the aggregation of CNTs was induced.

10.6.4.4 Ionic Liquids

MWCNTs were dispersed in hexagonal LLCs on the base of room temperature ionic liquid [255]. The hexagonal phase was obtained using the binary mixture of nonionic surfactants ($C_{16}EO_6$) and a room-temperature ionic liquid EAN. The strong integration of MWCNTs into the hexagonal LLC was confirmed by analysis of the texture of hexagonal phase as well by the data of FT-IR and Raman spectroscopy and SAXS results. It was shown that the LLCs can impose an alignment on the CNTs along a director of LLC. The potential application of these hydrid materials as lubricating materials was noted. In another study MWCNTs were dispersed in hexagonal (H1) LLCs on the base of a surfactant-like ionic liquid C14mimCl in an EAN [256]. The data of POM and SAXS evidenced that the incorporation of CNTs preserves the structure of H1 phase and increases the viscosity of H1 phase.

10.7 Dispersing of CNTs in LCS Assisted by Nanoplatelets

The effective approach for dispersion of MWCNTs into liquid crystals (LC), suitable for a wide range of concentrations ($C \leq 0.1$ wt%) was proposed [257]. It consists in doping of LC + MWCNT suspensions with organo-modified laponite (LapO). The LapO platelets have high affinity to the MWCNTs. They can cover the surface of MWCNTs and prevent their aggregation. This approach avoids problems of direct organo-modification of CNTs, associated with worsening of their unique properties. The increased dispersion degree results in drastic changes in dielectric and electro-optical characteristics of the suspensions. Dispersing and insulating effects of the LapO platelets on CNTs lead to the absence of classical percolation of conductivity, linear growth of dielectric constant with CNT concentration, as well as reduction of Freedericks threshold and significant growth of the contrast ratio of the E7 + MWCNTs suspensions.

10.8 Conclusions

Analysis of numerous experimental data, both taken from literature and obtained with participation of the authors, show that carbon nanotubes, as well as nanoparticles of other types, can be successfully dispersed in various liquid crystal media, yielding sufficiently homogeneous composite anisotropic systems with peculiar macroscopic properties. Suspensions doped by nanoparticles can be considered as promising for many practical applications. The LC phases included nematics, cholesterics, various smectics, as well as lyotropic or ionic liquid crystals. In many cases, the observed physical properties of LC + CNT suspension can be directly derived from the actual structural parameters of the dispersed particles in the LC media and reflect the formation of complex supramolecular structures. The observed dependences of thermal, electrophysical and optical properties of these systems upon the temperature, CNT concentration and influence of external factors (e.g., applied electric or magnetic fields) are in good agreement with each other and evidence the important role of mutual interactions between CNTs and LC molecules. Further possibilities for improvement of functional characteristics of LC + CNT suspension can be offered by incorporation into these systems the particles with different type of geometry (e.g., platelets), or organic molecules with specific properties (e.g., dyes). Basic description of such complex systems can be made using the same general approach.

The review mentioned possible practical applications of LC + CNT based materials in various electrooptic and optoelectronic devices. Many aspects, including ways of ensuring high time stability of these systems and further broadening of the range of involved components, were left outside the scope of our narration. Such works are now being actively carried out by many researchers worldwide, including studies in the authors' laboratories.

Acknowledgments This work was supported by the Program of fundamental research of Ukraine "Creating new nanomaterials and nanotechnologies". Authors also express our thanks to Dr. S.S. Minenko, Dr. A.I. Goncharuk, A.N. Samoilov and V.V. Ponevchinsky for participation in experimental measurements of LC + CNT suspensions.

References

1. N. Nakashima, Int. J. Nanosci. **4**, 119 (2005)
2. C. Zakri, Liquid Crystals Today **16**, 1 (2007)
3. S.E. San, M. Okutan, O. Köysal, Y. Yerli, Chin. Phys. Lett. **25**, 212 (2008)
4. S. Zhang, S. Kumar, Small **4**, 1270 (2008)
5. J.P.F. Lagerwall, G. Scalia, J. Mater. Chem. **18**(25), 2890 (2008)
6. M. Rahman, W. Lee, J. Phys. D Appl. Phys. **42**, 63001 (2009)
7. R. Stannarius, Nat. Mater. **8**, 617 (2009)
8. H.K. Bisoyi, S. Kumar, J. Indian Inst. Sci. **89**, 101 (2009)
9. I.C. Khoo, Phys. Rep. **471**, 221 (2009)

10. K. Iakoubovskii, Cent. Eur. J. Phys. **7**, 645 (2009)
11. G. Scalia, ChemPhysChem **11**, 333 (2010)
12. Y.A. Garbovskiy, A.V. Glushchenko, Solid State Phys. **62**, 1 (2010)
13. L. Dolgov, O. Kovalchuk, N. Lebovka, S. Tomylko, and Y. O., in *Carbon Nanotubes*, ed. by J.M. Marulanda (InTech, 2010)
14. H.K. Bisoyi, S. Kumar, Chem. Soc. Rev. **40**, 306 (2011)
15. V.A. Davis, J. Mater. Res. **26**, 140 (2011)
16. O. Stamatoiu, J. Mirzaei, X. Feng, T. Hegmann, Top. Curr. Chem. **318**, 331 (2012)
17. J.P.F. Lagerwall, G. Scalia, Curr. Appl. Phys. **12**, 1387 (2012)
18. G. Scalia, in *Liquid Crystals Beyond Displays: Chemistry, Physics, and Applications*, ed. by Q. Li (John Wiley & Sons, 2012), pp. 341–378
19. C. Blanc, D. Coursault, E. Lacaze, Liq. Cryst. Rev. **1**, 83 (2013)
20. T. Wilkinson, H. Butt, in *Carbon Nanotubes and Graphene for Photonic Applications*, ed. by S. Yamashita, Y. Saito, J.H. Choi (Woodhead Publishing, 2013), pp. 319–350e
21. S. Schymura, G. Scalia, Philos. Tran. R. Soc. A: Math. Phys. Eng. Sci. **371**(1988), 20120261 (2013)
22. S. Iijima, Nature **354**, 56 (1991)
23. A. Khare, S. Bose, J. Miner. Mater. Charact. Eng. **4**(1), 31 (2005)
24. M. Manilo, N. Lebovka, S. Barany, Colloids Surf., A **462**, 211 (2014)
25. C. Laurent, E. Flahaut, A. Peigney, Carbon **48**(10), 2994 (2010)
26. A. Peigney, C. Laurent, E. Flahaut, R.R. Bacsa, A. Rousset, Carbon **39**(4), 507 (2001)
27. W. Bauhofer, J.Z. Kovacs, Compos. Sci. Technol. **69**, 1486 (2009)
28. H.S. Lee, C.H. Yun, J. Phys. Chem. C **112**, 10653 (2008)
29. H.S. Lee, in *Syntheses and Applications of Carbon Nanotubes and Their Composites*, ed. by S. Suzuki (InTech, 2013)
30. B.I. Yakobson, L.S. Couchman, J. Nanopart. Res. **8**, 105 (2006)
31. K.N. Kudin, G.E. Scuseria, B.I. Yakobson, Phys. Rev. B: Condens. Matter Mater. Phys. **64**, 235 (2001)
32. N. Fakhri, D.A. Tsyboulski, L. Cognet, R.B. Weisman, M. Pasquali, Proc. Natl. Acad. Sci. U.S.A. **106**, 14219 (2009)
33. H.S. Lee, C.H. Yun, S.K. Kim, J.H. Choi, C.J. Lee, H.-J. Jin, H. Lee, S.J. Park, M. Park, Appl. Phys. Lett. **95**, 134104 (1) (2009)
34. W. Zhou, M.F. Islam, H. Wang, D.L. Ho, A.G. Yodh, K.I. Winey, J.E. Fischer, Chem. Phys. Lett. **384**, 185 (2004)
35. D.W. Schaefer, J. Zhao, J.M. Brown, D.P. Anderson, D.W. Tomlin, Chem. Phys. Lett. **375**, 369 (2003)
36. M. Endo, M.S. Strano, P.M. Ajayan, Top. Appl. Phys. **111**, 13 (2008)
37. T. Hasan, Z. Sun, F. Wang, F. Bonaccorso, P.H. Tan, A.G. Rozhin, A.C. Ferrari, Adv. Mater. **21**, 3874 (2009)
38. G. Pandey, E.T. Thostenson, Polym. Rev. **52**, 355 (2012)
39. F. Valentini, M. Carbone, G. Palleschi, Anal. Bioanal. Chem. **405**, 451 (2013)
40. A.S. Brady-Estevez, M.H. Schnoor, C.D. Vecitis, N.B. Saleh, M. Elimelech, Langmuir **26**, 14975 (2010)
41. A.V. Venediktova, A.Y. Vlasov, E.D. Obraztsova, D.A. Videnichev, I.M. Kislyakov, E.P. Sokolova, Appl. Phys. Lett. **100**, 251903 (2012)
42. A.Y. Vlasov, A.V. Venediktova, D.A. Videnichev, I.M. Kislyakov, E.D. Obraztsova, E.P. Sokolova, Physica Status Solidi (B) Basic Res. **249**, 2341 (2012)
43. B. Wang, W. Lou, X. Wang, J. Hao, New J. Chem. **36**, 1273 (2012)
44. S.D. Bergin, Z. Sun, P. Streich, J. Hamilton, J.N. Coleman, J. Phys. Chem. C **114**, 231 (2010)
45. J.M. Hughes, D. Aherne, S.D. Bergin, A. O'Neill, P.V. Streich, J.P. Hamilton, J.N. Coleman, Nanotechnology **23**, 265604 (2012)
46. J.Z. Kovacs, R.E. Mandjarov, T. Blisnjuk, K. Prehn, M. Sussiek, J. Müller, K. Schulte, W. Bauhofer, Nanotechnology **20**, 155703 (2009)

47. G.-X. Chen, Y. Li, H. Shimizu, Carbon **45**, 2334 (2007)
48. J.O. Aguilar, J.R. Bautista-Quijano, F. Avilés, Express Polymer Lett. **4**, 292 (2010)
49. N. Grossiord, J. Loos, O. Regev, C.E. Koning, Chem. Mater. **18**, 1089 (2006)
50. M. Poorteman, M. Traianidis, G. Bister, F. Cambier, J. Eur. Ceram. Soc. **29**, 669 (2009)
51. L. Jiang, L. Gao, J. Sun, J. Colloid Interface Sci. **260**, 89 (2003)
52. Q. Chen, C. Saltiel, S. Manickavasagam, L.S. Schadler, R.W. Siegel, H. Yang, J. Colloid Interface Sci. **280**, 91 (2004)
53. M.O. Lisunova, N.I. Lebovka, O.V. Melezhyk, Y.P. Boiko, J. Colloid Interface Sci. **299**, 740 (2006)
54. B. Lamas, B. Abreu, A. Fonseca, N. Martins, M. Oliveira, J. Colloid Interface Sci. **381**, 17 (2012)
55. B. Smith, K. Wepasnick, K.E. Schrote, H.-H. Cho, W.P. Ball, D.H. Fairbrother, Langmuir **25**, 9767 (2009)
56. M. Farbod, S.K. Tadavani, A. Kiasat, Colloids Surf., A **384**, 685 (2011)
57. Y. Maeda, S. Kimura, Y. Hirashima, M. Kanda, Y. Lian, T. Wakahara, T. Akasaka, T. Hasegawa, H. Tokumoto, T. Shimizu, H. Kataura, Y. Miyauchi, S. Maruyama, K. Kobayashi, S. Nagase, J. Phys. Chem. B **108**, 18395 (2004)
58. S.D. Bergin, V. Nicolosi, H. Cathcart, M. Lotya, D. Rickard, Z. Sun, W.J. Blau, J.N. Coleman, Phys. Chem. C **112**, 972 (2008)
59. M.S. Strano, V.C. Moore, M.K. Miller, M.J. Allen, E.H. Haroz, C. Kittrell, R.H. Hauge, R.E. Smalley, Nanosci. Nanotechnol **3**, 81 (2003)
60. L. Vaisman, H.D. Wagner, G. Marom, J. Colloid Interface Sci. **128–130**, 37 (2006)
61. M.D. Clark, S. Subramanian, R. Krishnamoorti, J. Colloid Interface Sci. **354**, 144 (2011)
62. B. Wang, Y. Han, K. Song, T. Zhang, J. Nanosci. Nanotechnol. **12**, 4664 (2012)
63. P.C. Ma, N.A. Siddiqui, E. Mader, J.K. Kim, Compos. Sci. Technol. **71**, 1644 (2011)
64. C.A. Furtado, U.J. Kim, H.R. Gutierrez, L. Pan, E.C. Dickey, P.C. Eklund, J. Am. Chem. Soc. **126**, 6095 (2004)
65. T. Gabor, D. Aranyi, K. Papp, F.H. Karman, E. Kalman, Mater. Sci. Forum **537–538**, 161 (2007)
66. S.D. Bergin, Z. Sun, D. Rickard, P.V. Streich, J.P. Hamilton, J.N. Coleman, ACS Nano **3**, 2340 (2009)
67. J.N. Coleman, Adv. Funct. Mater. **19**, 3680 (2009)
68. Z. Sun, I. O'Connor, S.D. Bergin, J.N. Coleman, J. Phys. Chem. C **113**, 1260 (2009)
69. S. Detriche, G. Zorzini, J.F. Colomer, A. Fonseca, J.B. Nagy, J. Nanosci. Nanotechnol. **8**, 6082 (2008)
70. H.T. Ham, Y.S. Choi, I.J. Chung, J. Colloid Interface Sci. **286**, 216 (2005)
71. C.M. Hansen, *Hansen Solubility parameters-A User's Handbook* (CRC Press, Boca Raton, 2007)
72. O. Deriabina, N. Lebovka, L. Bulavin, A. Goncharuk, Physica E **59**, 150 (2014)
73. B. Krause, G. Petzold, S. Pegel, P. Pötschke, Carbon **47**, 602 (2009)
74. Z. Wang, X. Meng, J. Li, D. X, S. Li, Z. Jiang, T. Tang, J. Phys. Chem. C **113**, 8058 (2009)
75. Y.F. Lan, J.J. Lin, Phys. Chem. A **113**, 8654 (2009)
76. Y.F. Lan, J.J. Lin, Dyes Pigm. **90**, 21 (2011)
77. M. Loginov, N. Lebovka, E. Vorobiev, J. Colloid Interface Sci. **365**, 127 (2012)
78. M. Loginov, N. Lebovka, E. Vorobiev, J. Colloid Interface Sci. **431**(1), 241 (2014)
79. L. Onsager, Ann. N.Y. Acad. Sci. **51**, 627 (1949)
80. H.N.W. Lekkerkerker, P. Coulon, R. Van Der Haegen, R. Deblieck, J. Chem. Phys. **80**(7), 3427 (1984)
81. W. Song, I.A. Kinloch, A.H. Windle, Science **302**, 1363 (2003)
82. W. Song, A.H. Windle, Macromolecules **38**, 6181 (2005)
83. V.A. Davis, L.M. Ericson, A.N.G. Parra-Vasquez, H. Fan, Y. Wang, V. Prieto, J.A. Longoria, S. Ramesh, R.K. Saini, C. Kittrell, W.E. Billups, W.W. Adams, R.H. Hauge, R.E. Smalley, M. Pasquali, Macromolecules **37**, 154 (2004)

84. V.A. Davis, A.N.G. Parra-Vasquez, M.J. Green, P.K. Rai, N. Behabtu, V. Prieto, R.D. Booker, J. Schmidt, E. Kesselman, W. Zhou, H. Fan, W.W. Adams, R.H. Hauge, J.E. Fischer, Y. Cohen, Y. Talmon, R.E. Smalley, M. Pasquali, Nat. Nanotechnol. **4**, 830 (2009)
85. C. Zakri, P. Poulin, J. Mater. Chem. **16**, 4095 (2006)
86. C. Zakri, C. Blanc, E. Grelet, C. Zamora-Ledezma, N. Puech, E. Anglaret, P. Poulin, Philos. Trans. R. Soc. A: Math. Phys. Eng. Sci. **371**(1988), 20120499 (2013)
87. N. Puech, C. Blanc, E. Grelet, C. Zamora-Ledezma, M. Maugey, C. Zakri, E. Anglaret, P. Poulin, J. Phys. Chem. C **115**, 3272 (2011)
88. Z. Dogic, S. Fraden, Curr. Opin. Colloid Interface Sci. **11**, 47 (2006)
89. N. Puech, M. Dennison, C. Blanc, P. van der Schoot, M. Dijkstra, R. van Roij, P. Poulin, E. Grelet, Phys. Rev. Lett. **108**, 247801 (2012)
90. P. Van Der Schoot, V. Popa-Nita, S. Kralj, J. Phys. Chem. B **112**, 4512 (2008)
91. R.W. Ruhwandl, E.M. Terentjev, Phys. Rev. E: Stat. Nonlinear Soft Matter Phys. **55**, 2958 (1997)
92. M. Tasinkevych, D. Andrienko, Condens. Matter Phys. **13**(33603), 1 (2010)
93. N. Lebovka, in *Polyelectrolyte Complexes in the Dispersed and Solid State I*, ed. by M. Müller (Springer Berlin Heidelberg, 2014), pp. 57–96
94. F. Mondiot, R. Botet, P. Snabre, O. Mondain-Monval, J.-C. Loudet, Proc. Natl. Acad. Sci. U.S.A. **111**, 5831 (2014)
95. D. Andrienko, M. Allen, G. Skačej, S. Žumer, Phys. Rev. E: Stat. Nonlinear Soft Matter Phys. **65**, 41702 (2002)
96. V. Popa-Nita, V. Barna, R. Repnik, S. Kralj, in *Syntheses and Applications of Carbon Nanotubes and Their Composites*, ed. by S. Suzuki (InTech, 2013)
97. V. Popa-Nita, J. Chem. Phys. **140**, 164905 (2014)
98. B.I. Lev, S.B. Chernyshuk, P.M. Tomchuk, H. Yokoyama, Phys. Rev. E: Stat. Nonlinear Soft Matter Phys. **65**, 21709 (2002)
99. F. Hung, Phys. Rev. E: Stat. Nonlinear Soft Matter Phys. **79**, 21705 (2009)
100. M. Tasinkevych, F. Mondiot, O. Mondain-Monval, J.-C. Loudet, Soft Matter **10**, 2047 (2014)
101. C. Lapointe, A. Hultgren, D.M. Silevitch, E.J. Felton, D.H. Reich, R.L. Leheny, Science **303**, 652 (2004)
102. Z. Dogic, J. Zhang, A. Lau, H. Aranda-Espinoza, P. Dalhaimer, D. Discher, P. Janmey, R. Kamien, T. Lubensky, A. Yodh, Phys. Rev. Lett. **92**, 125503 (2004)
103. U. Tkalec, M. Škarabot, I. Muševič, Soft Matter **4**, 2402 (2008)
104. F. Brochard, P.G. de Gennes, J. De Phys. **31**, 691 (1970)
105. S. Burylov, Y. Raikher, Phys. Rev. E: Stat. Nonlinear Soft Matter Phys. **50**, 358 (1994)
106. V. Popa-Nita, M. Cevko, S. Kralj, in *Electronic Properties of Carbon Nanotubes*, ed. by J.M. Marulanda (InTech, 2011), pp. 645–663
107. V. Popa-Nita, S. Kralj, J. Chem. Phys. **132**, 24902 (2010)
108. D.D. Nedyalkov, Ukrainian Phys. J. **50**, 662 (2005)
109. H. Duran, B. Gazdecki, A. Yamashita, T. Kyu, Liq. Cryst. **32**, 815 (2005)
110. L.N. Lisetski, S.S. Minenko, A.P. Fedoryako, N.I. Lebovka, Physica E **41**, 431 (2009)
111. N. Lebovka, T. Dadakova, L. Lysetskiy, O. Melezhyk, G. Puchkovska, T. Gavrilko, J. Baran, M. Drozd, J. Mol. Struct. **887**, 135 (2008)
112. W. Zhao, J. Wang, J. He, L. Zhang, X. Wang, R. Li, Appl. Surf. Sci. **255**, 6589 (2009)
113. R. Basu, G.S. Iannacchione, Appl. Phys. Lett. **95**, 173113 (2009)
114. R. Basu, G.S. Iannacchione, J. Appl. Phys. **106**, 124312 (2009)
115. I. Dierking, K. Casson, R. Hampson, Japn. J. Appl. Phys., Part 1: Regular Papers Short Notes Rev. Papers **47**, 6390 (2008)
116. R. Basu, G.S. Iannacchione, Phys. Rev. E: Stat. Nonlinear Soft Matter Phys. **80**, 10701 (2009)
117. L.N. Lisetski, S.S. Minenko, A.P. Fedoryako, N.I. Lebovka, M.S. Soskin, Funct. Mater. **20**, 153 (2013)

118. I.-S. Baik, S.Y. Jeon, S.H. Lee, K.A. Park, S.H. Jeong, K.H. An, Y.H. Lee, Appl. Phys. Lett. **87**, 1 (2005)
119. S.Y. Jeon, K.A. Park, I.-S. Baik, S.J. Jeong, S.H. Jeong, K.H. An, S.H. Lee, Y.H. Lee, NANO **02**, 41 (2007)
120. W. Gwizdala, K. Górny, Z. Gburski, J. Mol. Struct. **887**, 148 (2008)
121. J.J. Mallon, P.M. Adams, Mol. Cryst. Liq. Cryst. **213**, 173 (1992)
122. J.M. Russell, S. Oh, I. LaRue, O. Zhou, E.T. Samulski, Thin Solid Films **509**, 53 (2006)
123. F. Roussel, J.-F. Brun, A. Allart, L. Huang, S. O'Brien, AIP Adv. **2**, 12110 (2012)
124. V.V. Ponevchinsky, A.I. Goncharuk, V.I. Vasil'ev, N.I. Lebovka, M.S. Soskin, JETP Lett. **91**, 241 (2010)
125. V.V. Ponevchinsky, A.I. Goncharuk, V.I. Vasilev, N.I. Lebovka, M.S. Soskin, Proc. SPIE: Int. Soc. Opt. Eng. **7613**, 761306 (2010)
126. R. Basu, K.A. Boccuzzi, S. Ferjani, C. Rosenblatt, Appl. Phys. Lett. **97**, 121908 (2010)
127. R. Basu, C.-L. Chen, C. Rosenblatt, J. Appl. Phys. **109**, 83518 (2011)
128. R. Basu, R.G. Petschek, C. Rosenblatt, Phys. Rev. E: Stat. Nonlinear Soft Matter Phys. **83**, 41707 (2011)
129. R. Basu, C. Rosenblatt, R.P. Lemieux, Liq. Cryst. **39**, 199 (2012)
130. A. Harte, Caltech Undergraduate Res. J. **1**, 44 (2001)
131. M.D. Lynch, D.L. Patrick, Nano Lett. **2**, 1197 (2002)
132. C. Da Cruz, P. Launois, M. Veber, J. Nanosci. Nanotechnol. **4**, 86 (2004)
133. D.L. Patrick, F.S. Wilkinson, T.L. Fegurgur, Proc. SPIE: Int. Soc. Opt. Eng. **5936**, 59360A (2005)
134. M.D. Lynch, D.L. Patrick, Chem. Mater. **16**(5), 762 (2004)
135. I. Dierking, G. Scalia, P. Morales, D. LeClere, Adv. Mater. **16**, 865 (2004)
136. I. Dierking, G. Scalia, P. Morales, J. Appl. Phys. **97**, 44305 (2005)
137. G. Scalia, J.P.F. Lagerwall, M. Haluska, U. Dettlaff-Weglikowska, F. Giesselmann, S. Roth, Physica Status Solidi (B) Basic Research **243**, 3238 (2006)
138. S.J. Jeong, K.A. Park, S.H. Jeong, H.J. Jeong, K.H. An, C.W. Nah, D. Pribat, S.H. Lee, Y.H. Lee, Nano Lett. **7**, 2178 (2007)
139. A.I. Kocherzhin, M.P. Kuhtin, L.N. Lisetski, A.G. Neruh, E.I. Chernyakov, Appl. Electron. **8**, 209 (2009)
140. A.I. Goncharuk, N.I. Lebovka, L.N. Lisetski, S.S. Minenko, J. Phys. D Appl. Phys. **42**, 165411 (2009)
141. L.N. Lisetski, S.S. Minenko, V.V. Ponevchinsky, M.S. Soskin, A.I. Goncharuk, N.I. Lebovka, Materialwiss. Werkstofftech. **42**, 5 (2011)
142. O. Trushkevych, N. Collings, T. Hasan, V. Scardaci, A.C. Ferrari, T.D. Wilkinson, W.A. Crossland, W.I. Milne, J. Geng, B.F.G. Johnson, S. MacAulay, J. Phys. D Appl. Phys. **41**, 125106 (2008)
143. L. Lysetskiy, V. Panikarskaya, O. Sidletskiy, N. Kasian, S. Kositsyn, P. Shtifanyuk, N. Lebovka, M. Lisunova, O. Melezhyk, Mol. Cryst. Liq. Cryst. **478**, 127 (2007)
144. A.L. Efros, *Physics and Geometry of Disorder: Percolation Theory (Science for Everyone)* (Mir Publishers, 1987)
145. M. Sahimi, *Applications Of Percolation Theory* (CRC Press, 1994)
146. D. Stauffer, A. Aharony, *Introduction to Percolation Theory* (Taylor and Francis, London, 1992)
147. Y.Y. Tarasevich, *Percolation: Theory, Applications, Algorithms* (Editorial, Moscow, 2002)
148. N.I. Lebovka, L.N. Lisetskii, M.S. Soskin, A.I. Goncharuk, V.V Ponevchinsky, N.S. Pivovarova, in *Modeling of the Physical Properties of Disordered Systems: Self-organization, the Critical and Percolation Phenomena*, ed. by Y.Y. Tarasevich (Astrakhan University Publishing House, 2011), pp. 5–22
149. L. Berhan, A.M. Sastry, Phys. Rev. E: Stat. Nonlinear Soft Matter Phys. **75**, 41120 (2007)
150. F. Du, R.C. Scogna, W. Zhou, S. Brand, J.E. Fischer, K.I. Winey, Macromolecules **37**, 9048 (2004)
151. L. Berhan, A.M. Sastry, Phys. Rev. E: Stat. Nonlinear Soft Matter Phys. **75**, 41121 (2007)

152. A.V. Kyrylyuk, P. van der Schoot, Proc. Natl. Acad. Sci. U.S.A. **105**, 8221 (2008)
153. F. Du, J.E. Fischer, K.I. Winey, Phys. Rev. B: Condens. Matter Mater. Phys. **72**, 121404 (2005)
154. A. Behnam, A. Ural, Phys. Rev. B: Condens. Matter Mater. Phys. **75**, 125432 (2007)
155. A. Behnam, J. Guo, A. Ural, J. Appl. Phys. **102**, 44313 (2007)
156. S.I. White, B.A. DiDonna, M. Mu, T.C. Lubensky, K.I. Winey, Phys. Rev. B: Condens. Matter Mater. Phys. **79**, 24301 (2009)
157. W.S. Bao, S.A. Meguid, Z.H. Zhu, M.J. Meguid, Nanotechnology **22**, 485704 (2011)
158. S.K. Prasad, M.V. Kumar, C.V. Yelamaggad, Carbon **59**, 512 (2013)
159. A. Garcia-Garcia, X. Quintana, N. Bennis, D. Poudereux, J.M. Otón, in *VIII Reunión Española De Optoelectronica, OPTOEL'13* (Alcalá De Henares, Madrid, 2009), pp. 256–259
160. H.J. Shah, A.K. Fontecchio, D. Mattia, Y. Gogotsi, J. Appl. Phys. **103**, 64314 (2008)
161. Q. Wang, J. Dai, W. Li, Z. Wei, J. Jiang, Compos. Sci. Technol. **68**, 1644 (2008)
162. E.S. Snow, J.P. Novak, P.M. Campbell, D. Park, Appl. Phys. Lett. **82**, 2145 (2003)
163. A. Garcia-Garcia, R. Vergaz, J.F. Algorri, X. Quintana, J.M. Otón, Beilstein J. Nanotechnol. **6**, 396 (2015)
164. L.N. Lisetski, A.P. Fedoryako, A.N. Samoilov, S.S. Minenko, M.S. Soskin, N.I. Lebovka, Eur. Phys. J. E **37**, 68 (2014)
165. M. Gu, Y. Yin, S.V. Shiyanovskii, O.D. Lavrentovich, Phys. Rev. E: Stat. Nonlinear Soft Matter Phys. **76**, 061702 (1 (2007)
166. R. Basu, G.S. Iannacchione, Phys. Rev. E: Stat. Nonlinear Soft Matter Phys. **81**, 51705 (2010)
167. R. Basu, G.S. Iannacchione, Appl. Phys. Lett. **93**, 183105 (2008)
168. G.A. Rance, D.H. Marsh, R.J. Nicholas, A.N. Khlobystov, Chem. Phys. Lett. **493**, 19 (2010)
169. S.-T. Wu, J. Appl. Phys. **69**, 2080 (1991)
170. A.R.E. Bras, S. Henriques, T. Casimiro, A. Aguiar-Ricardo, J. Sotomayor, J. Caldeira, C. Santos, M. Dionísio, Liq. Cryst. **34**(5), 591 (2007)
171. H. Qi, J. O'Neil, T. Hegmann, J. Mater. Chem. **18**, 374 (2008)
172. L.N. Lisetski, N.I. Lebovka, O.T. Sidletskiy, V.D. Panikarskaya, N.A. Kasian, S.S. Kositsyn, M.O. Lisunova, O.V. Melezhyk, Funct. Mater. **14**(2), 233 (2007)
173. P.G. de Gennes, J. Prost, *The Physics of Liquid Crystals* (Clarendon Press, Oxford, 1993)
174. V.V. Ponevchinsky, A.I. Goncharuk, V.I. Vasilev, N.I. Lebovka, M.S. Soskin, Proc. SPIE Int. Soc. Opt. Eng. **7388**, 738802 (2009)
175. V.V. Ponevchinsky, A.I. Goncharuk, S.S. Minenko, L.N. Lisetskii, N.I. Lebovka, M.S. Soskin, Proc. SPIE Int. Soc. Opt. Eng. **8274**, 82740S (2012)
176. S.S. Minenko, A.I. Kocherzhyn, L.N. Lisetski, L. NI, Funct. Mater. **16**(3), 319 (2009)
177. C.-W. Lee, W.-P. Shih, Mater. Lett. **64**, 466 (2010)
178. H.-Y. Chen, W. Lee, Proc. SPIE Int. Soc. Opt. Eng. **5947**, 59470T (2005)
179. H.-Y. Chen, W. Lee, Appl. Phys. Lett. **88**, 222105 (2006)
180. W. Lee, H.-Y. Chen, Japn. J. Appl. Phys., Part 1: Regular Papers Short Notes Rev. Papers **46**, 2962 (2007)
181. K.-X. Yang, W. Lee, Mol. Cryst. Liq. Cryst. **475**, 201 (2007)
182. H.-Y. Chen, W. Lee, N.A. Clark, Appl. Phys. Lett. **90**, 33510 (2007)
183. W. Lee, H.-Y. Chen, Y.-C. Shih, J. Soc. Inform. Display **16**, 733 (2008)
184. T.-Y. Tsai, Y.-P. Huang, H.-Y. Chen, W. Lee, Y.-M. Chang, W.-K. Chin, Nanotechnology **16**, 1053 (2005)
185. W. Lee, C.-Y. Wang, Y.-C. Shih, Appl. Phys. Lett. **85**, 513 (2004)
186. R. Basu, Appl. Phys. Lett. **103**, 241906 (2013)
187. L. Dolgov, O. Yaroshchuk, M. Lebovka, Mol. Cryst. Liq. Cryst. **496**, 212 (2008)
188. L. Dolgov, O. Yaroshchuk, S. Tomylko, N. Lebovka, Condens. Matter Phys. **15(3)**, 33401: 1 (2012)
189. O. Yaroshchuk, S. Tomylko, L. Dolgov, T. Semikina, O. Kovalchuk, Diam. Relat. Mater. **19**, 567 (2010)

190. W. Lee, C.-S. Chiu, Opt. Lett. **26**, 521 (2001)
191. W. Lee, S.-L. Yeh, Appl. Phys. Lett. **79**, 4488 (2001)
192. W. Lee, S.-L. Yeh, C.C. Chang, C.C. Lee, Opt. Express **9**(13), 791 (2001)
193. W. Lee, H.-Y. Chen, S.-L. Yeh, Opt. Express **10**, 482 (2002)
194. S. Ghosh, G.O. Carlisle, J. Mater. Sci.: Mater. Electron. **16**, 753 (2005)
195. Y.S. Suleiman, S. Ghosh, M.E. Abbasov, G.O. Carlisle, J. Mater. Sci.: Mater. Electron. **19**, 662 (2008)
196. M.E. Abbasov, G.O. Carlisle, J. Nanophoton. **2**, 23510 (2008)
197. M.E. Abbasov, S. Ghosh, A. Quach, G.O. Carlisle, J. Mater. Sci.: Mater. Electron. **21**(8), 854 (2010)
198. M.E. Abbasov, G.O. Carlisle, J. Mater. Sci.: Mater. Electron. **23**(3), 712 (2011)
199. F. Akkurt, N. Kaya, A. Alicilar, Fullerenes, Nanotubes, Carbon Nanostruct. **17**, 616 (2009)
200. A. Alicilar, F. Akkurt, N. Kaya, Chin. J. Chem. Phys. **23**, 368 (2010)
201. N. Kaya, F. Akkurt, A. Alicilar, Fullerenes, Nanotubes, Carbon Nanostruct. **19**, 262 (2011)
202. N. Kaya, Liq. Cryst. **38**, 1 (2011)
203. E. Sen, N. Kaya, A. Alicilar, J. Mol. Liq. **186**, 33 (2013)
204. F. Akkurt, Liq. Cryst. **41**, 1269 (2014)
205. A. Akman, F. Akkurt, A. Alicilar, Liq. Cryst. **41**, 603 (2014)
206. I.C. Khoo, Opt. Photonics News **14**, 25 (2003)
207. I.C. Khoo, J. Ding, Y. Zhang, K. Chen, A. Diaz, Appl. Phys. Lett. **82**, 3587 (2003)
208. V.N. Vijayakumar, L.N.M. Madhu, Mohan. J. Dispersion Sci. Technol. **33**, 111 (2012)
209. V.N. Vijayakumar, M.L.N. Madhu, Mohan. Physica B: Condens. Matter **405**, 4418 (2010)
210. V.N. Vijayakumar, M.L.N. Madhu Mohan, T. Chitravel, J. Dispersion Sci. Technol. **33**, 623 (2012)
211. M. Petrov, B. Katranchev, P.M. Rafailov, H. Naradikian, U. Dettlaff-Weglikowska, E. Keskinova, T. Spassov, Phys. Rev. E: Stat. Nonlinear Soft Matter Phys. **88**, 42503 (2013)
212. D. Sharma, S. Tyagi, K.K. Raina, Adv. Mater. Res. **67**, 167 (2009)
213. S.K. Gupta, A. Kumar, A.K. Srivastava, R. Manohar, J. Non-Cryst. Solids **357**, 1822 (2011)
214. S.K. Gupta, D.P. Singh, R. Manohar, Curr. Appl. Phys. **13**, 684 (2013)
215. S.K. Gupta, D.P. Singh, R. Manohar, Ferroelectrics **468**, 84 (2014)
216. J.P.F. Lagerwall, R. Dabrowski, G. Scalia, J. Non-Cryst. Solids **353**, 4411 (2007)
217. Y. Zhao, Y. Xiao, S. Yang, J. Xu, W. Yang, M. Li, D. Wang, Y. Zhou, J. Phys. Chem. C **116**, 16694 (2012)
218. P. Arora, A. Mikulko, F. Podgornov, W. Haase, Mol. Cryst. Liq. Cryst. **502**, 1 (2009)
219. S. Khosla, A. Sharma, J. Inform. Display **14**, 127 (2013)
220. S. Kundu, S.K. Batabyal, P. Nayek, S.K. Roy, J. Nanosci. Nanotechnol. **8**, 1735 (2008)
221. P. Malik, A. Chaudhary, K.K. Raina, Asian J. Chem. **21**, S095 (2009)
222. P. Malik, A. Chaudhary, R. Mehra, K.K. Raina, J. Mol. Liq. **165**, 7 (2012)
223. Neeraj, K.K. Raina, Integr. Ferroelectr. **125**, 104 (2011)
224. Neeraj, P. Kumar, K.K. Raina, J. Mater. Sc. Techol. **27**, 1094 (2011)
225. R.K. Shukla, K.K. Raina, V. Hamplová, M. Kašpar, A. Bubnov, Phase Transitions **84**, 850 (2011)
226. M. Yakemseva, I. Dierking, N. Kapernaum, N. Usoltseva, F. Giesselmann, Eur. Phys. J. E **37**, 7 (2014)
227. P. Ganguly, A. Kumar, S. Tripathi, D. Haranath, A.M. Biradar, Liq. Cryst. **41**, 793 (2014)
228. A. Malik, J. Prakash, A. Kumar, A. Dhar, A.M. Biradar, J. Appl. Phys. **112**, 54309 (2012)
229. S. Tripathi, J. Prakash, A. Chandran, T. Joshi, A. Kumar, A. Dhar, A.M. Biradar, Liq. Cryst. **40**, 1255 (2013)
230. J. Prakash, A. Choudhary, D.S. Mehta, A.M. Biradar, Phys. Rev. E: Stat. Nonlinear Soft Matter Phys. **80**, 12701 (2009)
231. J. Prakash, A. Kumar, T. Joshi, D.S. Mehta, A.M. Biradar, W. Haase, Mol. Cryst. Liq. Cryst. **541**, 166 (2011)
232. N. Sood, S. Khosla, D. Singh, S.S. Bawa, Liq. Cryst. **39**, 1169 (2012)

233. F.V. Podgornov, A.M. Suvorova, A.V. Lapanik, W. Haase, Chem. Phys. Lett. **479**, 206 (2009)
234. A.K. Srivastava, E.P. Pozhidaev, V.G. Chigrinov, R. Manohar, Appl. Phys. Lett. **99**, 201106 (2011)
235. N.I. Lebovka, L.N. Lisetski, A.I. Goncharuk, S.S. Minenko, V.V. Ponevchinsky, M.S. Soskin, Phase Transitions **86**, 463 (2013)
236. S. Schymura, J. Lagerwall, in *Deutschen Bunsen-Gesellschaft e.V. 37. Arbeitstagung Flussigkristalle (2009, Stuttgart)* (2009)
237. N.I. Lebovka, L.N. Lisetski, M.I. Nesterenko, V.D. Panikarskaya, N.A. Kasian, S.S. Minenko, M.S. Soskin, Liq. Cryst. **40**, 968 (2013)
238. O. Koysal, Synth. Met. **160**, 1097 (2010)
239. G. Ao, D. Nepal, M. Aono, V.A. Davis, ACS Nano **5**, 1450 (2011)
240. C.-K. Chang, S.-W. Chiu, H.-L. Kuo, K.-T. Tang, Appl. Phys. Lett. **100**, 43501 (2012)
241. J.P.F. Lagerwall, G. Scalia, M. Haluska, U. Dettlaff-Weglikowska, F. Giesselmann, S. Roth, Physica Status Solidi (B) Basic Res. **243**, 3046 (2006)
242. J. Lagerwall, G. Scalia, M. Haluska, U. Dettlaff-Weglikowska, S. Roth, F. Giesselmann, Adv. Mater. **19**, 359 (2007)
243. V. Weiss, R. Thiruvengadathan, O. Regev, Langmuir **22**, 854 (2006)
244. G. Scalia, C. Von Bühler, C. Hägele, S. Roth, F. Giesselmann, J.P.F. Lagerwall, Soft Matter **4**, 570 (2008)
245. S. Schymura, S. Dölle, J. Yamamoto, J. Lagerwall, Soft Matter **7**, 2663 (2011)
246. S. Schymura, E. Enz, S. Roth, G. Scalia, J.P.F. Lagerwall, Synth. Met. **159**, 2177 (2009)
247. D. Vijayaraghavan, J. Mol. Liq. **199**, 128 (2014)
248. J. Lydon, Curr. Opin. Colloid Interface Sci. **3**, 458 (1998)
249. N. Ould-Moussa, C. Blanc, C. Zamora-Ledezma, O.D. Lavrentovich, I.I. Smalyukh, M.F. Islam, A.G. Yodh, M. Maugey, P. Poulin, E. Anglaret, M. Nobili, Liq. Cryst. **40**, 1628 (2013)
250. M.S. Mauter, M. Elimelech, C.O. Osuji, ACS Nano **4**, 6651 (2010)
251. Y.S. Kwon, B.M. Jung, H. Lee, J.Y. Chang, Macromolecules **43**, 5376 (2010)
252. X. Xin, H. Li, E. Kalwarczyk, A. Kelm, M. Fiałkowski, E. Gorecka, D. Pociecha, R. Hołyst, Langmuir **26**, 8821 (2010)
253. X. Xin, H. Li, S.A. Wieczorek, T. Szymborski, E. Kalwarczyk, N. Ziebacz, E. Gorecka, D. Pociecha, R. HoEyst, Langmuir **26**, 3562 (2010)
254. L. Wang, X. Xin, K. Guo, M. Yang, X. Ma, J. Yuan, J. Shen, S. Yuan, Phys. Chem. Chem. Phys. **16**, 14771 (2014)
255. W. Jiang, B. Yu, W. Liu, J. Hao, Langmuir **23**, 8549 (2007)
256. M. Zhao, Y. Gao, L. Zheng, Colloids Surf., A **369**, 95 (2010)
257. O. Yaroshchuk, S. Tomylko, O. Kovalchuk, N. Lebovka, Carbon **68**, 389 (2014)

Chapter 11
Structural Study of Star Polyelectrolytes and Their Porous Multilayer Assembly in Solution

Weinan Xu, Sidney T. Malak, Felix A. Plamper, Christopher V. Synatschke, Axel H.E. Müller, William T. Heller, Yuri B. Melnichenko and Vladimir V. Tsukruk

Abstract Star polyelectrolytes with responsive properties to external stimuli, such as pH, temperature and ionic condition, were utilized to fabricate layer-by-layer (LbL) microcapsules. The microstructure of star polyelectrolytes was first studied in semi-dilute solution by in situ small-angle neutron scattering (SANS). These measurements show that with the addition of salts, arms of strong cationic star polyelectrolytes will contract and the spatial ordering of the stars would be interrupted. SANS measurements were also performed on the microcapsules in order to study their internal structure and responsive properties in solution. The results show that with the increase of shell thickness, microcapsules undergo a change of fractal dimension. Microcapsules with thinner shell have a surface fractal structure with rough interface, while those with thicker shell generally have a mass fractal structure of 3D random network. With the change of surrounding environment (pH, temperature, or ionic condition), the morphology and permeability of microcapsules are changed concurrently, for example, with the addition of multivalent salt, there is

W. Xu · S.T. Malak · V.V. Tsukruk (✉)
School of Materials Science and Engineering, Georgia Institute of Technology, Atlanta, GA 30332, USA
e-mail: vladimir@mse.gatech.edu

F.A. Plamper
Institute of Physical Chemistry, RWTH Aachen University, 52056 Aachen, Germany

C.V. Synatschke · A.H.E. Müller
Makromolekulare Chemie II and Bayreuther Zentrum für Kolloide und Grenzflächen, Universität Bayreuth, 95440 Bayreuth, Germany

A.H.E. Müller
Institut für Organische Chemie, Johannes Gutenberg-Universität Mainz, 55099 Mainz, Germany

W.T. Heller · Y.B. Melnichenko
Biology and Soft Matter Science Division, Neutron Scattering Directorate, Oak Ridge National Laboratory, Oak Ridge, TN 37381, USA

L. Bulavin and N. Lebovka (eds.), *Physics of Liquid Matter: Modern Problems*, Springer Proceedings in Physics 171, DOI 10.1007/978-3-319-20875-6_11

a surface- to mass-fractal transition, with the correlation length decreasing by around 50 %. This study provides insight into the mechanism of the responsiveness of novel star polyelectrolytes and their assembled multilayer structures.

11.1 Introduction

Polyelectrolytes (PEs) are polymers with ionizable groups in their repeating units and they are usually charged in solution due to ionic dissociation. Due to the dramatic asymmetry in charge, mass and size between the polyelectrolyte backbone and counterions, co-ions and solvent molecules, polyelectrolyte solutions have rich and significantly different phase behaviors compared with that of neutral polymer solutions [1]. The structure and properties of linear polyelectrolytes have been extensity studied during the past several decades [2], but reports on the solution structure of branched polyelectrolytes are rare, and there are many important unresolved questions. Among many different kinds of branched polyelectrolytes, star polyelectrolytes constitute a particular class of macromolecules with high relevance in soft matter physics, chemistry, and materials science [3–6]. Due to the unique architecture of star polyelectrolytes, their conformational state can be easily affected by the degree of charging, the salt concentration, the valency of counterions and co-ions, as well as the temperature and pH of the solution [7, 8].

Theoretical studies have shown that in addition to the steric repulsion between star polymers, there are also a relatively short range attraction and a secondary repulsive barrier at longer distance [9]. Chen et al. [10] calculated the monomer-monomer structure factors of linear and star polyelectrolytes based on polymer reference interaction site model theory. They also showed that at lower polyelectrolyte concentration, the equilibrium distribution of counterions significantly depends on the polyelectrolyte conformation; but at higher concentration (well above the molecular overlap concentration) the association behavior of counterions has little correlation with polyelectrolyte geometry due to the steric effect.

One of the most important features of star polyelectrolytes is the strong localization of the counterions within the star interior, which leads to a stretching of the arms in salt-free condition. The degree of counterion confinement is stronger for multivalent ions than monovalent ones, therefore, in the presence of multivalent counterions, the release of a concomitant number of monovalent ions induces a significant drop in osmotic pressure within the stars, which results in the collapse of the star polyelectrolytes. Ballauff et al. [11] studied spherical polyelectrolyte brushes with a finite size polystyrene core, the structure of which is very similar to star polyelectrolytes, and thus expanded the theory of electrostatic repulsion of star polyelectrolyte to include particles with a finite core radius. They showed that most of the counterions are confined within the brush layer. The strong osmotic pressure thus created within the brush layer dominates the repulsive interaction between two such particles. Despite the similarity between star polyelectrolytes and spherical brushes, the main difference is the distribution of the counterions outside the intra-molecular volume. For polyelectrolyte stars the counterions are distributed uniformly in the solution, while for spherical brushes, the correlation of counterions with the macroion is even higher, so there is a cloud of counterions with thickness comparable to the brush layer, and only a small fraction of the counterions is uniformly distributed in the solution [12].

In order to utilize the branched polyelectrolytes in various applications such as drug delivery, smart surfaces and microreactors, it is necessary to assemble them into different nano- and micro-structures such as thin films, microcapsules, colloids and hydrogels. The main assembly techniques include solution assembly, interfacial assembly and layer-by-layer (LbL) assembly [13–15]. To elucidate the structure and properties of the assembled hierarchical structure, small angle neutron scattering (SANS) again is one of the most suitable experimental techniques, but there are actually very few reports on microstructures based on branched polyelectrolytes. For example, study on various star polymer networks composed of hydrophilic 2-(dimethylamino)ethyl methacrylate (DMAEMA) and hydrophobic methyl methacrylate (MMA) units, with identical topology but different molecular buildup, showed significant difference in structure when swollen with water [16]. SANS measurements showed that for the homopolymer and random copolymer star network, only relatively small structural units were observed; but for the heteroarm star co-networks, the presence of well-defined hydrophobic domains were observed indicating pronounced microphase separation in these systems.

Likos et al. [17] studied the complexation of oppositely charged colloidal particles and polyelectrolyte stars with the molecular dynamics simulations. The results indicated that besides electrostatic interaction, entropy also plays an important role in the adsorption of the stars on colloidal surface. The functionality (i.e. arm number) of the stars has significant influence on the adsorption process: higher functionality stars will not adsorb with all their arms on the colloid because of the internal Coulomb interactions between the arms. The maximum load of the PE stars clearly depends on all quantities, i.e. functionality, the length of the arms, and the overall charge of the PE stars, as well as the overall size and charge of the colloidal particles.

In this chapter, we discuss the morphology and responsive behavior of LbL microcapsules based on branched polyelectrolytes with responsive properties to external stimuli, such as pH, temperature and ionic condition. SANS is utilized to elucidate the internal structure of the microcapsule shells with different components and thickness. Moreover, the structural changes of the shell upon applying external stimuli, which are responsible for the permeability changes of microcapsules, have been examined by SANS. By fitting the experimental data with suitable models, we are able to quantitatively unveil the fractal dimensions and estimate the correlation length of the complex multilayered structures. We use deuterium labeling to contrast the inner nanoporous morphology of branched macromolecules in solution and in thin shell states.

11.2 Experimental Section

11.2.1 Materials

Poly(*N*,*N*-dimethylaminoethyl methacrylate) (PDMAEMA) star polymers were synthesized by atom transfer radical polymerization of 2-(*N*,*N*-dimethylamino)ethyl methacrylate employing a core-first route with functionalized polyhedral oligomeric silsesquioxane (POSS) core [18]. Sugar-based scaffolds as well as silsesquioxane nanoparticles were used as multifunctional initiators. Subsequent quaternization of the obtained PDMAEMA stars with methyl iodide yielded their star-shaped quaternized ammonium salts (qPDMAEMA). The rather low efficiency of the initiation sites (30–75 %) leads to a moderate arm number distribution of the prepared polyelectrolyte stars. Here, we used PDMAEMA star polymers with arm numbers of 9.5, 18 and 24 (number average), the number-average degrees of polymerization per arm are 170, 170 and 240, respectively. Therefore, they are named as $(PDMAEMA_{170})_{9.5}$, $(PDMAEMA_{170})_{18}$ and $(PDMAEMA_{240})_{24}$, with the number-average molecular weights of 250, 490 and 950 kg/mol, and polydispersity index of 1.20, 1.41 and 1.43, respectively.

Linear Poly(ethyleneimine) (PEI) was purchased from Polysciences. Poly (sodium 4-styrenesulfonate) (PSS, M_w = 70,000 g/mol) was purchased from Sigma-Aldrich. All commercial polyelectrolytes were used without further purification. Silica particles with a diameter of 4.0 ± 0.2 μm and 10 % dispersion in water were obtained from Polysciences. Hydrofluoric acid (48–51 %) was purchased from BDH Aristar. Nanopure water (Nanopure system, Barnstead) with a resistivity of 18.2 MΩ cm was used in all experiments. Tris-HCl (1.0 M) was purchased from Rockland and was diluted to 0.01 M in ultrapure pure water with pH adjusted by HCl or NaOH for use. To assure high scattering contrast, D_2O (99.9 %) was used to dissolve star polyelectrolytes for SANS experiments (Cambridge Isotope Laboratories). Sodium deuteroxide (40 wt% in D_2O, 99 at.% D) and deuterium chloride (99 at.% D) were purchased from Sigma-Aldrich, and used to adjust the pH of the samples.

11.2.2 Preparation of LbL Microcapsules

PSS and PDMAEMA star polyelectrolyte were each dissolved in 0.1 M NaCl solution with a concentration of 0.2 mg/mL. The preparation of LbL $(PSS/PDMAEMA)_n$ microcapsules have the following steps: the bare, negatively charged silica particles with average diameter of 4.0 μm were first coated with a PEI prelayer by incubating in 1.5 mL of PEI solution (1.0 mg/mL) for 15 min, followed by two centrifugation (3000 rpm for 3 min)/wash cycles. Subsequently, the silica particles were incubated in 1.5 mL PSS solution (0.2 mg/mL) for 15 min, followed by two centrifugation (3000 rpm for 3 min)/wash cycles. 1.5 mL of PDMAEMA star polyelectrolyte solution was then added and 15 min was allowed for adsorption, also followed by two centrifugation/wash cycles. The adsorption steps were repeated until the desired number of layers were built on silica particles. Hollow microcapsules were finally obtained by dissolving silica cores in 1 % HF solution for 2 h, followed by dialysis in Nanopure water for 2 days with repeated change of water. The preparation of other types of LbL microcapsules in this study follows similar procedure as described above.

11.2.3 SANS Experiments

The SANS measurements of PSS/qPDMAEMA$_{18}$ microcapsules were performed using the EQ-SANS instrument [19] at the Spallation Neutron Source of Oak Ridge National Laboratory (ORNL) employed a sample-to-detector distance of 4 m. The instrument was used in 30 Hz mode with a minimum wavelength setting of 2.5 Å, which gives a second band starting at 9.4 Å. The instrument configuration provides an effective q-range of ~0.005–0.45 Å^{-1}. Scattering data from the EQ-SANS were reduced using standard procedures with the MANTID software package [20].

SANS measurements of PSS/PDMAEMA$_{18}$ microcapsules were conducted at ORNL on the CG2 (GP-SANS) instrument [21] with a wavelength of $\lambda = 4.7$Å ($\Delta\lambda/\lambda \approx 0.14$). Polyelectrolyte or microcapsule solutions were loaded into 2 mm thick quartz cells. Quartz cells were mounted in a temperature-controlled sample holder (temperature stability and gradients are better than ±0.1 °C), and the samples were allowed to stabilize at a preset temperature for 10 min before each measurement. Polymer concentration in our experiments was chosen to be 1 wt% in order to keep high signal-to-noise ratio and minimize possible interactions between the stars and large scale aggregate formation. Two sample-detector distances were used (2.0 and 18.5 m with a 40 cm detector offset), which resulted in a range of scattering vectors q ($q = 4\pi \cdot \sin\theta/\lambda$, where 2θ is the scattering angle) covered in the experiment from 0.004 to 0.6 Å^{-1}. The data were corrected for instrumental background and detector efficiency and converted to an absolute scale (cross section I(q) in units of cm^{-1}) by means of a pre-calibrated secondary standard, Al-4 [22]. Scattering from the solvent was subsequently subtracted proportionally to its volume fraction.

11.3 Results and Discussion

11.3.1 Structure and Ion Responsive Behavior of Star Polyelectrolytes

The star polyelectrolytes discussed here are PDMAEMA stars and their quaternized salts, the structure of which is shown in Fig. 11.1a. PDMAEMA is a well-known

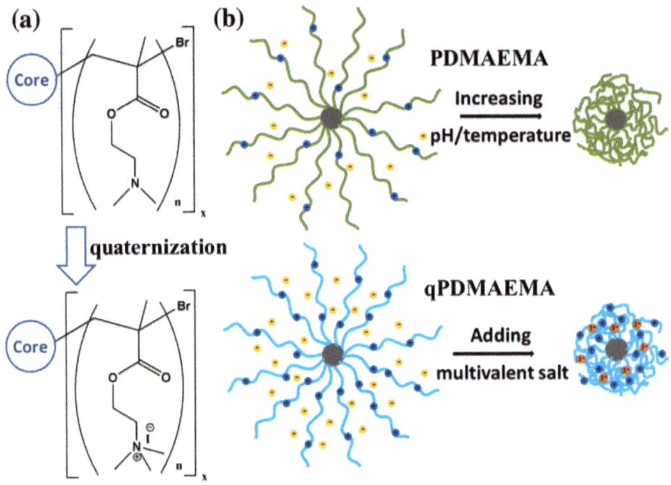

Fig. 11.1 **a** Chemical structure of PDMAEMA and qPDMAEMA star polyelectrolytes, **b** suggested structural changes of the two kinds of responsive star polyelectrolytes due to external condition variations

weak cationic polyelectrolyte with pH and temperature dual responsive properties [18, 23]. With the decrease of pH, the degree of protonation of the amine group in the repeating units increases, which leads to higher charge density and more extended chains due to electrostatic interaction. On the other hand, with the increase of temperature, the hydrogen bonding between the polymer chain and water weakens and the intramolecular hydrophobic interaction becomes stronger, therefore, the PDMAEMA chains would retract or collapse at temperature above the lower critical solution temperature (LCST) [24], as shown in Fig. 11.1b.

Quaternization of PDMAEMA introduces a permanent charge to the repeating unit, therefore the qPDMAEMA stars are strong cationic polyelectrolytes, and their charge density does not depend on pH or temperature. However, the conformation of the qPDMAEMA stars is very sensitive to the addition of multivalent salt. Plamper et al. [25] showed that adding trivalent hexacyanocobaltate(III) ions leads to a collapse of the polyelectrolyte star even at low concentrations (Fig. 11.1b); sufficiently high multivalent counterion concentration leads finally to the precipitation of the polymer from the solution.

The structure of qPDMAEMA stars in a semi-dilute solution is first studied by SANS, with three different types of stars having different arm numbers and arm lengths (Fig. 11.2a). The (number-average) arm numbers are 9.5, 18 and 24, while the degree of polymerization for each arm is either 170 or 240. In contrast to the weak star polyelectrolytes [42], the SANS for qPDMAEMA stars show sharper peaks indicating better intermolecular ordering with significant upturn at low q. The peaks are shifted to lower q indicating increased separation of star macromolecules as a result of increased repulsion. On the other hand, the excessive zero-angle

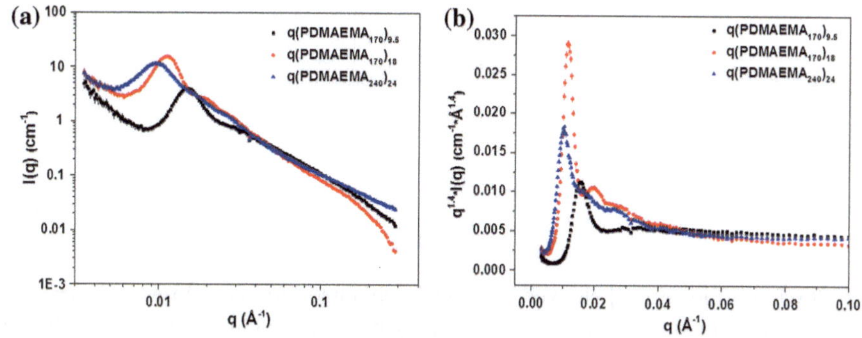

Fig. 11.2 **a** SANS data of qPDMAEMA star polyelectrolytes solution (1 wt% in D_2O, at 25 °C) with different number of arms or arm length. **b** Kratky plot for the corresponding quaternized PDMAEMA star polyelectrolytes (Reprinted from Tsukruk et al. [42] by permission of the American Chemical Society.)

scattering at very low q ($q < 0.008$ Å^{-1}) indicates large-scale concentration fluctuations and increasing osmotic pressure within solutions of highly charged stars [26].

In order to get the size information of the star polyelectrolytes from the scattering data, the generalized Kratky analysis can be used, which estimates the radius of gyration of the stars by plotting $I(q)q^{1/v}$ versus q, where v is the excluded volume parameter [27].

Briefly, in the region around the peak, the form factor can be approximated by the Gaussian star form factor of Benoit [28]:

$$P(q) = \frac{2}{(fv^4)\left(v^2 - [1 - \exp(-v^2)] + \frac{f-1}{2}[1 - \exp(-v^2)]^2\right)}, \tag{11.1}$$

where

$$v = \left(\frac{f}{3f - 2}\right)^{\frac{1}{2}} qR_g. \tag{11.2}$$

This form factor does, in principle, describe star polymers under θ-solvent conditions, but excluded volume effects for swollen chains are not affecting the q range around the peak [29].

The position of the maximum of the Benoit form factor, v_{max}, can be calculated from the first derivative of expression (11.2) with respect to v, and for star polymer with large number of arms ($f \gg 1$), $v_{max} \approx 1$. The radius of gyration can then be calculated by comparing v_{max} to the experimentally obtained q_{max}, after rearrangement of (11.2) we can get

$$R_g = \left(\frac{3f-2}{f}\right)^{\frac{1}{2}} \frac{v_{max}}{q_{max}}, \tag{11.3}$$

or

$$R_g = \frac{\sqrt{3}}{q_{max}}. \tag{11.4}$$

Previous studies [30] showed the values obtained are very close to those from the Zimm evaluation. Therefore, for star polymers whose Guinier regime could not be accessed easily, it is justified to take the R_g value from the Kratky approach.

Kratky analysis of the qPDMAEMA star polyelectrolytes shows a sharp peak in the low q range (Fig. 11.2b), which is significantly shifted to lower q indicating increase in effective molecular dimensions. Moreover, the excluded volume parameter ν in the Kratky analysis is found to be 0.7, while as known ν is 0.5 for ideal Gaussian chains, and 0.6 for fully swollen coils. This result indicates that the qPDMAEMA chains have a more expanded local blob structure due to the increased electrostatic repulsion and osmotic pressure within the star macromolecules [31].

The effect of concentration on the structure of qPDMAEMA stars was preliminarily studied as well, as shown in Fig. 11.3a. SANS curves from the 1 and 0.5 wt % $(qPDMAEMA_{240})_{24}$ solution have similar shape, but the characteristic peak shifts to lower q values was observed in the more dilute solution. The intermolecular distance calculated from the peak position increases from 65.8 to 82.8 nm, which is exactly the expected value, since dilution by a factor of 2 should increase the distance by $2^{1/3}$. Kratky analysis is used to estimate the radius of gyration, and shows that the R_g increases from 16.5 to 21.3 nm with the dilution. The result indicates that there is strong electrostatic repulsion between the cationic stars, at

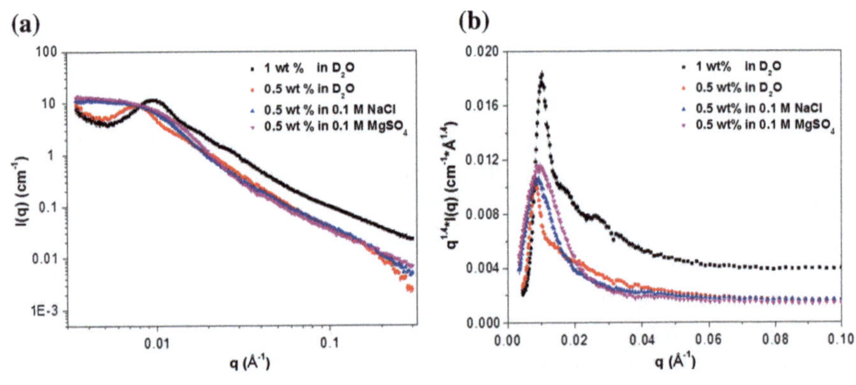

Fig. 11.3 **a** SANS data of $q(PDMAEMA_{240})_{24}$ star polyelectrolytes solution in D_2O with different concentrations and ionic conditions (at 25 °C). **b** Kratky plot for the corresponding samples from (**a**)

higher concentration the arms would be forced to retract from fully expanded conformation due to the strong intermolecular interaction. With the decrease of concentration, the arm chains recover to more extended conformation which is accompanied by the increase in the intermolecular distance.

As known, theoretical and experimental studies have already shown that salt concentration has a significant effect on the structure of polyelectrolytes. Here we also studied how the addition of mono- and divalent salts affects the conformation and organization of star polyelectrolytes in solution. As shown in Fig. 11.3, the most obvious change is that the characteristic peak at low q range disappears after adding 0.1 M NaCl or $MgSO_4$ to the solution. The R_g estimated from Kratky analysis (Fig. 11.3b) decreases from 21.3 nm in D_2O to 18.5 nm in 0.1 M NaCl and 17.7 nm in 0.1 M $MgSO_4$. The main reason is that the presence of salt can effectively screen the charge on qPDMAEMA chains, which significantly reduces the electrostatic repulsion between repeating units on the arm as well as between the stars; and the osmotic pressure within the stars also decreases. As a result, the arm chains contract and the intermolecular ordering is disrupted. SANS experiments on the effect of trivalent ions on the conformation and organization of the qPDMAEMA star polyelectrolytes is more challenging to do, because even a small amount of trivalent salt can lead to the precipitation of the stars from the solution [25]. Such experiments are planned for future studies.

11.3.2 Ion Responsive LbL Microcapsules

The responsive star polyelectrolytes were further utilized to fabricate LbL microcapsules via electrostatic interaction; the details of the fabrication and properties of the PSS/qPDMAEMA$_{18}$ microcapsules can be found in a previous report [32]. In order to characterize the structure of these thin shell microcapsules, confocal microscopy, AFM and TEM can be used. However, these techniques usually either only apply to samples in dry state or partially wetted state, which have significantly different structure as compared to that in solution. SANS, on the other hand, provides a powerful and non-destructive way to elucidate the porous morphology of the microcapsules in solution directly.

SANS measurements were first conducted for PSS/qPDMAEMA$_{18}$ microcapsules with different number of bilayers (Fig. 11.4). The q range of the scattering data in this experiment corresponds to a distance roughly from 1 to 100 nm, which actually covers several characteristic dimensions of the microcapsules including thickness of the shell, star polyelectrolyte size and the mesh size within the shell. However, considering the fact that the porous shell is filled with D_2O, and the scattering contrast is highest between D_2O and the hydrogenated polyelectrolyte matrix, while the contrast between the hydrogenated domain morphology is much smaller and can be neglected. Therefore, we suggest that the scattering in this region is likely dominated by the porous structure filled with D_2O within the LbL shell.

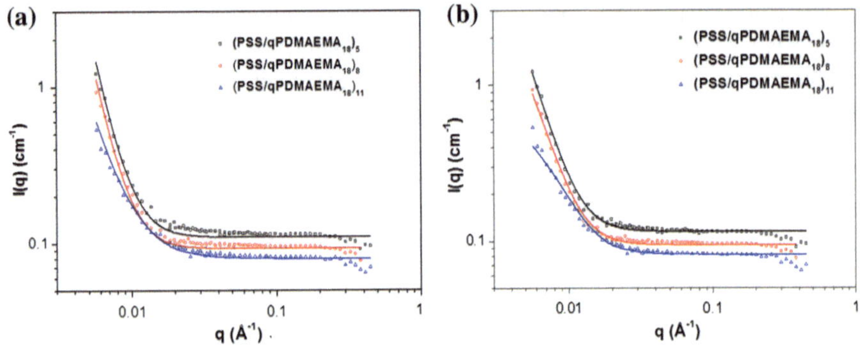

Fig. 11.4 SANS data of PSS/qPDMAEMA$_{18}$ microcapsules with (*square*) 5 bilayers, (*circle*) 8 bilayers, and (*triangle*) 11 bilayers in D_2O solution (at 25 °C) that have been fitted with a power law model (**a** *solid lines*) to determine the evolution of the fractal dimension, and Debye-Anderson-Brumberger (DAB) model (**b** *solid lines*) to obtain the correlation length

Our initial attempt to fit the data utilized a shape-dependent model to describe the pores within the shell, with the shapes (spherical, ellipsoidal, cylindrical etc.), dimensions, and their polydispersity to be varied. However, no such model can provide satisfactory fitting of the entire q range. Therefore, the assumption of a porous shell with isolated and well-defined closed pores can be excluded from further consideration. Next, we suggest that the structure of the microcapsule shell can be more accurately described as randomized interconnected network.

Two shape-independent models which are suitable for weakly-contrasted inhomogeneities at multiple length scales were applied to the SANS data of PSS/qPDMAEMA$_{18}$ microcapsules with different number of bilayers. The first one to be used is a power law or Porod model, which corresponds to a probed range smaller than the scattering object, so that the scattering is related to the local structure [33]. The scattering intensity can be expressed as $I(q) = A/q^n + B$, where a power law exponent n between 3 and 4 characterizes rough interfaces, which is called surface fractal, and the surface fractal dimension $D_f = 6 - n$. A power law exponent between 2 and 3 is for “mass fractals” such as branched systems (gels) or networks [34], the mass fractal dimension $D_m = n$. It can be seen that the power law model yields relatively good fitting for the scattering data over the entire q range for microcapsules with different number of bilayers (Fig. 11.4a), and the fractal dimensions obtained are summarized in Table 11.1.

Importantly, the fitting results from PSS/qPDMAEMA$_{18}$ microcapsules with different number of bilayers have obvious differences. The 5 and 8 bilayer microcapsules have power law exponents of 3.98 and 3.92, which strongly suggests surface fractal morphology, in other words, a thin shell network of pores and with rough surface. Measurements from other techniques such as AFM confirm that the 5 and 8 bilayers microcapsules have thin porous shells and rough surfaces, with a thickness of 12.8 and 16.1 nm in the dry state, respectively [32]. This result is consistent with SANS characterization of particles like microgels, where a power

Table 11.1 SANS fitting results for from $(PSS/qPDMAEMA_{18})_n$ microcapsules using power-law model and DAB model

Sample	Power-law model		DAB model
	Power law exponent	Fractal type	Correlation length (nm)
$(PSS/qPDMAEMA_{18})_5$	3.98	surface	31.4
$(PSS/qPDMAEMA_{18})_8$	3.92	surface	26.7
$(PSS/qPDMAEMA_{18})_8$ with salt	2.63	mass	13.5
$(PSS/qPDMAEMA_{18})_{11}$	2.84	mass	13.0
$(PSS/qPDMAEMA_{18})_{11}$ with salt	2.12	mass	9.0

The first index indicates the number of arms of the star and the second one the number of bilayers. The salt added is $K_3[Co(CN)_6]$

law exponent of 4 indicating smooth surfaces, and a power law exponent between 3 and 4 indicating a microporous system with rough surfaces [35].

In contrast, the shells with 11 bilayers have a fractal dimension of 2.84, which suggests a mass fractal structure. The shell structure corresponds to a network-like porous morphology with network elements randomly oriented within the shell. This combination is likely indicative of a denser shell with major elements of high contrast formed by the swollen hydrogenated polyelectrolyte matrix and the nanopores filled with deuterated water.

Such a transition from the surface fractal to the mass fractal structure with increasing shell thickness corresponds to general trends in morphological changes based on microscopic observations and expected for the growth of LbL structures [36, 37]. The gradual filling of the initial two-dimensional thin shell by subsequent polymer layers results in the formation of more uniform films with diminishing through-pores and decreasing pore dimensions. The occurrence of such a reorganization is further supported by the results from confocal microscopy and AFM [32], which demonstrate a densification of the surface morphology and a consistent decrease in the permeability.

Another model to calculate the scattering from a randomly distributed, two-phase system, the Debye-Anderson-Brumberger (DAB) model, was used to fit the SANS data of the LbL microcapsules. The two-phase system is characterized by a single correlation length, which is a measure of the average spacing between regions of phase 1 and phase 2 [38]. The model also assumes a smooth interface between the phases and hence exhibits Porod behavior at large q. The scattering intensity can be expressed as

$$I(q) = \frac{scale \cdot \xi^3}{(1 + (q\xi)^2)^2} + bck, \tag{11.5}$$

where ξ is the correlation length, bck is background [39]. The DAB model fits the SANS data quite well over the entire q-range (Fig. 11.4b) and provides correlation lengths which can be interpreted as the characteristic dimensions of density inhomogeneities represented by pores. The results are shown in Table 11.1. The correlation length for PSS/qPDMAEMA$_{18}$ microcapsules with 5, 8 and 11 bilayers are 31.4, 26.7 and 13.0 nm, respectively, which is in good agreement with the results estimated from permeability measurements [32].

As we discussed before, adding multivalent salt to the solution of qPDMAEMA star polyelectrolytes would induce the collapse of the arm chains. Taking advantage of this unique salt-responsive behavior, it is possible to achieve salt controlled permeability changes in the PSS/qPDMAEMA$_{18}$ microcapsules. Our previous study showed that by adding a small amount of $K_3[Co(CN)_6]$ salt to the microcapsule suspension, the permeability of the microcapsules can be dramatically reduced, which was proven by the permeability test with FITC-dextran with different molecular weights [32]. To get direct evidence of the structure and mesh size changes of the microcapsules after adding the multivalent salt, SANS was performed on the microcapsule solution before and after the addition, as shown in Fig. 11.5. The scattering intensity in the low q range significantly decreased after adding $K_3[Co(CN)_6]$ salt, which indicates the characteristic dimension which related to the structure of the shell decreases.

By fitting with the DAB model, the results in Table 11.1 show that after adding 0.8 mM trivalent salt, the correlation length ξ decreases from 26.7 to 13.5 nm for (PSS/qPDMAEMA$_{18}$)$_8$ microcapsules, and from 13.0 to 9.0 nm for (PSS/qPDMAEMA$_{18}$)$_{11}$ ones. Moreover, there is a surface- to mass-fractal transition upon adding salt for (PSS/qPDMAEMA$_{18}$)$_8$ microcapsules, and the mass fractal dimension also significantly decreases for (PSS/qPDMAEMA$_{18}$)$_{11}$ microcapsules. The decrease in fractal dimension is generally related to the increased aggregation and roughness in the local structure [40].

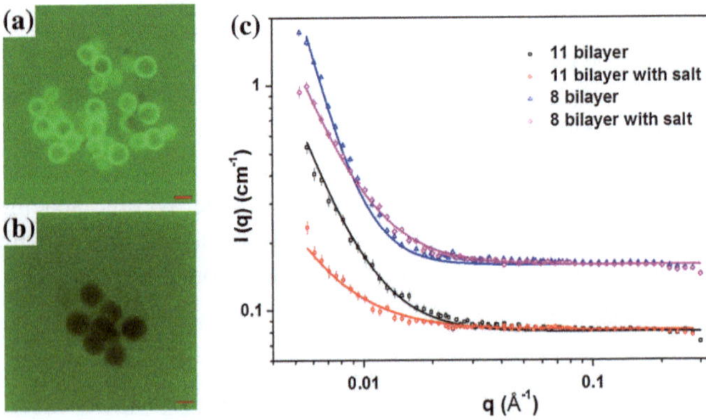

Fig. 11.5 Permeability of (PSS/qPDMAEMA$_{18}$)$_8$ microcapsules to 500 kDa FITC-dextran before (**a**) and after (**b**) adding 0.8 mM $K_3[Co(CN)_6]$ (Reprinted from Tsukruk et al. [32] by permission of the American Chemical Society.). **c** SANS data of (PSS/qPDMAEMA$_{18}$)$_8$ and (PSS/qPDMAEMA$_{18}$)$_{11}$ microcapsules before and after adding 0.8 mM $K_3[Co(CN)_6]$ (at 25 °C); *solids lines* are fitting by the power law model, the *two curves* for (PSS/qPDMAEMA$_{18}$)$_8$ were shifted upward for clarity

With the addition of multivalent salt, the residual, entrapped monovalent counterions are replaced by $[Co(CN)_6]^{3-}$; on average three I^- ions are replaced by one $[Co(CN)_6]^{3-}$, with ion exchange process controlled by Donnan effect [41]. Thus the osmotic pressure inside $qPDMAEMA_{18}$ is reduced by a factor of 3 and strong contraction of the arm or collapse of the stars occurs. On the other hand, due to the presence of $[Co(CN)_6]^{3-}$ ions between $qPDMAEMA_{18}$ stars, there is an attraction force between the star polyelectrolytes, which induces the decrease of intermolecular distance. As a result, the mesh size as well as the overall size of the microcapsule decreases. The results from SANS data analysis well supported the proposed mechanism of structural organization.

11.3.3 Temperature Responsive LbL Microcapsules

As discussed before, the PDMAEMA star polyelectrolytes are responsive to temperature, which also depends on the pH conditions [24]. Our previous SANS study [42] on the phase behavior of PDMAEMA stars in semi-dilute solution showed that at pH values close to the pK_a, all PDMAEMA stars studied here are partially charged and show a core-shell quasi-micellar morphology caused by microphase separation with the collapsed core region possessing high monomer density and the hydrated loose brush shell region. Upon increasing the temperature, the PDMAEMA star polyelectrolytes first experience a contraction in the loose shell region while the core size remains almost unchanged, and then start to form intermolecular aggregates within narrow temperature range. With decreasing pH value, the transition temperature increases and the size of the aggregates decreases. We suggest that these changes are triggered by the decrease in solvent quality with increasing temperature, which leads to the transition from an electrostatically dominated regime to a regime dominated by hydrophobic interactions. The observed phenomenon is in striking contrast with the behavior of linear PDMAEMA polyelectrolytes, which show macrophase separation with increasing temperature under the same conditions.

By applying LbL assembly of PDMAEMA stars and PSS on spherical substrates, hollow microcapsules can be fabricated after dissolving the core. The microcapsules are dual responsive to pH and temperature, as was demonstrated in our recent study [43]. Basically, the overall size and permeability of the microcapsules decrease with increasing temperature (with a shrinkage of 54 % in diameter at 60 °C as compared to the value at room temperature), thus allowing to reversibly load and unload the microcapsules with high efficiency. The organization and interaction of star polyelectrolytes within confined multilayers are the main driving forces for the responsiveness to external stimuli. In order to elucidate the detailed structural changes of $(PSS/PDMAEMA)_n$ microcapsules during external stimuli, we also conducted SANS measurements on the microcapsule solutions.

Figure 11.6a shows the SANS data from microcapsules with different number of bilayers. It can be seen that for thinner shells, the scattering curve has a

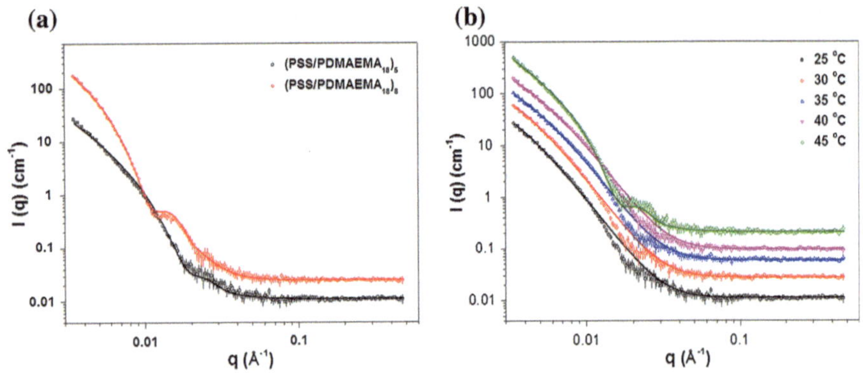

Fig. 11.6 **a** SANS data of $(PSS/PDMAEMA_{18})_n$ microcapsules with 5 and 8 bilayers at 25 °C and pH 7 condition, *solid curves* are fitting from lamellar model. **b** SANS curves of $(PSS/PDMAEMA_{18})_5$ microcapsules at pH 7 condition with increasing temperature, *solid curves* are fitting from DAB model (25–40 °C) and lamellar model (45 °C). The *curves* are mutually offset by a factor of 2 for better visualization

monotonically decreasing trend with increasing q, while for microcapsules with 8 bilayers, there are obviously characteristic humps in the q range from 0.01 to 0.05 $Å^{-1}$. The reason for such a difference is that the shell consisting of 5 bilayers is thin enough to be considered as a simple two phase system composed of a hydrogenated polyelectrolyte shell and D_2O inside. Fitting from the power law model gives a surface fractal dimension of 2.60, which corresponds to surface fractal structure; the DAB model fitting gives a correlation length of 35.6 nm (Fig. 11.6b), which also matches with the result from the permeability studies.

Microcapsules with a much thicker shell of 8 bilayers have complex hierarchical internal structures, which prevents a simple power law model or DAB model to give a satisfactory fitting. In fact, such characteristic humps in the middle q range are an indication of lamellar-like structures [44]. Although well-defined lamellar layering is probably not the accurate description of the shell structure, the result indicates the thicker shell may have reorganized to microphase separated internal structures.

The lamellar model provides the scattering intensity for a lamellar phase where a uniform scattering length density and random distribution in solution are assumed [45]. The scattering intensity is expressed as:

$$I(q) = 2\pi \frac{P(q)}{\delta q^2}, \tag{11.6}$$

and the form factor is

$$P(q) = \frac{2\Delta\rho^2}{q^2}(1 - \cos(q\delta)), \tag{11.7}$$

where δ is the lamellar thickness. Fitting by the lamellar model gives a thickness of 57.8 nm for the 8 bilayer shells and 33.0 nm for the 5 bilayer shell, which confirms the increase in shell thickness with increasing number of layers, the results also match well with thickness from AFM measurements.

The interdiffusion of polyelectrolyte chains as well as the increased thickness and roughness with bilayer number are the probable driving forces for the appearance of lamellar-like shell structure.

On the other hand, in situ SANS measurements were conducted for $(PSS/PDMAEMA_{18})_5$ microcapsules with increasing temperature, which are shown in Fig. 11.6b. It can be seen that from 25 to 40 °C, the overall shapes of the scattering curves are similar, but fitting from power law and DAB models is able to provide insightful information. Power law model fitting shows that the surface fractal dimension gradually increases from 2.60 (25 °C) to 2.63 (30 °C), 2.80 (35 °C) and 2.84 (40 °C), which means that the shell structure has a densification trend with temperature, although still in the surface fractal range. Accordingly, DAB model fitting shows that the correlation length decreases from 35.6 nm (25 °C) to 33.8 nm (30 °C), 28.2 nm (35 °C) and 26.3 (40 °C), which provides direct evidence about the permeability decrease of the microcapsule with increasing temperature.

Moreover, when the temperature further increases to 45 °C, the scattering curve undergoes a significant change in shape, which is similar to the thicker, lamellar-like structure as we discussed for the 8 bilayer microcapsules above. Fitting by the lamellar model for the 45 °C scattering curve gives a thickness of 38.3 nm. Such a transition provides another strong evidence that increasing temperature leads to densification of the shell, which has both a thickness increase and mesh size decrease.

With the increase of temperature, water becomes a bad solvent for PDMAEMA, the hydrogen bonding between PDMAEMA chains and water weakens, and the hydrophobic interaction increases [46], so that the arms of PDMAEMA stars shrink to a more collapsed conformation [47], which leads to changes in the structure and permeability of the microcapsule. The decreasing correlation length and increasing surface fractal dimension of the shell with temperature from SANS analysis is a strong proof to this hypothesis.

11.4 Conclusions

In conclusion, cationic star polyelectrolytes with responsiveness to pH, temperature or ionic conditions, were utilized to fabricate LbL microcapsules via electrostatic interactions. SANS measurements were conducted for the microcapsules with different composition and thickness. The results indicate that increasing thickness of the shell leads to a surface- to mass-fractal transition; the correlation length which represents the mesh size in the shell decreases. Taking advantage of the responsive properties of the branched polyelectrolytes, the microcapsules also exhibit responsiveness in shell structure and permeability to pH, temperature or ionic conditions. SANS experiments showed the detailed structural changes of the polyelectrolyte shell with the change of surrounding conditions. Adding multivalent ions leads to a decrease in correlation length of $(PSS/qPDMAEMA_{18})_n$ microcapsules of around 50 %. For the non-quaternized $(PSS/PDMAEMA_{18})_n$ microcapsules a temperature increase from 25 to 40 °C leads to an increase in surface fractal dimension from 2.60 to 2.84, and correlation length from 35.6 to 26.3 nm. This study is one of the first to reveal the internal structures of multilayer microcapsules and their evolution upon applying external stimuli, which is critical for their applications in drug delivery, as microreactors and self-healing materials. Nevertheless, there is still much work to be done in this direction, for example, the adsorption process of branched polyelectrolytes on spherical templates, and the kinetics of the structure reorganization in the multilayered thin films or shells.

Acknowledgments This work is supported by the NSF-DMR 1002810 grant. Research conducted at ORNL's High Flux Isotope Reactor and Spallation Neutron Source was sponsored by the Scientific User Facilities Division, Office of Basic Energy Sciences, US Department of Energy.

References

1. A.V. Dobrynin, M. Rubinstein, Prog. Polym. Sci. **30**, 1049 (2005)
2. G.D. Wignall, Y.B. Melnichenko, Rep. Prog. Phys. **68**, 1761 (2005)
3. S. Peleshanko, V.V. Tsukruk, Progr. Polym. Sci. **33**, 523 (2008)
4. S. Peleshanko, V.V. Tsukruk, J. Polym. Sci., Part B: Polym. Phys. **50**, 83 (2012)
5. Y. Xu, F.A. Plamper, M. Ballauff, A.H.E. Müller, Adv. Polym. Sci. **228**, 1 (2010)
6. O.V. Borisov, E.B. Zhulina, F.A.M. Leermakers, M. Ballauff, A.H.E. Müller, Adv. Polym. Sci. **241**, 1 (2011)
7. M.C. Stuart, W. Huck, J. Genzer, M. Müller, C. Ober, M. Stamm, G. Sukhorukov, I. Szleifer, V.V. Tsukruk, M. Urban, F. Winnik, S. Zauscher, I. Luzinov, S. Minko, Nat. Mater. **9**, 101 (2010)
8. F.A. Plamper, A.P. Gelissen, J. Timper, A. Wolf, A.B. Zezin, W. Richtering, H. Tenhu, U. Simon, J. Mayer, O.V. Borisov, D.V. Pergushov, Macromol. Rapid Commun. **34**, 855 (2013)
9. F.L. Verso, C.N. Likos, L. Reatto, Progr. Colloid Polym. Sci. **133**, 78 (2006)
10. C.Y. Shew, C. Do, K. Hong, Y. Liu, L. Porcar, G.S. Smith, W.R. Chen, J. Chem. Phys. **137**, 024907 (2012)
11. A. Jusufi, C.N. Likos, M. Ballauff, Colloid. Polym. Sci. **282**, 910 (2004)
12. M. Ballauff, M. Patel, S. Rosenfeldt, N. Dingenouts, T. Narayanan, A.H.E. Müller, F.A. Plamper, Polym. Mater. Sci. Eng. **93**, 232 (2005)
13. I. Choi, D.D. Kulkarni, W. Xu, C. Tsitsilianis, V.V. Tsukruk, Langmuir **29**, 9761 (2013)
14. B.S. Kim, H.F. Gao, A.A. Argun, K. Matyjaszewski, P.T. Hammond, Macromolecules **42**, 368 (2009)
15. I. Choi, S.T. Malak, W. Xu, W.T. Heller, C. Tsitsilianis, V.V. Tsukruk, Macromolecules **46**, 1425 (2013)
16. M. Vamvakaki, C.S. Patrickios, P. Lindner, M. Gradzielski, Langmuir **23**, 10433 (2007)
17. R. Blaak, C.N. Likos, J. Phys.: Condens. Matter **24**, 322101 (2012)
18. F.A. Plamper, A. Schmalz, E. Penott-Chang, M. Drechsler, A. Jusufi, M. Ballauff, A.H.E. Müller, Macromolecules **40**, 5689 (2007)
19. J.K. Zhao, C.Y. Gao, D. Liu, J. Appl. Crystallogr. **43**, 1068 (2010)

20. O. Arnold, J. C. Bilheux, J.M. Borreguero, A. Buts, S.I. Campbell, L. Chapon, M. Doucet, N. Draper, R. Ferraz Leal, M.A. Gigg, V.E. Lynch, A. Markvardsen, D.J. Mikkelson, R.L. Mikkelson, R. Miller, K. Palmen, P. Parker, G. Passos, T.G. Perring, P.F. Peterson, S. Ren, M. A. Reuter, A.T. Savici, J.W. Taylor, R.J. Taylor, R. Tolchenov, W. Zhou, J. Zikovsky, Nucl. Instrum. Methods Phys. Res. Sect. A **764**, 156 (2014)
21. G.D. Wignall, K.C. Littrell, W.T. Heller, Y.B. Melnichenko, K.M. Bailey, G.W. Lynn, D.A. Myles, V.S. Urban, M.V. Buchanan, D.L. Selby, P.D. Butler, J. Appl. Crystallogr. **45**, 990 (2012)
22. G.D. Wignall, F.S. Bates, J. Appl. Cryst. **20**, 28 (1987)
23. Z. Iatridi, C. Tsitsilianis, Polymers **3**, 1911 (2011)
24. F.A. Plamper, M. Ruppel, A. Schmalz, O. Borisov, M. Ballauff, A.H.E. Müller, Macromolecules **40**, 8361 (2007)
25. F.A. Plamper, A. Walther, A.H.E. Müller, M. Ballauff, Nano Lett. **7**, 167 (2007)
26. D. Wang, D. Moses, G. C. Bazan, A.J. Heeger, J. Lal, J. Macromol. Sci. A Pure Appl. Chem. **38**, 1175 (2001)
27. M.K. Crawford, R.J. Smallew, G. Cogen, B. Hogan, B. Wood, S.K. Kumar, Y.B. Melnichenko, L. He, W. Guise, B. Hammouoda, Phys. Rev. Lett. **110**, 196001 (2013)
28. H. Benoit, J. Polym. Sci. **11**, 507 (1953)
29. M. Laurati, J. Stellbrink, R. Lund, L. Willner, E. Zaccarelli, D. Richter, Phys. Rev. E **76**, 041503 (2007)
30. L. Willner, O. Jucknischke, D. Richter, J. Roovers, L.L. Zhou, P.M. Toporowski, L.J. Fetters, J.S. Huang, M.Y. Lin, N. Hadjichristidis, Macromolecules **27**, 3821 (1994)
31. H. Hsu, W. Paul, K. Binder, J. Chem. Phys. **137**, 174902 (2012)
32. W. Xu, I. Choi, F.A. Plamper, C.V. Synatschke, A.H.E. Müller, V.V. Tsukruk, ACS Nano **7**, 598 (2013)
33. T. Freltoft, J.K. Kjems, S.K. Sinha, Phys. Rev. B **33**, 269 (1986)
34. P.W. Schmidt, J. Appl. Cryst. **24**, 414 (1991)
35. K. Kratz, T. Hellweg, W. Eimer, Polymer **42**, 6631 (2001)
36. V.V. Tsukruk, V.N. Bliznyuk, D.W. Visser, A.L. Campbell, T. Bunning, W.W. Adams, Macromolecules **30**, 6615 (1997)
37. V.V. Tsukruk, Prog. Polym. Sci. **22**, 247 (1997)
38. P. Debye, A.M. Bueche, J. Appl. Phys. **20**, 518 (1949)
39. P. Debye, R. Anderson, H. Brumberger, J. Appl. Phys. **28**, 679 (1957)
40. S. Roldán-Vargas, R. Barnadas-Rodríguez, A. Martín-Molina, M. Quesada-Pérez, J. Estelrich, J. Callejas-Fernández, Phys. Rev. E **78**, 010902 (2008)
41. Y. Mei, K. Lauterbach, M. Hoffmann, O.V. Borisov, M. Ballauff, A. Jusufi, Phys. Rev. Lett. **97**, 158301 (2006)
42. W. Xu, I. Choi, F.A. Plamper, C.V. Synatschke, A.H.E. Müller, Y.B. Melnichenko, V.V. Tsukruk, Macromolecules **47**, 2112 (2014)
43. W. Xu, P.A. Ledin, F.A. Plamper, C.V. Synatschke, A.H.E. Müller, Y.B. Melnichenko, V.V. Tsukruk, Macromolecules **47**, 7858 (2014)
44. M. Bergström, J.S. Pedersen, P. Schurtenberger, S.U. Egelhaaf, J. Phys. Chem. B **103**, 9888 (1999)
45. F. Nallet, R. Laversanne, D. Roux, J. Phys. II **3**, 487 (1993)
46. T. Thavanesan, C. Herbert, F.A. Plamper, Langmuir **30**, 5609 (2014)
47. W. Yuan, H. Zou, W. Guo, A. Wang, J. Ren, J. Mater. Chem. **22**, 24783 (2012)

Chapter 12
Thermodynamic and Phase Behavior of Nanofluids

Sergey Artemenko, Victor Mazur and Olena Vasilieva

Abstract The importance of thermodynamic and phase behavior of working fluids embedded with nanostructured materials is fundamental to new nanotechnology applications. Considering the extremely large number of different both nanoparticle types and reference fluids, it is obvious that there is need for developing theoretically sound methods of the prompt estimation thermodynamic properties and phase equlibria for emerging working media. The effect of nanoparticles on the critical point shift for classical fluids doped by nanoparticles is examined. Global phase diagrams of two-component fluids with nanoparticles are analyzed. The global phase diagram studies of binary mixtures provide some basic ideas of how the required methods can be developed to visualize the phase behavior of nanofluid blends. The mapping of the global equilibrium surface in the parameter space of the equation of state (EoS) model provides the most comprehensive system of criteria for predicting binary mixture phase behavior. Results of calculations of phase equilibria for some nanofluids are described.

12.1 Introduction

Nanofluids (NF), i.e. fluids embedded with nanostructured materials, have recently became a subject of growing scientific interest due to reports of greatly enhanced thermal properties [1–4]. Key features of NF include thermal conductivity

---

S. Artemenko · V. Mazur (✉)
Institute of Refrigeration, Cryotechnologies, and Eco-Power Engineering (Former Academy of Refrigeration), Odessa National Academy of Food Technologies, 1/3, Dvoryanskaya Street, Odessa 65082, Ukraine
e-mail: victor.mazur@gmail.com

S. Artemenko
e-mail: sergey.artemenko@gmail.com

O. Vasilieva
Kiev National University of Trade and Economics, 19, Kyoto Street, Kiev 02156, Ukraine
e-mail: vasulieva35@mail.ru

L. Bulavin and N. Lebovka (eds.), *Physics of Liquid Matter: Modern Problems*, Springer Proceedings in Physics 171, DOI 10.1007/978-3-319-20875-6_12

exceeding those of conventional suspensions, a nonlinear relationship between thermophysical properties and concentration for NF containing carbon nanotubes, and a significant increase in critical heat flux in boiling heat transfer [5–7]. NF phenomena will allow to create a new class of efficient working and bring such benefits like energy efficiency (e.g., improving heat transfer, reducing pumping power), lower operating costs, smaller/lighter systems (small heat exchangers) and cleaner environment (e.g., reducing heat transfer fluid inventory) [3, 5]. The heat transfer in nanofluids the Thomson Reuters rating agency associates with "Research Fronts 2013" 100 top-ranked specialties in the sciences and social sciences [8].

The key values that define thermodynamic and phase behavior of fluids are the critical point for pure substance and critical lines for binary mixtures. The mapping of the global equilibrium surface on the critical parameter space of components provides the most comprehensive system of criteria for predicting of the binary mixture phase behavior. These singular properties of nanofluids have not been studied yet. At present time, neither experimental data nor theoretical assessments about changes in the singularities of conventional fluids after nanoparticle doping are available in literature. The presence of nanostructured materials should displace the singularity allocation and change the phase behavior of mixtures due to critical point shift of components.

The principal aim of this work is to study the effect of nanoparticle doping on thermodynamic and phase behavior of conventional fluids and their mixtures.

This work is organized as follows. In the first part, we study the influence of nanoparticle adding on critical point location in the one-component fluids. We suggest that regular and singular parts of thermodynamic surface of reference fluid and nanofluid with small nanoparticle volume concentration (<5 %) are coincided in reduced form. The shift of critical point for CO_2 doped with graphene genealogic tree nanoparticles (carbon nanotubes, fullerenes, grapheme flakes) and metal oxides (TiO_2) nanoparticles is theoretically predicted. In the second part, we consider possible changes in the phase behavior of two component fluids under nanoparticle doping. Analysis of phase behavior is based on global phase diagram of binary mixtures. These diagrams are not represented in pressure–temperature variables; instead, they are represented in the space of parameters of an equation of state. For example, in terms of the van der Waals equation of state the constants a and b directly related with the critical parameters of pure components. The critical parameters change for nanofluids leads to a change both combination rules and equation of state parameters for binary mixture. It can shift the position of characteristic point on the global phase diagram and change the type of phase behavior. It is demonstrated that carbon nanotube (CNT) doping can provide the transition from zeotropic to azeotropic state for R1234yf/R161 mixture. Some examples of the nanoparticle adding effect on liquid–vapor and liquid–liquid equilibria are given.

12.2 Influence of Nanoparticle Adding on Critical Point Location. One-Component Fluids

Critical point identifies thermodynamic behavior of pure substance in wide range of parameters of state. Adding of nanoparticles changes the intermolecular interactions between fluid components embedded with nanostructured materials and shifts phase equilibria in the nanofluids. Here we suggest the fluids with small impurities obey the corresponding state principle. It is hypothesized that the regular and singular parts of thermodynamic surface of base fluid and nanofluid with small nanoparticle volume concentration are coincided in reduced form. The compressibility factor (Z) of nanofluid is defined via scaled pure reference fluid properties

$$Z = Z(\rho_{Cnf}/\rho, T/T_{Cnf}), \tag{12.1}$$

where critical parameters of nanofluid (ρ_{Cnf}, T_{Cnf}) are calculated from available equation of state in vicinity of critical point. To estimate the critical parameters of reference fluid—nanoparticles system the fundamental equations of state in reduced form for industrial fluids [9] are used. To compute thermodynamic properties of nanofluids under investigation in the range 0–5 % volume concentrations of nanoparticles (*np*) the density of nanofluid (*nf*) calculated via reference fluid density (*rf*) by standard relation [10]:

$$\rho_{nf} = (1-\varphi)\rho_{rf} + \varphi\rho_{np}. \tag{12.2}$$

The search algorithm of nanofluid critical parameters is as follows. From the fundamental EoS of given substance the p–ρ–T data are generated in vicinity of critical point to establish the power law equations [9]

$$\frac{\rho_\sigma}{\rho_c} - 1 = N_1\left(1 - \frac{T_\sigma}{T_c}\right) \pm N_2\left(1 - \frac{T_\sigma}{T_c}\right)^{\beta}, \tag{12.3}$$

where ρ_σ is the saturation density for the liquid or the vapor; T_σ is the temperature along saturation curve; ρ_c, T_c are critical density and temperature; N_1, N_2, β are fitting parameters.

This equation is valid only in the critical region. The critical pressures for each fluid are determined from calculations with the equation of state at the critical temperature and density. It is assumed that (12.3) is valid both reference substance and nanofluids. The application of universality to dissimilar critical points is based on the isomorphism principle, which formulates the conditions for expressing the theoretical scaling fields through physical field variables. The N_1, N_2, β are calculated from the ρ–T data set for pure substances and then used to estimate nanofluid critical parameters at different nanoparticle concentrations. The algorithm accuracy is checked at limit $\varphi = 0$ to reproduce the critical point data for pure

Table 12.1 Critical temperature ($\Delta T_c = T_{cnf} - T_{cCO2}$) and density ($\Delta\rho_c = \rho_{cnf} - \rho_{cCO2}$) shifts in carbon dioxide embedded with different nanoparticles

Nanoparticle types	$\Delta\rho_c$, kg/m^3 at $\varphi = 1$ %	$\Delta\rho_c$, kg/m^3 at $\varphi = 5$ %	ΔTc, K at $\varphi = 1$ %	ΔTc, K at $\varphi = 5$ %
CNT	5.74	22.6	0.01	0.11
C_{60}	7.47	33.2	0.01	0.13
Graphene	12.47	55.7	0.02	0.19
SiO_2	12.93	58.0	0.02	0.20
TiO_2	22.99	107.1	0.05	0.42
ZnO	34.40	187.4	0.07	0.71
CuO	39.69	162.0	0.08	0.64

components. The results of our calculations reproduce the EoS data from [9, 11] within experimental accuracy of density measurements for given substance.

Here we consider as an example the critical point shift for CO_2 embedded with different types of nanoparticles: graphene genealogic tree (CNT, fullerenes, and graphene flakes) and some oxides (TiO_2, SiO_2, ZnO, CuO) which have bulk density ρ_{CNT} = 1330 kg/m^3, ρ_{C60} = 1650 kg/m^3, ρ_G = 2230 kg/m^3, ρ_{TiO2} = 3900 kg/m^3, ρ_{SiO2} = 2400 kg/m^3, ρ_{ZnO} = 5600 kg/m^3, ρ_{CuO} = 5400 kg/m^3, correspondingly.

Thermodynamic behavior for selected fluids near critical point is generated via the fundamental equations of state from [9]. The predicted critical temperature shifts under adding of different nanoparticles are presented in Table 12.1 and Figs. 12.1 and 12.2. The growth of volume nanoparticle concentration tends to increase slightly the CO_2 nanofluid critical temperature. Changes that are more significant observed for critical density (Figs. 12.3 and 12.4).

The critical parameters for nanofluids also give an opportunity to calculate their thermodynamic properties from the reduced EoS (12.1). The speed of sound is most effective estimator of thermodynamic surface description since includes main thermodynamic derivatives. In addition, speed of sound is related with different mechanisms of thermal conductivity enhancement in nanofluids. Phonon transport speed is related the sound speed as function of the compressibility and the density of the fluid. We have evaluated the speed of sound via equation of state by conventional thermodynamic relationships.

Figure 12.5 presents the calculation results for sound speed of carbon dioxide along isobar at different volume fractions of the graphene genealogic tree nanoparticles.

The change of thermodynamic properties due to the compressibility of the fluid with nanoparticles suspended is negligible in the low volume fraction limit. A significant growth of sound speed is appeared at the higher concentrations and corresponds to similar picture of sound speed increasing at pressure rising. The shift of critical point parameters can change the landscape of phase behavior in binary mixtures and requires more detailed analysis which unavailable in literature.

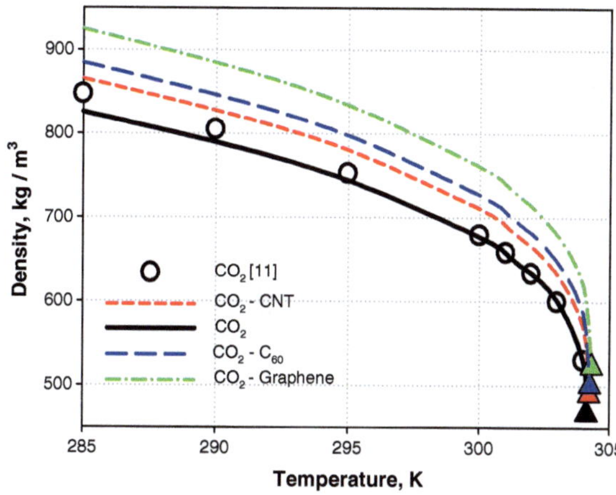

Fig. 12.1 Saturation curve and critical point CO_2 shift under graphene genealogic tree nanoparticle doping

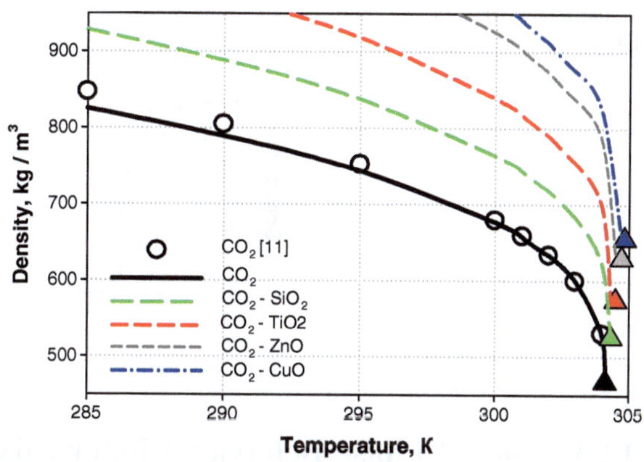

Fig. 12.2 Saturation curve and critical point CO_2 shift under different oxides nanoparticle doping

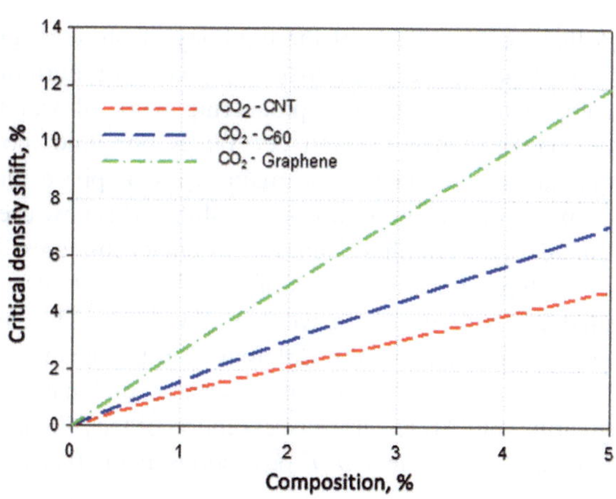

Fig. 12.3 Critical density shift for CO_2 under graphene genealogic tree nanoparticle doping

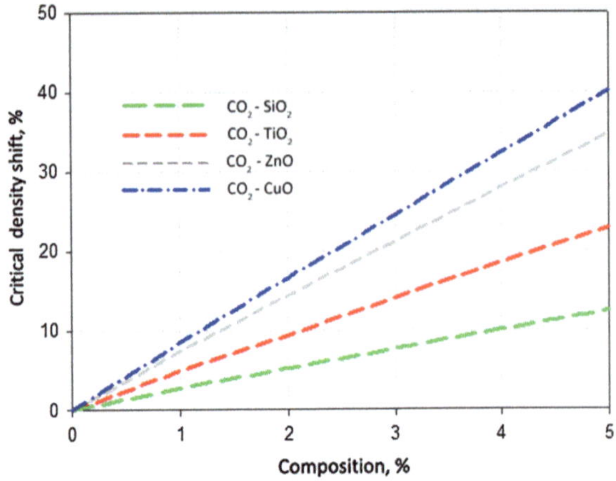

Fig. 12.4 Critical density shift for CO_2 under different oxides nanoparticle doping

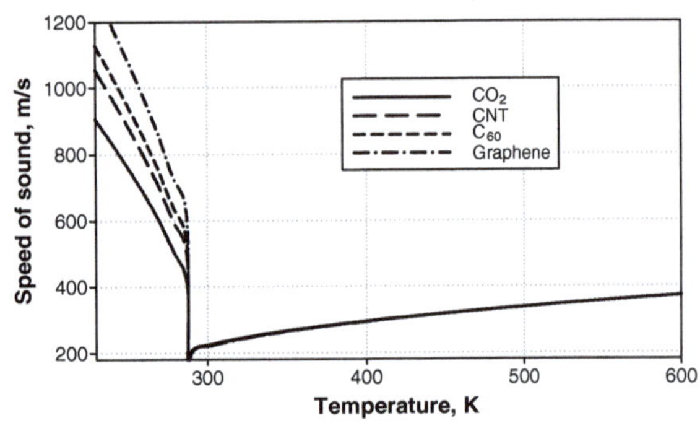

Fig. 12.5 Sound speed along isobar $p = 5$ MPa for CO_2 and CO_2 embedded with CNT, fullerenes, and graphene

12.3 Global Phase Behavior Binary Mixtures with Nanostructured Materials

A theoretical analysis of the topology of phase diagrams is a very useful tool for understanding the phenomena of phase equilibrium that are observed in multi-component systems. The pioneering work of van Konynenburg and Scott [12] demonstrated that the van der Waals one-fluid model has wide possibilities of qualitative reproducing the main types of phase diagrams of binary fluids. The proposed classification was successful, and is now used as a basis for describing the different types of phase behavior in binary mixtures. A more rigorous classification of the typical characteristics of equilibrium surfaces and phase diagrams of binary mixtures is given in the work of Varchenko [13], in which it is proved that the number of topologically different rearrangements is eight. At present, the topological analysis of equilibrium surfaces of binary fluid systems contains 26 singularities and 56 scenarios of evolution of the p–T diagrams [14]. The various phase diagrams classes and p–T projections main types of phase diagrams have been described in literature. Here for the reader's convenience we list the main types of phase behavior in binary mixtures.

12.3.1 Main Types of Phase Behavior in Binary Mixtures

The mapping of the global surface of a thermodynamic equilibrium onto the space of parameters of an equation of state gives the possibility to obtain the most extensive and sequential system of criteria for predicting the phase behavior of a binary mixtures. The effect of model parameters of mixture components on the topology of phase behavior is visualized using global phase diagrams. Conventional phase diagrams are a visual representation of the state of a substance as a function of temperature T, pressure p, and component concentration x. Therefore, they are used as a tool for visual analysis of the physical picture of the solubility phenomena. These variables are inherently different.

Pressure and temperature are the "field" variables that are the same for all phases coexisting in equilibrium. The molar fraction is the "density" that is in principle different for different phases. Global phase diagrams of binary mixtures represent boundaries between different types of phase behavior in a dimensionless space of equation of state parameters. For the first time, the idea of mapping the surface of phase equilibria onto the space of field variables, i.e., parameters of an equation of state, was proposed by van der Waals.

The boundaries of the global phase diagrams (tricritical points (TCPs), double critical end points (DCEPs), azeotropic line, etc.) divide the space of model parameters into the regions that correspond to the different types of phase behavior. The mapping of the global surface of a thermodynamic equilibrium onto the space of parameters of an equation of state is the most extensive and sequential system of criteria for predicting the phase behavior of a binary mixture. The types of phase behavior within the Van Konynenburg and Scott [12] classification scheme of interest are characterized as follows (Fig. 12.6).

- Type I: a single permanent critical line between $\mathbf{C_l}$ and $\mathbf{C_2}$;
- Type II: one critical line connecting $\mathbf{C_l}$ and $\mathbf{C_2}$, another line going from $\mathbf{C_m}$ to a critical endpoint;
- Type III: one critical line going from $\mathbf{C_l}$ to an upper critical endpoint, another line going from $\mathbf{C_2}$ to $\mathbf{C_m}$;
- Type III-H: a subclass of III having hetero-azeotropic three-phase curve.
- Type III-A: a subclass of III with a genuine positive azeotropic line.
- Type IV: one critical line going from $\mathbf{C_l}$ to an upper critical endpoint, a second critical line going from $\mathbf{C_2}$ to a lower critical endpoint, a third line going from $\mathbf{C_m}$ to an upper critical endpoint.
- Type V: similar to IV, but without the low temperature critical curve going to $\mathbf{C_m}$.
- Type V-A: a subclass of V with a genuine negative azeotropic line.

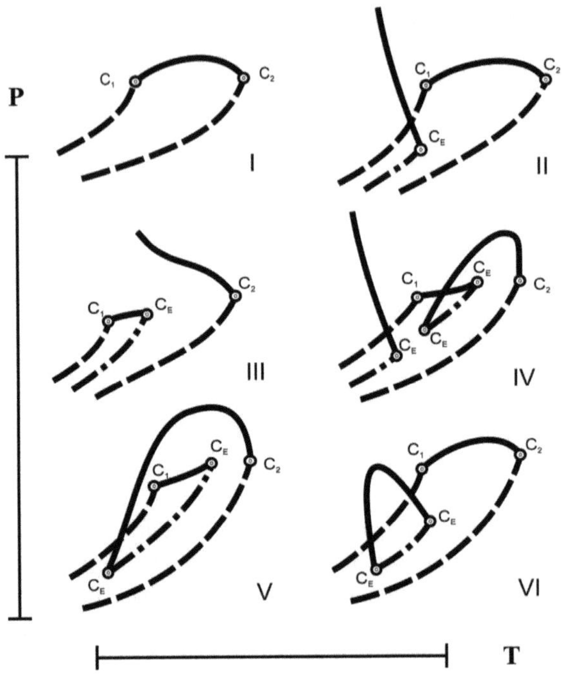

Fig. 12.6 Main types of phase behavior binary mixtures. Notations: *solid lines* are vapor pressure curves; *dashed lines* are critical curves; *dash-and-dot lines* are three-phase curves

- Type VI: involving closed-loop liquid-liquid immiscibility at low temperatures and practically impossible for supercritical conditions.

Here $\mathbf{C_1}$ and $\mathbf{C_2}$ are critical points of components 1 and 2; $\mathbf{C_m}$ is hypothetic critical point beyond solidification line.

12.3.2 Equation of State Model and Mixing Rules

To describe the thermodynamic properties and phase equilibria in the mixtures we use the one-fluid model of the Redlich-Kwong equation of state (RK EoS) [15, 16]. The RK EoS is used in its classical, non-modified, form:

$$p = \frac{RT}{(V-b)} - \frac{a}{T^{0.5}V(V-b)}, \tag{12.4}$$

where R is the universal gas constant.

The van der Waals co-volume parameter b as well as the attraction parameter a of the EoS (12.4) are given by mixing rules as function of the mole fractions x_i and x_j:

$$a = \sum_{i=1}^{2} \sum_{j=1}^{2} x_i x_j a_{ij}, b = \sum_{i=1}^{2} \sum_{j=1}^{2} x_i x_j b_{ij}. \tag{12.5}$$

The convenient set of dimensionless parameters for the Redlich-Kwong model is as follows [17]:

$$\begin{aligned} Z_1 &= \frac{d_{22} - d_{11}}{d_{22} + d_{11}}, \\ Z_2 &= \frac{d_{22} - 2d_{12} + d_{11}}{d_{22} + d_{11}}, \\ Z_3 &= \frac{b_{22} - b_{11}}{b_{22} + b_{11}}, \\ Z_4 &= \frac{b_{22} - 2b_{12} + b_{11}}{b_{22} + b_{11}}, \end{aligned} \tag{12.6}$$

where

$$d_{ij} = \frac{T_{ij}^* b_{ij}}{b_{ii} b_{jj}}, \ T_{ij}^* = \left(\frac{\Omega_b a_{ij}}{R \Omega_a b_{ij}}\right)^{2/3}, \ \Omega_a = \left[9\left(2^{1/3} - 1\right)\right]^{-1}, \ \Omega_b = \frac{2^{1/3}}{3}.$$

The combining rules for the binary interaction parameters are

$$a_{ij} = (1 - k_{ij})\sqrt{a_{ii} a_{jj}}, \ b_{ij} = (1 - l_{ij})\frac{b_{ii} + b_{jj}}{2}. \tag{12.7}$$

where k_{ij} and l_{ij} are fitting coefficients in the Lorentz-Berthelot combining rule ($k_{ij} = l_{ij} = 0$).

It should be brought to the attention that the dimensionless parameter Z_1 represents the difference of the critical pressures of the components, and that the dimensionless parameter Z_3 represents the difference of the critical volumes (co-volumes in EoS (12.4)). Therefore, there is a direct correlation of the global phase behaviour between mixtures and critical properties, i.e. geometry and energy parameters of real binary fluids.

Global phase diagrams for all realistic models have an extremely similar structure, particularly for the case of molecules of the same size. For example, the global phase diagrams of such different models as the Redlich–Kwong [17] or the Lennard-Jones binary fluid [18] are almost identical, including such sensitive phenomena as the presence of closed immiscibility regions. Accordingly, most of assumptions and conclusions based on the above-mentioned models of phase behavior can be transferred to other cases.

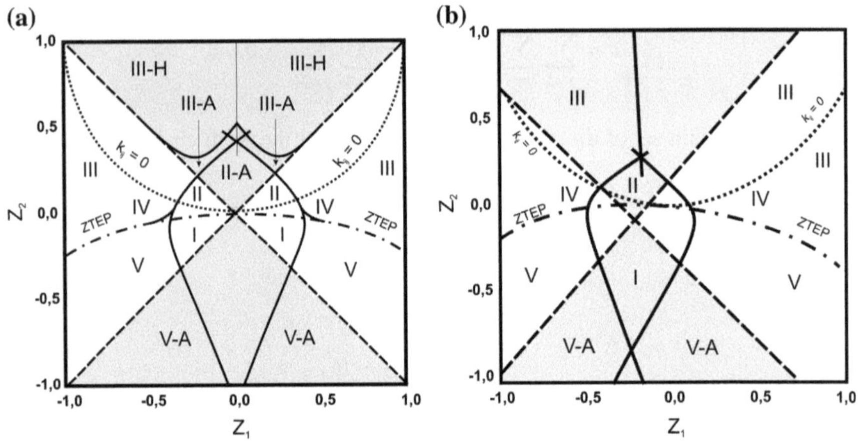

Fig. 12.7 Global phase diagram of the Redlich-Kwong model **a** equal size molecules ($Z_3 = 0$); **b** systems with nonadditive co-volumes ($Z_3 = 0.15$). The phase boundaries are represented by different lines: tricritical (*solid*), azeotropic (*dashed*), ZTEP (*dashed-dotted*) and the Lorentz–Berthelot combining rule (12.7) (*dotted*)

Figure 12.7 shows the global phase diagram for binary mixtures of equal-sized molecules, plotted in the two-dimensional (Z_1–Z_2) space. The tricritical points (TCP) visualize one of the most important boundaries. This boundary divides the classes I and V, II and IV, or III and IV. The tricritical state is a state, where the regions of the liquid-liquid-gas immiscibility shrink to one point, which is named the TCP. Three phases become identical at a TCP. Another important boundary in the global phase diagrams is the locus of double critical end-points (DCEP) that delimits types III and IV, or II and IV. Type IV is characterised by two liquid–liquid–gas curves. One is at high temperatures and is restricted by two critical end points [lower critical end point (LCEP) and an upper critical end point (UCEP)]. At the upper critical end point, the solution becomes immiscible as the temperature is lowered. At the lower critical end point, the solution separates into two phases as the temperature is increased. The DCEP occurs in a type IV when LCEP high-temperature three-phase region joins the UCEP of low-temperature three-phase region. The DCEP is produced in a type III system when the critical curve cuts tangentially the three-phase line in a pressure–temperature diagram. The types I and II, or IV and V, differ in the existence of a three-phase line, which goes from high pressures to an UCEP. For this case, the boundary situation is defined by the zero-temperature end point (ZTEP). An appearance of type VI involving closed-loop liquid-liquid immiscibility at low temperatures has topological origin rather than more sophisticated explanation via the association interactions in SAFT or CPA models. Thermodynamic expressions and mathematical tools are given in the literature [19, 20]. To predict the conventional phase diagrams the computational schemes of phase equilibria calculations were realized in MATLAB.

12.3.3 Boundary States Among Phase Diagram Classes

Our aim is to recognise a wide variety of phase diagrams from analysis of variations in geometry and energy characteristics (e.g., in critical density and critical temperature) of mixture components. The influence of these two parameters on the phase diagram topology could be conveniently visualised on the master diagram, called a "*global phase diagram*". Such a diagram shows the different areas of occurrence of the possible phase diagrams as a function of the geometry and energy factors of the compounds used. Ever since the work of van Konynenburg and Scott [12], numerous studies have been carried out on other, more realistic EoS [17, 18]. The boundaries, between the various types in the global phase diagram, can be calculated directly using the thermodynamic description of the boundary states (tri-critical line, double critical end-points, etc.). The dimensionless co-ordinates that are used for the representation of the boundary states depend on the equation of state model, but normally they are designed similar to those proposed by van Konynenburg and Scott for the van der Waals model [12]. In this case, the global phase diagrams of all realistic models have a very similar structure, in particular for the case of equal sized molecules.

The simplest boundary is a normal critical point when two fluid phases are becoming identical. Critical conditions are expressed in terms of the molar Gibbs energy derivatives in the following way:

$$\left(\frac{\partial^2 G}{\partial x^2}\right)_{p,T} = \left(\frac{\partial^3 G}{\partial x^3}\right)_{p,T} = 0. \tag{12.8}$$

Corresponding critical conditions for the composition—temperature—volume variables are:

$$\begin{aligned} &A_{xx} - WA_{xV} = 0, \\ &A_{xxx} - 3WA_{xxV} + 3W^2A_{xVV} - 3W^3A_{VVV} = 0, \end{aligned} \tag{12.9}$$

where A is the molar Helmholtz energy, $W = \frac{A_{xx}}{A_{VV}}$, $A_{mVnx} = \left(\frac{\partial^{n+m}A}{\partial x^n \partial V^m}\right)_T$ are the contracted notations for differentiation operation which can be solved for V_C and T_C at given concentration x.

At present time, Patel and Sunol [19] developed an automated and reliable procedure for systematic generation of global phase diagrams for binary systems. The approach utilizes equation of state, incorporates solid phase and is successful in generation of type VI phase diagram. The procedure enables automatic generation of GPD which incorporates calculations of all important landmarks such as critical endpoints (*CEP*), quadruple point (*QP*, if any), critical azeotropic points (*CAP*), azeotropic endpoints (*AEP*), pure azeotropic points (*PAP*), critical line, liquid–liquid–vapor line (*L*1*L*2 *V*, if any), solid–liquid–liquid line (*SL*1*L*2, if any), solid–liquid–vapor line (*SLV*) and azeotropic line. The proposed strategy is completely general in that it does not require any knowledge about the type of phase diagram and can be applied to any pressure explicit equation of state model. Recently, Cismondi and Michelsen [20] introduced a procedure to generate different type of phase diagrams classified by van Konynenburg and Scott. Their strategy does not take into account an existence of solid phase. Figure 12.7 shows the global phase diagrams for binary mixtures of equal-sized molecules (Fig. 12.7a) and systems with nonadditive co-volumes (Fig. 12.7b), plotted in the two-dimensional $(Z_1–Z_2)$ space.

Nanoparticle adding has different influence on the global phase diagram variables Z_i (12.6). Critical temperature variation of nanofluids is not significant to change the phase behavior types on global phase diagram. Hence, the cohesive energy density for reference fluid and nanofluid are approximately the same. Corresponding variables Z_1, Z_2 which define the type of phase behavior should remain without changes after nanoparticle doping. Effect of nanoparticles can appear due to changes in critical density and correspondingly in co-volumes b_{ii} for equation of state model, e.g. for the RK EoS (12.4). The Z_3 variation depends on different scenario of nanofluid preparing. If nanoparticles are added to the first pure component, the co-volume b_{11} is changed considerably together with variable Z_3. It can change the type of phase behavior as shown in Fig. 12.7b. An opportunity to change phase diagrams types allows considering the nanoparticle doping as a smart tool for phase equilibria control. One of these possible approaches is illustrated for zeotropic–azeotropic transitions.

12.4 Azeotropy in Binary Mixtures with Embedded Nanomaterials

The conditions of azeotropic state are

$$\mu_i^l = \mu_i^g, \quad i = 1, 2, \quad x_i^l = x_i^g. \tag{12.10}$$

Azeotropy in binary fluids can be easily predicted in the framework of global phase diagrams. The corresponding boundary state is called the degenerated critical azeotropic point (CAP) and represents the limit of the critical azeotropy at $x_i \to 0$ or at $x_i \to 1$. This results in solving the system of thermodynamic equations for a degenerated critical azeotrope. One may obtain the relationships for azeotropy boundaries from the global phase diagram (shaded **A**(Azeotropy) and **H** (Hetero-azeotropy)) regions in Fig. 12.7. The above azeotropic borders are straight lines in the (Z_1, Z_2)-plane that cross at a single point near the centre for equal sized molecules. It opens the opportunity for obtaining the series of inequalities to separate azeotropic and non-azeotropic regions of the global phase diagram. For the Redlich–Kwong EoS a corresponding relationship was obtained in the analytical form [21]:

$$Z_2 = \mp Z_1 - 0.67(1 \pm Z_1)\left(\frac{1 - Z_4}{1 \pm Z_3} - 1\right). \tag{12.11}$$

Global phase diagrams of binary fluids represent the boundaries between different types of phase behaviour in a dimensionless parameter space. In a real p–T–x space, two relatively similar components usually have an uninterrupted critical curve between the two critical points of the pure components.

Here we consider phase behavior of the R1234yf–R161zeotropic blend and R1234yf–R161–Fe_3O_4 nanoparticles as most likely azeotropic system that are recognized as low global warming potential (GWP) refrigerant to replace the R134a refrigerant. The equation of state parameter for low-boiling component R1234yf were taken from [22, 23]: $T_C = 367.85$ K, $p_C = 3.382$ MPa and $\omega = 0.280$. For the R161 equation of state parameters, corresponding values are as follows: $T_C = 375.35$K, $p_C = 4.7$ MPa and $\omega = 0.210$. The binary interaction parameters were fitted by the Lorentz-Berthelot combination rule ($k_{ij} = l_{ij} = 0$). The results of phase equilibria calculations for different temperatures in the pressure–composition and pressure–temperature diagrams are shown in Figs. 12.8 and 12.9. The R1234yf doped with the Fe_3O_4 nanoparticles increases the critical temperature of pure low-boiling component until 371 K. This shift conduces to transformation from zeotropic state (Fig. 12.8) to azeotropic state (Fig. 12.10).

The topological predictions based on a global phase diagram will become a convenient method in the analysis of nanofluid mixtures of scientific and industrial interest. Topologically, there is no difference in the isoproperty behaviour for any pure fluid. This fact allows us to find the parameters of the equation of state model, which can reproduce thermodynamic properties of an arbitrary substance in a local region of a phase diagram.

If combination rules are known, then it is possible to determine the global phase diagram via the ratio of critical parameters of pure components only.

The traditional classification of fluid phase behavior can easily be discussed with the aid of the p–T projections of fluid phase diagrams. There are two kinds of phase

Fig. 12.8 p–x,y diagram of the R1234yf–R161 binary blend at different temperatures

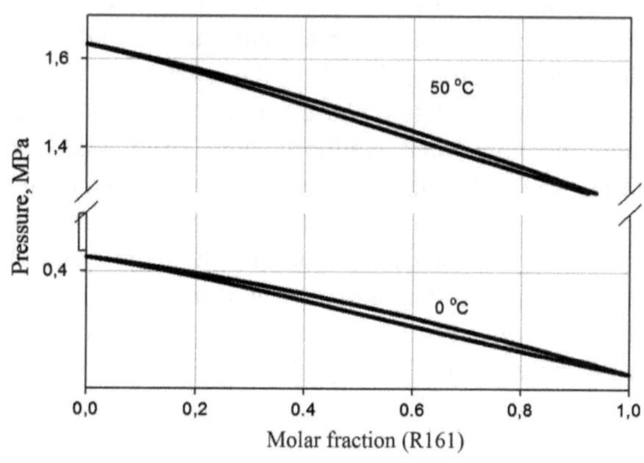

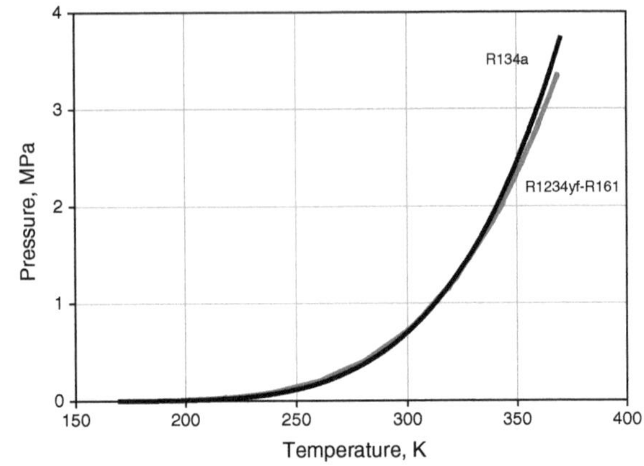

Fig. 12.9 *p–T* diagram of the R134a and the R1234yf–R161 binary blend

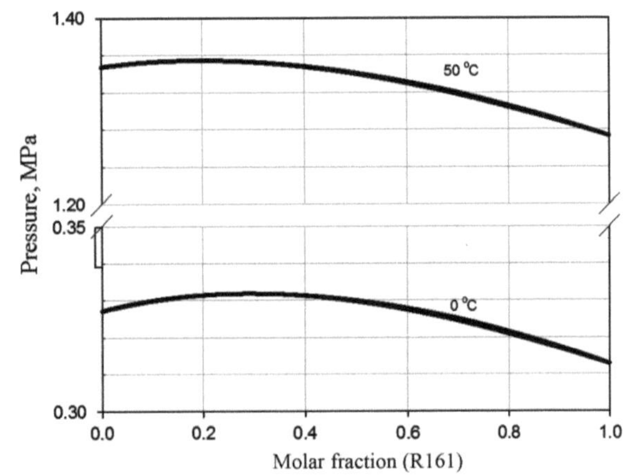

Fig. 12.10 *p–x,y* diagram of the R1234yf–R161–Fe_3O_4 nanoparticles blend at different temperatures

diagrams. Phase diagrams of types I, V, and VI have the vapor pressure curves that are started and ended in nonvariant points with equilibria where the no solid phase exist. In the case of types II, III and IV, some critical curves, starting in high-temperature nonvariant points, are not ended by the nonvariant points from the lower temperature side, where the solid phase should exist. A solid phase is absent in calculations of fluid phase diagrams using previously discussed equations of state and the nonvariant equilibria with solid could not be obtained even at 0 K. Therefore, the monovariant curves remain incomplete on the theoretical *p–T* projections. As a result, these diagrams can be considered as the '*derivative*' *versions.* It demonstrates not only the main types of fluid phase behavior but also the fluid phase diagrams that appear when the heterogeneous fluid equilibria are bounded not only by another fluid equilibrium but also by the equilibrium with solid phase that is usually observed in the most real systems.

12.5 Influence of Nanoparticles on the Shift of Liquid–Liquid Equilibria

Influence of nanoparticles on the shift of liquid–liquid equilibria we have studied for the liquid–liquid coexistence curve of the binary fluids nitrobenzene-heptane and nitrobenzene–heptane–TiO_2 nanoparticles. Fitting coefficients k_{ij} and l_{ij} have been restored from experimental data Borzenkov, Zhelezny [24]. The RK or other cubic EoS belong to the mean field models that cannot simultaneously describe experimental data near critical point and at low temperatures. Here we preferred the more exact description of near-critical area. Coefficients k_{ij} (l_{ij}) are 0.01824 (0.01392) for nitrobenzene-heptane mixture and 0.01794 (0.0148) for nitrobenzene–heptane–TiO_2, correspondingly. Deviations of mean field model from experimental data for binary mixtures of interest are shown in Fig. 12.11. The changes of coefficients k_{ij} and l_{ij} displace the liquid–liquid coexistence curve of binary mixture and location of upper critical end point. Experimental data treatment show that upper critical end point for the liquid–liquid coexistence curve of the binary fluids nitrobenzene–heptane and nitrobenzene–heptane–nanoparticles TiO_2 are as follows: T_{UCEP} = 292.998 K, x_{UCEP} = 0.519 and T_{UCEP} = 292.46 K,

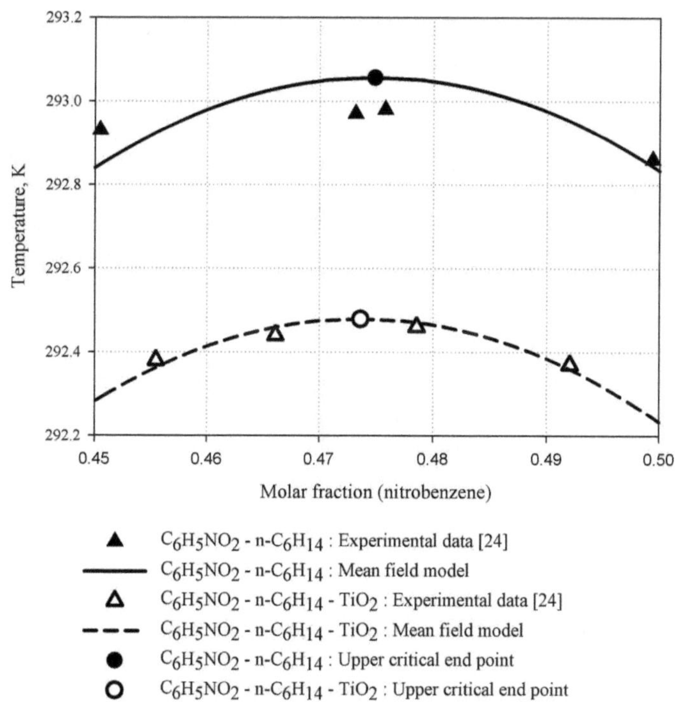

Fig. 12.11 Liquid–liquid coexistence curves of the binary fluids nitrobenzene–heptane and nitrobenzene–heptane–nanoparticles TiO_2

$x_{UCEP} = 0.531$, correspondingly. There is a small divergence between thermodynamic data Borzenkov, Zhelezny [24] and Shelton, Balzarini [25], which measured the liquid–liquid coexistence curve by interferometric means.

The value for the critical exponent β is higher than other liquid–liquid equilibria systems. The value 0.3328 is near classic value 1/3 for pure components. Recent evidence indicated that critical exponent β for liquid–liquid equilibria is lower than for pure substance. Nanoparticle adding decreases value of critical exponent ($\beta = 0.3252$) in comparison with liquid–liquid equilibria value.

12.6 Conclusion

This study is one of the first attempts to establish and demonstrate multiple links existing between the critical point shift in classical fluids and phase equilibria phenomena in mixtures embedded with nanostructured materials. From the very beginning of these efforts, the obtained results serve very useful information for scientists and engineers working in the field of emerging nanotechnology applications. The examples the critical point shift for CO_2 with different types of nanoparticle doping: graphene genealogic tree (CNT, fullerenes, and graphene flakes) and some oxides (TiO_2, SiO_2, ZnO, CuO) are given. A wide variety of phase diagrams from analysis of variations in geometry and energy characteristics of mixture components are analyzed via global phase diagrams. As illustration, phase behavior of systems R1234yf–R161 and R1234yf–R161–nanoparticles that are recognized as low GWP refrigerants has been studied. Computer modeling has shown a possible azeotropy appearance in the zeotropic blend with nanoparticle doping that can pretend to be replacement of conventional refrigerant R134a. Impact of nanoparticles on the shift of liquid–liquid equilibria are discussed for the liquid–liquid coexistence curve of the binary fluids nitrobenzene-heptane and nitrobenzene–heptane–TiO_2 nanoparticles. We have shown that the presence of nanoparticles in a binary critical mixture of two liquids changes the location of upper critical end point.

There is no doubt that extension of our knowledge about thermodynamic and phase behavior of nanofluids will lead to the creation of reliable engineering recipes for solving the actual problems of nanotechnologies.

References

1. S. Choi, Z. Zhang, W. Yu, F. Lockwood, E. Grulke, Appl. Phys. Lett. **79**, 2252 (2001)
2. J. Eastman, S. Choi, S. Li, W. Yu, L. Thompson, Appl. Phys. Lett. **78**, 718 (2001)
3. R. Saidura, K. Leong, H. Mohammad, Renew. Sustain. Energy Rev. **15**, 1646 (2011)
4. W. Daungthongsuk, S. Wongwises, Renew. Sustain. Energy Rev. **11**(5), 797 (2007)
5. P. Keblinski, J. Eastman, D. Cahill, Mater. Today **8**(6), 36 (2005)
6. S. Das, S. Choi, H. Patel, Heat Transfer Eng. **27**(10), 3 (2006)

7. S. Murshed, K. Leong, C. Yang, Appl. Therm. Eng. **28**(17–18), 2109 (2008)
8. C. King, D. Pendlebury, WEB of knowledge—research fronts. http://sciencewatch.com/sites/sw/files/sw-article/media/research-fronts-2013.pdf (2013)
9. E. Lemmon, R. Span, J. Chem. Eng. Data **51**, 785 (2006)
10. J. Avsec, M. Oblak, Int. J. Heat Mass Transfer **50**(21–22), 4331 (2007)
11. R. Span, W. Wagner, J. Phys. Chem. Ref. Data **25**(6), 1509 (1996)
12. P. van Konynenburg, R. Scott, Philos. Trans. R. Soc. Lond. Ser. A **298**, 495 (1980)
13. A. Varchenko, J. Sov. Math **52**(4), 305 (1990)
14. F. Aicardi, P. Valentin, E. Ferrand, Phys. Chem. Chem. Phys. **4**, 884 (2002)
15. O. Redlich, J. Kwong, Chem. Rev. **44**, 233 (1949)
16. G. Soave, Chem. Eng. Sci. **27**, 1197 (1972)
17. V. Mazur, L. Boshkov, V. Murakhovsky, Phys. Lett. **104A**, 415 (1984)
18. U. Deiters, J. Pegg, Journ. Chem. Phys. **90**, 6632 (1989)
19. K. Patel, A. Sunol, Comput. Chem. Eng. **3**, 1793 (2009)
20. M. Cismondi, M. Michelsen, J. of Supercritical Fluids **39**, 287 (2007)
21. S. Artemenko, V. Mazur, Int. J. Refrig. **30**, 831 (2007)
22. M. Richter, M. McLinden, E. Lemmon, J. Chem. Eng. Data **56**, 3254 (2011)
23. J. Calm, G. Hourahan, Eng. Syst. **18**(11), 74 (2001)
24. P. Borzenkov, V. Zheleznyj, Refrig. Eng. Technol. **152**, 4–9 (2014)
25. J. Shelton, D. Balzarini, Can. J. Phys. **59**, 334 (1981)

Part IV
Medical and Biological Aspects

Chapter 13
RNA Nanostructures in Physiological Solutions: Multiscale Modeling and Applications

Shyam Badu, Roderick Melnik and Sanjay Prabhakar

Abstract In this review chapter we focus on the nucleic acid nanotechnology research and its application in the biomedical field. We also describe some of our most recent results on the modeling of ribonucleic acid (RNA) nanotubes and their characteristics in physiological solutions. This includes the properties that can be characterised by root mean square deviation (RMSD), radius of gyration and radial distribution function (RDF) for the RNA nanoclusters, paying special attention to RNA nanotubes. We describe the distribution of $^{23}Na^{+}$ and $^{35}Cl^{-}$ ions around the tube as a function of time within a distance of 5 Å from the surface of the tube. The results obtained from our computational studies are compared with available experimental results in the literature. The current developments in the coarse grain modeling of the RNA nanoclusters and other biomolecules are also highlighted.

13.1 Introduction

Extensive studies have been done on nucleic acid nanotechnology research in the biomedical applications. Among others, we recall that the bacteriophage ϕ 29 motor has been constructed to package the deoxyribonucleic acid (DNA) and the X ray crystallography has been used to determine the process of DNA packaging [1]. To construct any kind of nanostructure involving biomolecules the self assembly of small building blocks is very important. One of the notable experimental works has been the development of the crystalline bacterial cell surface using the streptavidin protein [2] which has the tendency of self assembling to construct the building blocks for the nanoclusters that can be used for biomedical applications. The self

S. Badu (✉) · R. Melnik (✉) · S. Prabhakar
MS2 Discovery Interdisciplinary Research Institute, M2NeT Laboratory,
Wilfrid Laurier University, Waterloo, ON N2L 3C5, Canada
e-mail: sbadu@wlu.ca

R. Melnik
e-mail: rmelnik@wlu.ca

L. Bulavin and N. Lebovka (eds.), *Physics of Liquid Matter: Modern Problems*,
Springer Proceedings in Physics 171, DOI 10.1007/978-3-319-20875-6_13

assembly of alkylated peptides can assist in creating the nanobelts in solutions which can be used for various therapeutic applications [3]. Due to the importance of therapeutic applications several kinds of polygon-shaped self assemblies have been developed by using the deoxyribonucleic acid [4–7].

Furthermore, the self assembly of biomolecules, including DNA [8–14], can be used for various bionanodevices in nanobiotechnology. The DNA nanotubes can be used for the alignment of the membrane protein, in order to take the NMR spectrum to determine its crystal structure [15–17]. It has been shown that the stability of the RNA assemblies is higher than that of the DNA self assembled nanoparticles in solutions [18–23]. DNA relaxation under internal viscosity was studied for DNA [24, 25] but no similar study is available for RNA. The thermodynamic stability of any system is determined by calculating its free energy. The system with smaller free energy is more stable, i.e., the RNA has smaller free energy in solutions compared to DNA nanoparticles. The differently shaped structures of the RNA molecules have been formed from the RNA building blocks as well as from their complexes with other biomolecules [26–31].

In the current age of scientific discoveries in nanobiotechnology, the RNA plays a vital role in drug delivery applications. The use of this important molecular system as a drug delivery object to the human body is due to its flexibility in structure. One of the most important building blocks of RNA nanoclusters is the RNAI/II complex. The RNAI and RNAII are defined as the sense and anti sense plasmids that control the replication of COLE1 plasmid [32, 33]. COLE1 is a DNA molecule separated from chromosomal DNA that is found in the cell of bacteria. The sequence for the RNAI is (GGCAACGGAUGGUUCGUUGCC) and the sequence for the RNAII is (GCACCGAACCAUCCGGUGC) [34]. The schematic diagram for the RNAI/II complex is shown is Fig. 13.1. Using these building blocks, the experimental work has been performed to see the varieties of self assemblies of this complex. Experimentally it has been found that there are varieties of self assemblies formed by the end of the experiment. Notably, it has been found that the hexagonal ring is one of the most abundant self assemblies [35] formed during the experiment performed at the solution phase. The formation of the hexagonal nanoring from the pRNA strand has been studied experimentally, and has

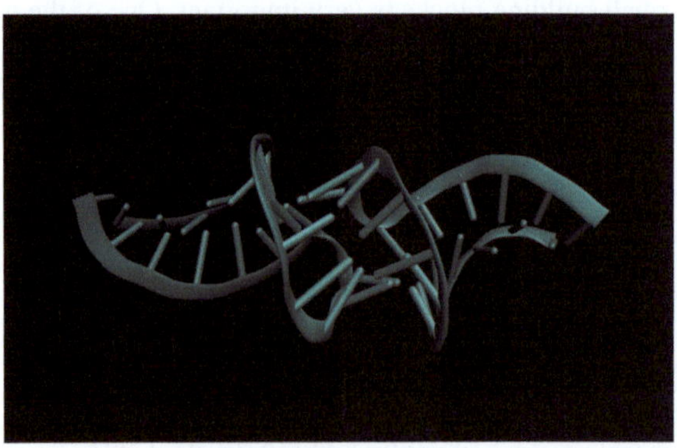

Fig. 13.1 Formation of the RNAI/II complex via base pairing between two segments

been later verified theoretically. During the experiment the RNAI/RNAII complex was put into the vessel and let molecules to assemble following the specific experimental protocol. The system was heated at first to the 95 °C for 2 min, then cooled to 4 °C and then warmed to 30 °C and supplied to the buffer solution. Then the system is supplied to the polycrystalamide gel experiment (PAGE) and TGGE experiment using a specific experimental setup which is described in [35].

It has been found that the majority of the assemblies are hexagonal rings with very few assemblies with square and octahedral structures. The low abundance of the tetramer and the pentamer assemblies of the RNAI/II complexes is due to the fact that they are less thermostable. Similar situation has been found in the case of higher order species like septamers and octamers. This result is found to be similar to the results obtained in [36]. An increasing interest to the field of RNA nanoclusters is due to their potential use in the drug delivery, nanodesign, therapy, among other fields. The ribonucleic acid is the polymer containing four kind of nucleobases; adenine, uracil, cytosine and guanine. These four kinds of the nucleobases are connected via sugar ring and the phosphate backbone to form the long chain polymer. A combination of the sugar ring and the nucleobase is known as the nucleotide. The size of the RNA strand is defined by the number of the nucleotides present in the polymer which eventually gives the number of possible structures that can be made from the given set of the nucleotides. The difference between the RNA and the DNA is that the uracil will be replaced by the thymine in DNA. In their double strand structure the stability is well described on the basis of the base pairing between them.

The development of bionanotechnology has facilitated the varieties of the techniques to detect and diagnose the cancer and other diseases. The use of bionanoparticles consisting of the RNA is due to their small size which let them to have assess to most of the parts of the body to interact with the infected or damaged cells. Ultimately, the models for RNA should be coupled with nonlinear dynamics models of cells [37]. In order to make the modeling of RNA nanoclusters successful, it is very important to have the suitable RNA building blocks. The self assemblies of RNA nanoclusters are performed in two ways. One is templated and the other is non-templated. If the self assemblies of RNA building blocks are done by the interaction between them from the external influence then this kind of self assemblies are known as templated, whereas if there is no external influence, during the interaction between the building blocks, then that kind of self assemblies are known as the non-templated self assemblies. Furthermore, the building blocks of nanorings are engineered in such a way that the RNAI and RNAII ends are complementary to each other. The complexation of these two RNA fragments via sticky ends is an important feature for constructing nanoclusters like RNA nanorings and nanotubes. In short, by using six helical building blocks of either one or two types (RNAI/RNAII) the nanoring is formed by self-assembling them via base pairing hydrogen bonds. The structure of the nanoring including the links that are used to form the nanotube is presented in Fig. 13.2. The stability of the nanoring depends on RNAI/RNAII interactions. The design of the sticking ends helps to assemble the nanorings to build the nanotubes. The starting structures of the RNAIi/RNAIIi

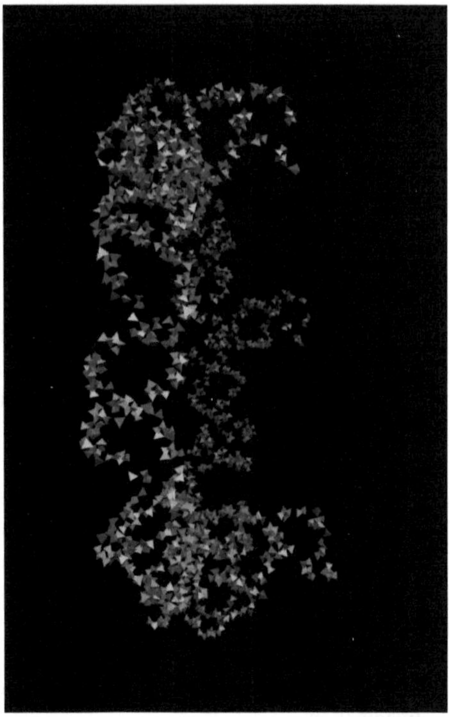

Fig. 13.2 RNA nanoring including three tails to link them to form the RNA nanotube

complex are taken from the protein data bank with the pdb code (2bj2.pdb) [34]. Recently, RNA has been self-assembled to build nanoscale scaffolds [38] using computational techniques and experiments. In the literature it is reported that several forms of RNA motifs can be constructed to provide a proper multifunctional RNA nanocluster; however only a few of them are found to be useful for drug delivery [39]. Assembling should be used to build the RNA nanocluster from the RNA building blocks [40].

RNA has been used to build the higher order self assembly in vitro [41] and in vivo [42] by assembling the multidimensional RNA structures. It has been used for the bacterial metabolism. RNA nanotechnology research is still in progress to achieve complete benefit from it. Furthermore, the RNA molecules can survive at low pH values that makes RNA to be compatible for the drug delivery and therapy in vivo. These RNA molecules can produce the self assemblies in vivo. They are transcribable using the DNA as a template [43–50]. The important factor that needs to be considered during the drug delivery process is the toxicity and the safety issues related to nanomaterials used for this process [51, 52].

There are two possible ways to deliver therapeutic drugs into the human body. Firstly, it can be done by directly including it into the RNA building blocks. Secondly, it can be done by attaching the drug at some particular ends of the RNA nanotube. The RNA nanotube can be useful for many applications, in particular for the delivery of drugs into the human body due to its stable condition at all temperatures as experiments show [53]. The target delivery vehicle, i.e. the RNA nanoclusters built for the drug delivery, should be stable, so that the modeling of such kinds of structures is critical. We note that the hairpin-like structure of the nanocluster has already been modeled [54] from RNA interference polymers. The discovery of a small interfering RNA (siRNA) [55, 56] is one of the most important research achievements in this field. The siRNA is basically a synthetic double stranded RNA with 21 basepairs. The function of siRNA is to suppress the problematic genes by RNA interference [57]. It also can be used for safe and efficient delivery of siRNA to cells. The use of the small interfering RNA is somewhat incomplete until its safe way of the delivery to the human body is determined. For this purpose the DNA packaging of bacteriophase ϕ 29 has been modified and used to package the siRNA for safe delivery to the human body [58, 59]. The delivery of oligonucleotides has also been studied to understand the effectiveness of models of the cancer therapy in humans [60]. The RNA molecules of the size with 18–30 nucleotides are also important in regulating the gene expression in the cytoplasm and nucleus [58, 61–63].

Previously, our group [64, 65] studied the mechanical and thermodynamical properties of RNA nanorings [36] using the molecular dynamics technique, and such studies on the RNA nanotubes have also been under way. The results obtained for the nanoring were later supported by experimental results using biochemical and biophysical techniques [35]. Specifically, the issues addressed in the previous papers have been the stability of the nanoring versus temperature, effect of the environment (i.e. solvent and counteractions) on its stability, as well as the conformations and dynamics under external forces. Some anomalous behaviour has been observed with the variation of temperature of the simulation box containing the nanoring. In a recent study, the properties of human immunodeficiency virus on hairpin-like subtype-A and subtype-B at different salt concentrations and magnesium bindings have been explored using molecular dynamics techniques [66]. Also an experimental study of the concentration dependence of NaCl and KCl on the free energy of RNA hairpin folding has been done [67]. Such studies provide additional motivations to do the molecular dynamics simulations of RNA nanoclusters. Furthermore, the first coarse-grained model for RNA nanorings has been developed [68] by utilizing the molecular dynamics simulations method.

13.2 Molecular Dynamics Simulation of Biomolecules

Molecular dynamics simulation involving biomolecules has become very important to understand the dynamics of biological systems. Specifically, the molecular dynamics simulation gives the idea about the dynamic behaviour of the biomolecules in physiological solutions, it gives the average of the thermal properties of the biomolecules under study and prediction of the thermally compatible conformation of the molecules [69, 70]. Theoretical basis of the molecular dynamics lies with statistical thermodynamics models and solving the Newton's equations of motion of the many particle system involving atoms. Recently, several molecular dynamics studies have been done on the RNA molecules and their derivatives [71–73]. Some of the most important structures derived from RNA molecules are the RNA hairpin loops for which the molecular dynamics simulation has been performed to calculate potential of mean force as a function of distance between two ends of the loop. The stability of the hairpin loops has been calculated and compared to the experimental results which were found to be in close agreement [73]. Furthermore, the molecular dynamics simulation has been performed on the viral RNA dependent polymers to understand their function and structure [72]. In our group the molecular dynamics simulation has been done on the varieties of the RNA nanoclusters. Some of the most recent results related to the RNA nanoclusters [64, 65] will be briefly presented in the following sections.

13.3 Multiscale Modeling

In multiscale modeling calculation of the properties of the system at one level is done using the models from a different level. There are a number of multiscale modeling methodologies used in applications. Here we describe three of them, typical for biological applications [74–77], namely:

1. The Boltzmann inversion method
2. The force matching method for developing the coarse-grained modeling
3. Multiscale coupling method for direct transfer of the information from mesoscopic and atomic scales during the simulation.

The first method commonly used in the multiscale modeling is the Boltzmann inversion method which has also been used in the coarse-grained modeling of the RNA nanoclusters. This will be discussed in details in the following section.

In the force matching process [78], the objective function, depending upon the parameter α, is defined follows

$$Z(\alpha) = Z_F(\alpha) + Z_c(\alpha), \tag{13.1}$$

$$Z_F(\alpha) = \left(3\sum_{k=1}^{M} N_k\right)^{-1} \sum_{k=1}^{M}\sum_{i=1}^{N_k} |F_{ki}(\alpha) - F_{ki}^0|^2, \tag{13.2}$$

$$Z_C(\alpha) = \sum_{r=1}^{N_C} W_r |A_r(\alpha) - A_r^0|^2. \tag{13.3}$$

The integer M in Z_F (the force objective function) is the number of configurations, N_k is the number of atoms in the kth configuration and $F_{ki}(\alpha)$ is the force on the ith atom in the kth configuration which is obtained from the parametrization of α, and the F_{ki}^0 is the corresponding reference force obtained from the first principles calculations. In the constraint objective function Z_C the quantities $A_r(\alpha)$ are also physical parameters obtained from parametrization, A_r^0 are experimental values or the values calculated from the first principles methods and W_r is the weight factor. The force objective function defined in (13.1) is minimized for given α to calculate the classical force parameters by using the force and physical quantities obtained from ab initio calculations. Therefore, in order to calculate the force objective function, and constraint objective function it is necessary to do the ab initio calculations. The parameters α defined in the above (13.1, 13.2, 13.3) are calculated by matching the forces obtained by using the first-principles calculations of the several configurations of the molecular system and the classical potentials [78–80].

At the mesoscopic level, there exist effective field theories to apply the continuum mechanics. Also there are particle-based methodologies that have been developed to give more accurate results in the study than it is possible to deduce from the mesoscopic scale. Further details on particle-based methods can be found in [81–83].

13.4 Developments in Coarse-Grained Modeling of RNAs

Although due to the current development of the computational techniques and feasible computational resources, the theoretical study of the bimolecular systems has become much easier, preserving the physical information of the molecular system in the model remains very crucial. In view of this, the coarse-grained models are designed to explain information about the system at larger scales from the smaller scale that are modeled from the atomistic classical approach. The developed coarse-grained models should be easy enough to simulate accurately enough the physical characteristics of the system. In the coarse-grained modeling we represent the sum of atoms as a pseudo atom and then define an effective energy function U_{CG} that determines the thermodynamical properties, which should be identical to the system's properties once the proper energy function is predicted. Using coarse-grained models, the research has been done to study the structural and physical properties of DNA [84]. There are many other DNA studies found in the literature to describe the six helical systems [85] using atomic force microscopy. Also the study has been done for the improved angle potential [86]. In this study the array of hexagonal six helix bundles are described in 1D and 2D cases. Furthermore, several studies [87–89] of other bimolecular systems using the coarse-grained modeling have been done. Recently, the modeling of the coarse-grained structure of RNA and RNA-Protein using the fluctuation matching method has been performed [90], in which the authors also followed the assumptions used in [68]. There are a number of other investigations done on coarse grained modeling of RNA for the prediction of the tertiary structures [91–94]. In one of the earlier studies on the coarse-grained modeling of the RNA 3D structure, a single nucleobase has been approximated by five pseudo atoms [95]. In order to determine the forcefield parameters the 688 experimentally determined structures of

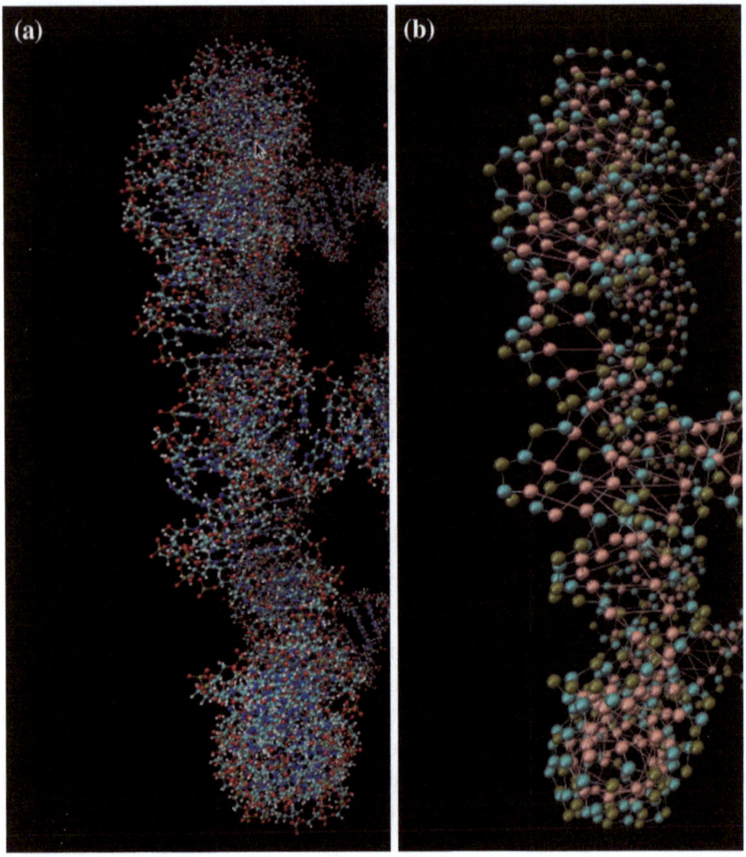

Fig. 13.3 **a** Part of the RNA nanocluster in all atom representation. **b** Part of the RNA nanocluster in coarse grained representation

RNA have been used. The transformation of the all atom model to the coarse-grained model in our typical three bead approximation is demonstrated in Fig. 13.3 where a part of the RNA nanoring is presented in both all atom and coarse-grained representations.

13.5 Computational Details

In molecular dynamics simulation the classical equations of motion of a molecular system are solved by their time dependent integration. The potential of the system used during the molecular dynamics simulation using CHARMM force field can be expressed as follows.

$$V_{total} = \sum_{bond} K_b(r - r_0)^2 + \sum_{angle} K_\theta(\theta - \theta_0)^2 + \sum_{dihedral} K_\phi(1 + cos(n\phi - \gamma))$$
$$+ \sum_{Hbond} \left(\frac{C_{ij}}{r_{ij}^{12}} - \frac{D_{ij}}{r_{ij}^{10}} \right) + \sum_{impropers} K_w(w - w_0)^2$$
$$+ \sum_{Urey-Bradley} K_u(u - u_0)^2 + \sum_{Vanderwaals} \left(\frac{A_{ij}}{r_{ij}^{12}} - \frac{B_{ij}}{r_{ij}^{10}} \right) + \sum \frac{q_i q_j}{\varepsilon r_{ij}}. \tag{13.4}$$

In (13.4), the first term corresponds to bonds, second corresponding to angle parameters, the third term corresponds to the potential energy and interactions arised from the dihedral angles in the molecular system, the fourth term defines the interaction coming from the hydrogen bonds which includes the base pairing as well as the hydrogen bonding between the RNA and the water molecules. The fifth term known as the improper term that arised due to out-of-plane bending of molecular system and the sixth term is the Urey-Bradley contribution. The improper term is included in the potential energy expression to maintain the planarity of the molecule. Finally, the last term in the potential expression represents the long distance interactions known as the van der Waals' interactions. We have performed all-atom molecular dynamics simulations of RNA nanotubes by using the CHARMM27 force field [96] implemented in the NAMD package [97] as it was done for the nanoring [64, 65].

The CHARMM is the force field widely used for the molecular dynamics simulation that is implemented in several molecular dynamics packages like LAMMP, NAMD and GROMACS. The CHARMM27 force field is one of the most important force fields which is developed for the nucleic acid through the empirical force field determinations. This is the most recent force field which is obtained from the reoptimization of the earlier force field CHARMM22 [97–99]. During the optimization of the force field the importance is given to the balancing of the properties of the local small molecules to the global system. In the studies [98, 99] the development of the CHARMM force field was carried out and its compatibility for the DNA and the RNA has been tested. It was found that the results are close to the experimental results.

The modeling of the nanotube, visualization and the analysis of the simulation outputs have been performed using the software visual molecular dynamics (VMD) [100]. The VMD is a molecular graphics software developed to display the biomolecular systems like biopolymers and proteins interactively. In this program the molecules can be viewed in several colors and several kind of representations. This can allow us to modify a particular protein structure as needed. In particular, one can do mutation, delation or addition of bonds between the atoms using VMD tool. Furthermore, we can display several structures at the same time using VMD. The trajectory of the molecular dynamics simulation can be displayed as well as analysed to study the molecular properties. The VMD program is written in C++

and is provided with the complete documentation with instructions to use it. This program is compatible with several kinds of molecular dynamics simulation packages including NAMD, LAMMP, GROMACS etc.

In our typical runs, the RNA-nanotube has been solvated in a water box. The size of the box is taken in such a way that the distance from the surface of the nanocluster to the wall is slightly larger than the cut off radius used in the molecular dynamics simulation. In order to make the system neutral we have added 924, and 1254 $^{23}Na^{+}$ ions for three ring and four ring nanotubes, respectively. Furthermore, to make the solution equivalent to physiological solutions we have added extra 924 and 1254 $^{23}Na^{+}$ and $^{35}Cl^{-}$ ions to the three ring and four ring nanotube, respectively. The resulting system has been first simulated at constant temperature and pressure using the NAMD software package. The temperature in the system has been controlled by using Langevin's method with damping $\eta = 5\ ps^{-1}$. For adding chemical bonds between the segments in the nanoclusters we have used the topotools available in the VMD.

13.6 Results and Discussions

Advancing further multiscale models for RNA nanoclusters, we have modeled the RNA nanotubes with multiple nanorings. Some of the most recent results have been presented in our papers [65, 101]. For the modeling of RNA nanotubes the hexagonal nanorings were connected to each other by using the links between them as described in our earlier studies [64, 65, 101]. Typical sample structures of the three ring nanotube without water and with water are presented in Fig. 13.4a, b respectively. The six helical segments are constructed from RNAI and RNAII building blocks. Also, the tails used to connect the RNA nanorings are the double strand RNAs with the length of 22 nucleotides.

As we discussed earlier, the RNAIi and RNAIIi are the double strand RNAs. By using the VMD tools we were able to connect multiple rings via three links at junctions presented in Fig. 13.4b. The links used in connecting the multiple numbers of nanorings (to build the RNA nanotube) are composed of helical double strand RNAs with 22 nucleobases. Three links are used in between two consecutive rings to connect them to form the nanotubes as shown in Fig. 13.4b. The chemical bonds between the ring and the links are mediated through the phosphorous of the phosphate group in the ring and the oxygen in the sugar ring of the corresponding link or vice versa. Using NAMD, we optimized the chemical bonds added between different segments of the RNA nanoclusters.

Here at first, we present the results for the RNA nanotube of different sizes obtained from the molecular dynamics simulation. The results for the simulation of three the ring nanotube are summarized in Figs. 13.5 and 13.6. Figure 13.5 describes the variation of the energy and temperature as a function of simulation time. At the beginning of the simulation the energy of the system varies and then becomes stable once the system becomes stabilized. Here in our results we have presented only the latter part of the simulation, which is also known as the

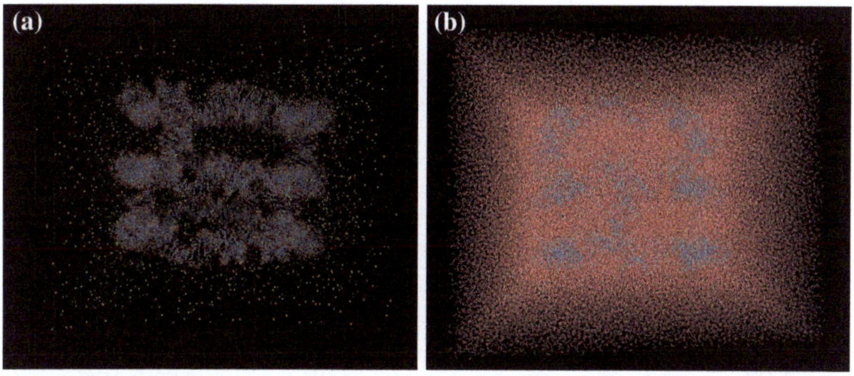

Fig. 13.4 **a** Three ring RNA nanotube and ions without water and **b** three ring RNA nanotube in a physiological solution

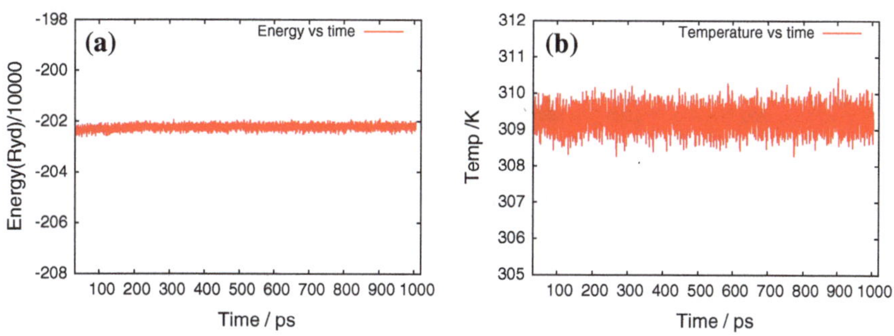

Fig. 13.5 **a** Energy and **b** temperature versus simulation time for the all-atom molecular dynamics simulation of three ring RNA nanotube in water and salt

production region of simulation. All our results and analysis of the properties are presented from this production region of molecular dynamics simulation. The temperature of the system remains almost stable with some fluctuations. In Fig. 13.6 we present the calculated properties such as the number of ions around the RNA nanotube within the distance of 5 Å at different temperatures, the number of bonds per basepairs, the radius of gyration and the root mean square deviation at two temperatures, 310 and 510 K. The results corresponding to variations of the parameters are similar to the results obtained for the other nanoclusters described in our earlier studies [64, 65, 101].

The nature of the radial distribution function plots calculated for the three ring RNA nanotube is revealed in Fig. 13.7. In particular, four subplots in this figure present the RDF plots for phosphorous-phosphorous, phosphorous-water, phosphorous-sodium and phosphorous-chlorine, respectively. From the P–P RDF plots presented in Fig. 13.7a, we see that there are three well-pronounced peaks around the same positions at it was observed for other nanoclusters studied in our

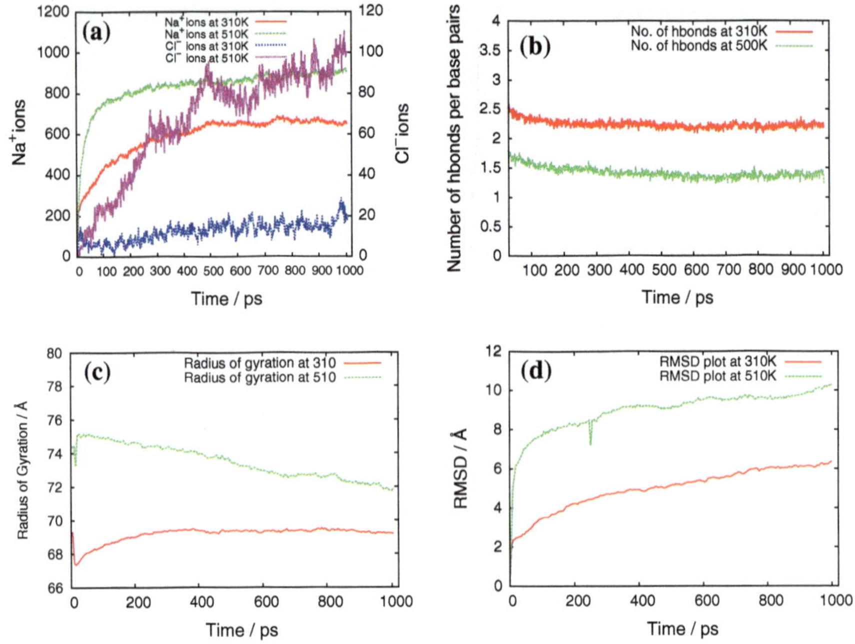

Fig. 13.6 **a** Number of ions with the range of 5 Å. **b** Number of bonds per base pairs. **c** Radius of gyration and **d** RMSD of three ring RNA nanotube obtained from all atom molecular dynamics simulation

paper [65]. These peaks actually show the first, second and third nearest neighbours of the phosphorous atom respectively. The intensity of the peaks is increased on going from 310 to 510 K. The position of the first peak is at the same position, whereas the second and third peaks are shifted slightly to the lower distances at 510 K in comparison to their positions at 310 K.

From the P–OH2 RDF plots presented in Fig. 13.7b calculated at temperatures 310 and 510 K, it is clear that for each RDF there is a peak around the distance of 4 Å. This first peak indicates the first solvation shell around the phosphorous atom taken from the surface of the RNA nanotube. Similarly, the second small peak shows the second solvation peak for the phosphorous atom in the phosphate backbone of RNA strands that builds the RNA nanotubes. In the rest of the range, the nature of the P–OH2 RDF plots remained more or less stable showing that the water molecules are distributed uniformly after certain distance from the surface of the RNA nanotube. In spite of showing a similar trend at both temperatures, the height of the first peak is significantly dropped on going from 310 to 510 K. This indicates that a significant amount of water molecules are expelled out from the surface of the RNA nanocluster at higher temperatures as demonstrated in our earlier papers [64, 65].

The P–Na RDF plots for the three ring RNA nanotube at both temperatures are presented in Fig. 13.7c. The RDF plot for P–Na shows the first peak at around 3.5 Å

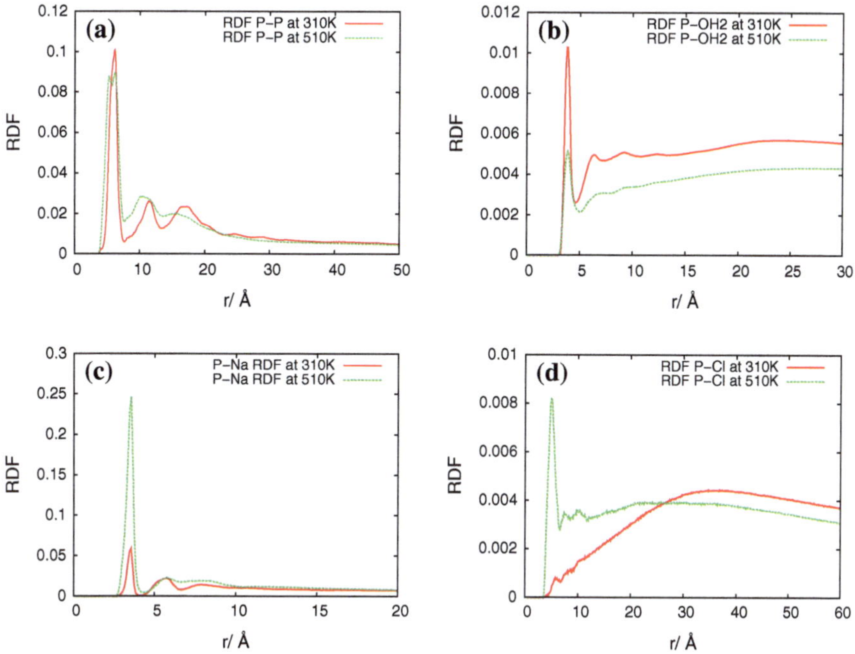

Fig. 13.7 Radial distribution function for three ring RNA nanotube. **a** P–P. **b** P–OH2. **c** P–Na. **d** P–Cl

at both temperatures, 310 and 510 K. This shows that most of the sodium ions are around this distance at the final step of all-atom molecular dynamics simulation. One significant difference between the RDF plots at 310 and 510 K is that the first peak of the radial distribution function is significantly increased on going from 310 to 510 K. This feature supports the conclusion that we made from the ionic distribution plots presented in Fig. 13.6a. Furthermore, the P–Cl RDF plots presented in Fig. 13.6d show that the chloride ions are far away from the surface of the RNA nanotube at 310 K. When the temperature of the system is increased from 310 to 510 K the $^{35}Cl^-$ ions are also aggregated significantly closer to the surface of the RNA nanotube as observed from the first peak of RDF plot at 510 K. The height of the peak at a particular distance from the surface of RNA nanotube in the RDF plot is proportional to the number of $^{35}Cl^-$ ions at that distance. This means that the number of $^{35}Cl^-$ ions around the first peak at 510 K are larger in comparison to the number of $^{35}Cl^-$ ions at 310 K as observed from the plots for the number of $^{35}Cl^-$ ions at 310 and 510 K presented in Fig. 13.6a.

The results for the four ring RNA nanotube are presented in Figs. 13.8, 13.9 and 13.10. The nature of solvation and the ionic distribution during the molecular dynamics simulation have been found to be similar to those found in the case of the three ring nanotube as well as to the results described in our earlier work [64, 65]. For all of these systems we see that the peaks for the P–P RDF remain almost the

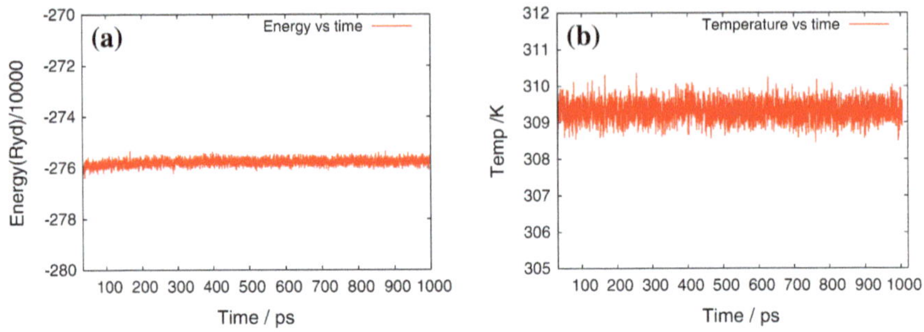

Fig. 13.8 **a** Energy and **b** temperature versus simulation time for the all-atom molecular dynamics simulation of four ring RNA nanotube in water and salt

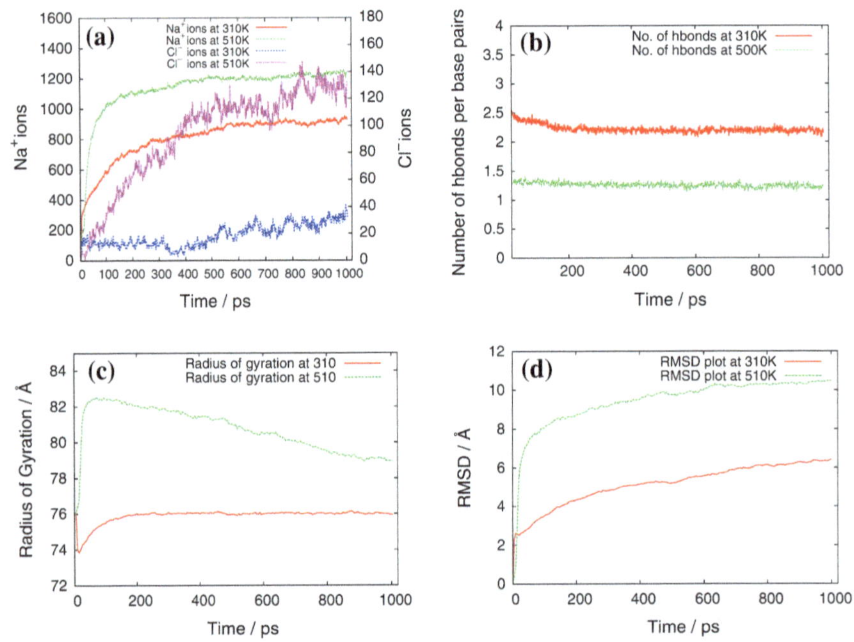

Fig. 13.9 **a** Number of ions with the range of 5 Å. **b** Number of bonds per base pairs. **c** Radius of gyration and **d** RMSD of four ring RNA nanotube obtained from all atom molecular dynamics simulation

same, for P–OH2 the intensity at the peak is decreased on increasing the temperature, but in the P–Na and P–Cl RDF plots the intensity of the first peaks is significantly increased on going from 310 to 510 K temperature. From these observations we can conclude that the $^{23}Na^{+}$ and $^{35}Cl^{-}$ ions are attracted toward the surface of the nanocluster and the water molecules are pushed away from the surface as the temperature is increased. In short, we observe that the ions are being

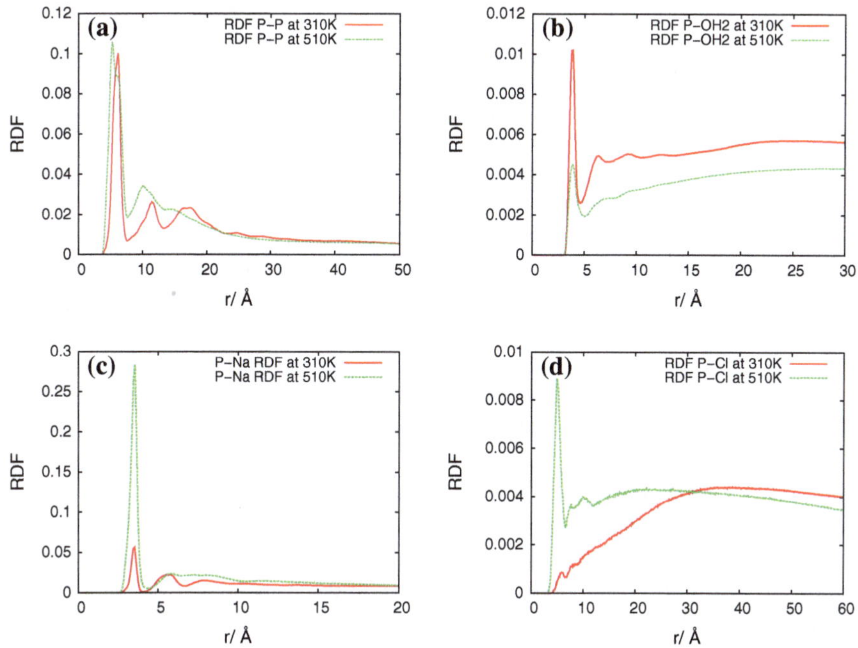

Fig. 13.10 Radial distribution function for four ring RNA nanotube. **a** P–P. b P–OH2. **c** P–Na. **d** P–Cl

precipitated around the surface of the RNA nanocluster as the temperature of the system is increased. This phenomenon of self stabilization was first discovered in [64] and has recently been observed for a larger class of RNA nanoclusters in our earlier work [65].

In spite of existing difficulties in the modeling of the nanostructures from the building blocks consisting of nucleic acids and their derivatives structures, significant achievements have been observed in the field of RNA nanotechnology. Clearly, there is a scope for further improvements. Notably that although there are several state-of-the-art computational software packages developed to predict RNA nanostructures, only 70 % accuracy has been found in the prediction and folding of the RNA nanoclusters [102, 103].

13.7 Conclusions

In this chapter we have reviewed some of the aspects of the nucleic acid research and their potential applications in nanobiomedicine and other related fields. From the review of the literature, we conclude that the RNA is more versatile and compatible for the biomedical applications in the human body, compared to other alternatives, due to its flexibility in structure as well as the absence of toxicity issues to be addressed. In view of this, we have studied the structure and properties of RNA self assemblies in physiological solutions. In our earlier investigations [64, 65], the analysis of the physical properties of the nanoring has been reported. The RNA nanoring is a small system in comparison to the nanotube and has limited practical applications in bionanotechnology. However, it provides an excellent testing ground for further studies. In our most recent studies [65, 101]. This included the optimized structures of nanotubes up to the size of 40 nm have been analyzed for the first time. The individual RNA nanorings were connected via double-helical rings mediated by the bonds between the phosphate group and sugar ring. The newly added bond lengths have been optimized by using algorithms available in NAMD. Then, starting from nanorings, the results for the RNA-nanotubes of different sizes have been exemplified in this chapter for three and four nanoring structures and discussed in details. Similar to our earlier study [65] we presented some of the results for these RNA nanotube structures via calculations of the root mean square deviation, radius of gyration, number of hydrogen bonds per basepair, ion accumulation around the tube, and the radial distribution functions. From our present analysis it is clear that the quality of the results are likely to be improved further by doing the molecular dynamics simulation for longer time ranges. Another way to improve the quality of the results lies with further developing of the multiscale models methodologies specifically for these structures e.g. coarse-grained modeling methods that would allow one to perform molecular dynamics simulations for longer ranges of time, closer to the time scales of real biological phenomena. These kinds of developments are currently under way in our lab. Furthermore, these new developments will help further progress in theoretical bionanotechnology, as well as guide the experimentalists working in this filed. Also, the use of reduced order modeling [104] would increase the compatibility of the results with the time range better comparable to real biological processes. So far we have been using the SHARCNET parallel computing facilities for calculating of the properties of the RNA nanoclusters using the 64 processors for each simulation. Using the clusters such as GPGPUs or Phi co-processors may make these computations more efficient for larger RNA nanoclusters as well as for longer time ranges. For drug delivery applications computational efforts will also be dependent on the shape of the nanocluster as well as on the drug particle to be delivered [105]. In our modeling of the RNA nanoclusters, we have control on the size of the nanocluster which gives better potentiality of their use in the nanomedicine.

Our progress in the development of RNA nanoclusters will boost further their applications in the therapy, nanodesign and drug delivery, among other fields.

Acknowledgments Authors are grateful to the NSERC and CRC Programs for their support and Shared Hierarchical Academic Research Computing Network (SHARCNET: www.sharcnet.ca) for providing the computational facilities. Finally, we would like to thank Dr. P.J. Douglas Roberts for helping with technical SHARCNET computational aspects.

References

1. A.A. Simpson, Y. Tao, P.G. Leiman, M.O. Badasso, Y. He, P.J. Jardine, N.H. Olson, M.C. Morais, S. Grimes, D.L. Anderson, T.S. Baker, M.G. Rossmann, Nature **408**(6813), 745 (2000)
2. D. Moll, C. Huber, B. Schlegel, D. Pum, U.B. Sleytr, M. Sra, PNAS **99**(23), 14646 (2002)
3. H. Cui, T. Muraoka, A.G. Cheetham, S.I. Stupp, Nano Lett. **9**(3), 945 (2009)
4. D. Liu, M. Wang, Z. Deng, R. Walulu, C. Mao, J. Am. Chem. Soc. **126**(8), 2324 (2004)
5. F.A. Aldaye, H.F. Sleiman, J. Am. Chem. Soc. **129**(44), 13376 (2007)
6. P.W.K. Rothemund, N. Papadakis, E. Winfree, PLoS Biol. **2**(12), e424 (2004)
7. S.H. Park, R. Barish, H. Li, J.H. Reif, G. Finkelstein, H. Yan, T.H. LaBean, Nano Lett. **5**(4), 693 (2005)
8. T.P.J. Knowles, T.W. Oppenheim, A.K. Buell, D.Y. Chirgadze, M.E. Welland, Nat. Nanotechnol. **5**(3), 204 (2010)
9. N.C. Seeman, Annu. Rev. Biochem. **79**, 65 (2010)
10. C. Lin, Y. Liu, H. Yan, Biochemistry **48**(8), 1663 (2009)
11. F.A. Aldaye, A.L. Palmer, H.F. Sleiman, Science **321**(5897), 1795 (2008)
12. K. Tanaka, A. Tengeiji, T. Kato, N. Toyama, M. Shionoya, Science **299**(5610), 1212 (2003)
13. M. Endo, T. Sugita, Y. Katsuda, K. Hidaka, H. Sugiyama, Chem. Eur. J. **16**(18), 5362 (2010)
14. Y. Ke, S. Lindsay, Y. Chang, Y. Liu, H. Yan, Science **319**(5860), 180 (2008)
15. B. Yurke, A.J. Turberfield, A.P. Mills, F.C. Simmel, J.L. Neumann, Nature **406**(6796), 605 (2000)
16. S.M. Douglas, J.J. Chou, W.M. Shih, PNAS **104**(16), 6644 (2007)
17. M. Endo, N.C. Seeman, T. Majima, Angew. Chem. Int. Ed. **44**(37), 6074 (2005)
18. N. Sugimoto, S.I. Nakano, M. Katoh, A. Matsumura, H. Nakamuta, T. Ohmichi, M. Yoneyama, M. Sasaki, Biochemistry **34**(35), 11211 (1995)
19. A. Kitamura, P.J. Jardine, D.L. Anderson, S. Grimes, H. Matsuo, Nucl. Acids Res. **36**(3), 839 (2008)
20. P. Guo, Nat. Nano. **5**(12), 833 (2010)
21. W.W. Grabow, L. Jaeger, Acc. Chem. Res. **47**(6), 1871 (2014)
22. M.S. Searle, D.H. Williams, Nucl. Acids Res. **21**(9), 2051 (1993)
23. K.A. Afonin, W. Kasprzak, E. Bindewald, P.S. Puppala, A.R. Diehl, K.T. Hall, T.J. Kim, M.T. Zimmermann, R.L. Jernigan, L. Jaeger, B.A. Shapiro, Methods **67**(2), 256 (2014)
24. J. Yang, R. Melnik, Discrete and continuous dynamical systems, supplement pp. 1052–1060 (2007)
25. X. Yang, R. Melnik, Comput. Biol. Chem. **31**, 110 (2007)
26. M. Anokhina, S. Bessonov, Z. Miao, E. Westhof, K. Hartmuth, R. Lührmann, EMBO J. **32**(21), 2804 (2013)
27. I. Severcan, C. Geary, E. Verzemnieks, A. Chworos, L. Jaeger, Nano Lett. **9**(3), 1270 (2009)
28. I. Severcan, C. Geary, A. Chworos, N. Voss, E. Jacovetty, L. Jaeger, Nat. Chem. **2**(9), 772 (2010)
29. E. Osada, Y. Suzuki, K. Hidaka, H. Ohno, H. Sugiyama, M. Endo, H. Saito, ACS Nano **8**(8), 8130 (2014)
30. L. Jaeger, A. Chworos, Curr. Opin. Struct. Biol. **16**(4), 531 (2006)
31. N.B. Leontis, E. Westhof, Science **345**(6198), 732 (2014)
32. J.I. Tomizawa, Cell **38**(3), 861 (1984)
33. J.I. Tomizawa, Cell **47**(1), 89 (1986)
34. A.J. Lee, D.M. Crothers, Structure **6**(8), 993 (1998)
35. W.W. Grabow, P. Zakrevsky, K.A. Afonin, A. Chworos, B.A. Shapiro, L. Jaeger, Nano Lett. **11**(2), 878 (2011)
36. Y.G. Yingling, B.A. Shapiro, Nano Lett. **7**(8), 2328 (2007)
37. R.V.N. Melnik, X. Wei, G. MorenoHagelsieb, J. Biol. Syst. **17**(03), 425 (2009)

38. K.A. Afonin, E. Bindewald, A.J. Yaghoubian, N. Voss, E. Jacovetty, B.A. Shapiro, L. Jaeger, Nat. Nanotechnol. **5**(9), 676 (2010)
39. N.B. Leontis, A. Lescoute, E. Westhof, Curr. Opin. Struct. Biol. **16**(3), 279 (2006)
40. D. Shu, Y. Shu, F. Haque, S. Abdelmawla, P. Guo, Nat. Nanotechnol. **6**(10), 658 (2011)
41. B. Cayrol, C. Nogues, A. Dawid, I. Sagi, P. Silberzan, H. Isambert, J. Am. Chem. Soc. **131** (47), 17270 (2009)
42. C.J. Delebecque, A.B. Lindner, P.A. Silver, F.A. Aldaye, Science **333**(6041), 470 (2011)
43. E. Bindewald, R. Hayes, Y.G. Yingling, W. Kasprzak, B.A. Shapiro, Nucleic Acid Res. **36** (suppl 1), D392 (2008)
44. A. Fire, S. Xu, M.K. Montgomery, S.A. Kostas, S.E. Driver, C.C. Mello, Nature **391**(6669), 806 (1998)
45. C. Wagner, C. Ehresmann, B. Ehresmann, C. Brunel, J. Biol. Chem. **279**(6), 4560 (2004)
46. E. Laurenti, I. Barde, S. Verp, S. Offner, A. Wilson, S. Quenneville, M. Wiznerowicz, H.R. MacDonald, D. Trono, A. Trumpp, Stem Cells **28**(8), 1390 (2010)
47. K.Y. Chang, I. Tinoco, PNAS **91**(18), 8705 (1994)
48. S. Hoeprich, Q. Zhou, S. Guo, D. Shu, G. Qi, Y. Wang, P. Guo, Gene Ther. **10**(15), 1258 (2003)
49. L. Ponchon, G. Beauvais, S. Nonin-Lecomte, F. Dardel, Nat. Protocols **4**(6), 947 (2009)
50. C. Chen, S. Sheng, Z. Shao, P. Guo, J. Biol. Chem. **275**(23), 17510 (2000)
51. S.T. Stern, S.E. McNeil, Toxicol. Sci. **101**(1), 4 (2008)
52. S. Singh, A. Sharma, G.P. Robertson, Cancer Res. **72**(22), 5663 (2012)
53. Y. Shu, M. Cinier, S.R. Fox, N. Ben-Johnathan, P. Guo, Mol. Ther. **19**(7), 1304 (2011)
54. J.B. Lee, J. Hong, D.K. Bonner, Z. Poon, P.T. Hammond, Nat. Mater. **11**(4), 316 (2012)
55. J.B. Bramsen, J. Kjems, Front Genet. **3**, 154 (2012)
56. R. Kanasty, J.R. Dorkin, A. Vegas, D. Anderson, Nat. Mater. **12**(11), 967 (2013)
57. X.D. Yang, D.R. Mahapatra, R.V.N. Melnik, in *AIP Conference Proceedings*, vol. 952 (AIP Publishing, 2007), pp. 229–237
58. S. Guo, N. Tschammer, S. Mohammed, P. Guo, Hum. Gene Ther. **16**(9), 1097 (2005)
59. A. Khaled, S. Guo, F. Li, P. Guo, Nano Lett. **5**(9), 1797 (2005)
60. J. Yano, K. Hirabayashi, S.I. Nakagawa, T. Yamaguchi, M. Nogawa, I. Kashimori, H. Naito, H. Kitagawa, K. Ishiyama, T. Ohgi, T. Irimura, Clin. Cancer Res. **10**(22), 7721 (2004)
61. H. Li, W.X. Li, S.W. Ding, Science **296**(5571), 1319 (2002)
62. C. Zhang, Curr. Opin. Mol. Ther. **11**(6), 641 (2009)
63. M.R. Fabian, N. Sonenberg, W. Filipowicz, Annu. Rev. Biochem. **79**(1), 351 (2010)
64. M. Paliy, R. Melnik, B.A. Shapiro, Phys. Biol. **6**(4), 046003 (2009)
65. S.R. Badu, R. Melnik, M. Paliy, S. Prabhakar, A. Sebetci, B.A. Shapiro, Eur. Biophys. J. **43** (10–11), 555 (2014)
66. T. Kim, B.A. Shapiro, J. Biomol. Struct. Dyn. **31**(5), 495 (2013)
67. J. Vieregg, W. Cheng, C. Bustamante, I. Tinoco, J. Am. Chem. Soc. **129**(48), 14966–14973 (2007)
68. M. Paliy, R. Melnik, B.A. Shapiro, Phys. Biol. **7**(3), 036001 (2010)
69. T. Hansson, C. Oostenbrink, W. van Gunsteren, Curr. Opin. Struct. Biol. **12**(2), 190 (2002)
70. W. van Gunsteren, D. Bakowies, R. Brgi, I. Chandrasekhar, M. Christen, X. Daura, P. Gee, A. Glttli, T. Hansson, C. Oostenbrink, C. Peter, J. Pitera, L. Schuler, T. Soares, H. Yu, CHIMIA Int. J. Chem. **55**(10), 856 (2001)
71. X. Wang, Y. Wang, L. Zheng, J. Chen, Curr. Med. Chem. **21**(17), 1968 (2014)
72. I.M. Moustafa, H. Shen, B. Morton, C.M. Colina, C.E. Cameron, J. Mol. Biol. **410**(1), 159 (2011)
73. N.J. Deng, P. Cieplak, Biophys. J. **98**(4), 627 (2010)
74. J.W. Chu, G.S. Ayton, S. Izvekov, G.A. Voth, Mol. Phys. **105**(2–3), 167 (2007)
75. J.W. Chu, S. Izveko, G.A. Voth, Mol. Simul. **32**(3–4), 211 (2006)
76. G.S. Ayton, W.G. Noid, G.A. Voth, Curr. Opin. Struct. Biol. **17**(2), 192 (2007)
77. J. Zhou, I.F. Thorpe, S. Izvekov, G.A. Voth, Biophys. J. **92**(12), 4289 (2007)
78. F. Ercolessi, J.B. Adams, Euro. Phys. Lett. **26**(8), 583 (1994)

79. M.S. Daw, M.I. Baskes, Phys. Rev. B **29**, 6443 (1984)
80. F. Ercolessi, E. Tosatti, M. Parrinello, Phys. Rev. Lett. **57**, 719 (1986)
81. G.S. Ayton, J.L. McWhirter, P. McMurtry, G.A. Voth, Biophys. J. **88**(6), 3855 (2005)
82. G.S. Ayton, G.A. Voth, Biophys. J. **87**(5), 3299 (2004)
83. G.S. Ayton, G.A. Voth, Int. J. Multiscale Comput. Eng. **2**(2), 291 (2004)
84. T.E. Ouldridge, A.A. Louis, J.P.K. Doye (2010)
85. F. Mathieu, S. Liao, J. Kopatsch, T. Wang, C. Mao, N.C. Seeman, Nano Lett. **5**(4), 661 (2005)
86. M. Bulacu, N. Goga, W. Zhao, G. Rossi, L. Monticelli, X. Periole, D.P. Tieleman, S.J. Marrink, J. Chem. Theory Comput. **9**(8), 3282 (2013)
87. I. Bahar, A. Rader, Curr. Opin. Struct. Biol. **15**(5), 586 (2005)
88. F. Tama, M. Valle, J. Frank, C.L. Brooks, PNAS **100**(16), 9319 (2003)
89. V. Tozzini, Curr. Opin. Struct. Biol. **15**(2), 144 (2005)
90. N. Hori, S. Takada, J. Chem. Theory Comput. **8**(9), 3384 (2012)
91. R. Das, D. Baker, PNAS **104**(37), 14664 (2007)
92. J. Bernauer, X. Huang, A.Y.L. Sim, M. Levitt, RNA **17**(6), 1066 (2011)
93. F. Ding, S. Sharma, P. Chalasani, V.V. Demidov, N.E. Broude, N.V. Dokholyan, RNA **14** (6), 1164 (2008)
94. J. Bida, R. Das, Curr. Opin. Struct. Biol. **22**(4), 457 (2012)
95. Z. Xia, D.P. Gardner, R.R. Gutell, P. Ren, J. Phys. Chem. B **114**(42), 13497 (2010)
96. J.C. Phillips, R. Braun, W. Wang, J. Gumbart, E. Tajkhorshid, E. Villa, C. Chipot, R. D. Skeel, L. Kal, K. Schulten, J. Comput. Chem. **26**(16), 1781 (2005)
97. D. MacKerell, D. Bashford, M. Bellott, R.L. Dunbrack, J.D. Evanseck, M.J. Field, S. Fischer, J. Gao, H. Guo, S. Ha, D. Joseph-McCarthy, L. Kuchnir, K. Kuczera, F.T.K. Lau, C. Mattos, S. Michnick, T. Ngo, D.T. Nguyen, B. Prodhom, W.E. Reiher, B. Roux, M. Schlenkrich, J.C. Smith, R. Stote, J. Straub, M. Watanabe, J. Wirkiewicz-Kuczera, D. Yin, M. Karplus, J. Phys. Chem. B **102**(18), 3586 (1998)
98. N. Foloppe, A.D. MacKerell Jr, J. Comput. Chem. **21**(2), 86 (2000)
99. A.D. MacKerell, N.K. Banavali, J. Comput. Chem. **21**(2), 105 (2000)
100. W. Humphrey, A. Dalke, K. Schulten, J. Mol. Graph. **14**(1), 33 (1996)
101. S.R. Badu, R. Melnik, M. Paliy, S. Prabhakar, A. Sebetci, B.A. Shapiro, in *Proceedings of IWBBIO-2014* (International Work-Conference on Bioinformatics and Biomedical Engineering, 2014), pp. 601–607
102. M. Zuker, Nucl. Acids Res. **31**(13), 3406 (2003)
103. N.R. Markham, M. Zuker, in *Bioinformatics*, ed. by J.M. Keith, no. 453 in Methods in Molecular Biology (Humana Press, 2008), pp. 3–31
104. Z. Bai, Appl. Numer. Math. **43**(12), 9 (2002)
105. P. Guo, O. Coban, N.M. Snead, J. Trebley, S. Hoeprich, S. Guo, Y. Shu, Adv. Drug Deliv. Rev. **62**(6), 650 (2010)

Chapter 14
Nuclear Spin Catalysis: From Molecular Liquids to Biomolecular Nanoreactors

Vitaly K. Koltover

Abstract All chemical reactions obey the law of conservation of spin angular momentum ('spin'): any reaction is only allowed when the total spin of reactants is identical to the total spin of products. Correspondingly, free-radical reactions in molecular liquids can be accelerated by changing in electron spin of the reactants via magnetic fields of magnetic nuclei, the so-called "magnetic isotope effect" (MIE). In molecular liquids, the magnetic isotope effects have been discovered for a number of magnetic isotopes, among them H–D, ^{13}C, ^{17}O, ^{29}Si, ^{33}S, ^{73}Ge, $^{117,119}Sn$, $^{199,201}Hg$, and ^{235}U. Recently MIE has been discovered in living cells. It was revealed that the rate constant of post-radiation recovery of yeast cells is twice higher for the cells enriched with the magnetic ^{25}Mg when compared to the cells with the nonmagnetic ^{24}Mg. Furthermore, it has been revealed that ^{25}Mg essentially accelerates, 2–2.5 times by comparison to the spin-less ^{24}Mg and ^{26}Mg, the reaction of ATP hydrolysis catalyzed by myosin, the enzyme isolated from muscle cells. Although detailed mechanisms of the ability of biomolecular nanoreactors to perceive the nuclear magnetism require further investigations, the recent developments in this new field highlight promising venues for future research of the magnetic isotope effects ('nuclear spin catalysis') in molecular liquids and biopolymer nanoreactors with possible application of the stable magnetic isotopes for control over efficiency and reliability of molecular nanoreactors in engineering and biomedicine.

List of Abbreviations

MIE Magnetic isotope effect
ATP Adenosine 5'-triphosphate
ADP Adenosine 5'-diphosphate
P_i Inorganic phosphate

V.K. Koltover (✉)
Institute of Problems of Chemical Physics, Russian Academy of Sciences, Chernogolovka, Moscow Region 142432, Russian Federation
e-mail: koltover@icp.ac.ru

L. Bulavin and N. Lebovka (eds.), *Physics of Liquid Matter: Modern Problems*, Springer Proceedings in Physics 171, DOI 10.1007/978-3-319-20875-6_14

14.1 Introduction

Apart from the energy control (the law of conservation of energy), any chemical reaction as electron-nuclear rearrangement of reactants into products is controlled by angular momentum, spin, of reactants. The total spin of products must be identical to that of reactants. This law of spin conservation immediately follows from quantum mechanics, from the fundamental and universal Pauli principle: no two electrons may occupy the same quantum state simultaneously [1].

Correspondingly, acceleration of the free-radical reactions can be achieved through changes in the total electron spin of reactants by interaction with the magnetic fields of magnetic nuclei. It is known as magnetic isotope effect (MIE), a new trend in chemical and biochemical physics within last years (see [2–9] and references therein). This paper is a brief review of recent developments in this field highlighting promising venues for future research of magnetic isotope effects ('nuclear spin catalysis') in molecular liquids and biomolecular nanoreactors.

14.2 Nuclear Spin Catalysis in Molecular Liquids

Figure 14.1 illustrates how the law of conservation of the spin gives control over reactivity of free radicals ($R^{\bullet}$) in molecular liquids. For example, a pair of free radicals, each with the electron spin $S = 1/2$, is to form a chemical bond and the resultant diamagnetic molecule, the total electron spin of which $S = 0$ (Fig. 14.1a). From the law of conservation of spin, it follows that the chemical bond between these two radicals may happen only if the spin state of the pair at collision is singlet, i.e., the spins of two electrons are subtracted to give the net $S = 0$ (spin multiplicity, $2S + 1 = 1$). If the spin state of the radical pair is triplet, i.e., the electron spins are added up to give the net $S = 1$ (spin multiplicity, $2S + 1 = 3$), then the radicals cannot react immediately. As a result, only one-quarter of encounters, when the total spin state of the radical pair is singlet, gives the recombination product while three-quarters of the radical pairs are inhibited from the reaction. Another example is presented on Fig. 14.1b. Namely, it is the reaction of a radical $R^{\bullet}$ with oxygen, the molecules of which are normally in the triplet spin state. The total spin of this reagent pair can be 1/2 when the individual spins are subtracted (spin multiplicity is 2) or 3/2

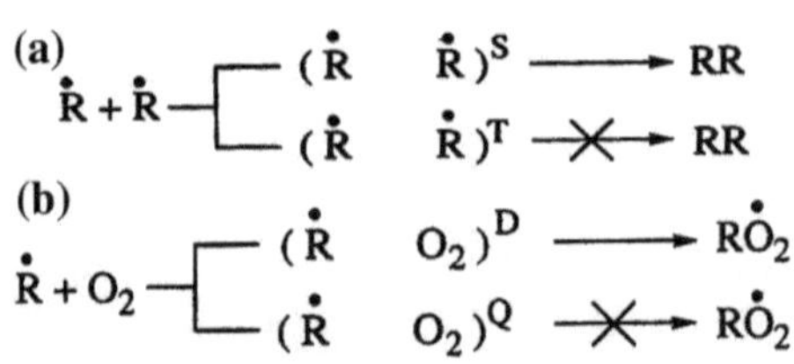

Fig. 14.1 Spin control over chemical reactions in molecular liquids

when the individual spins are added up (spin multiplicity is 4). Meanwhile, the reaction product (peroxyl radical $RO_2^{\bullet}$) has the total electron spin 1/2. Hence, from six possible spin states of the reactants, only two states do not require the change in the total electron spin of the reactants and, therefore, are permitted for formation of $RO_2^{\bullet}$; other four states are forbidden for the reaction. To lift the ban forced by the spin conservation law, the spins of the reactants must be changed.

In reality, owing to spin-spin and spin-lattice interactions, the spin state of the radical pair is an occasional superposition of the singlet and triplet states. As a consequence, the probability that two radicals during their diffusion meet each other in the proper spin state and, hence, the probability that the collision results in the chemical reaction is quite high [10]. In molecular liquids, however, with the time allotted for any collisions of radicals of order of nanoseconds, neither spin-spin interaction nor spin-lattice relaxation have time to fit the spin orientation. Moreover, in organic free radicals, spin-orbit coupling is small. So, magnetic fields are the only means to change the spin state and, thereby, switch the reaction over the spin-forbidden and spin-allowed channels.

Accordingly, the probability of chemical reaction is a function of parameters of the magnetic interactions [2–5]:

$$P = f(H; \omega; H_1; J; a; I; m_I; \mu_n).$$

In this equation H is an applied magnetic field; ω and H_1 are frequency and amplitude of microwave magnetic fields (interaction that produces chemically detected magnetic resonance and stimulates nuclear polarization); J is exchange energy (exchange interaction). Correspondingly, acceleration of the free-radical reaction can be achieved through changes in the total electron spin of reactants by their interaction with external, applied, magnetic fields. The equation also contains parameters of hyperfine coupling a, nuclear spin I, nuclear spin projection m_I, and nuclear magnetic moment μ_n, i.e. the parameters of interactions of electron spins with magnetic nuclei. Correspondingly, the spin ban can be lift through the relevant changes in the total electron spin of reactants owing to interactions of the electron spins with magnetic fields of nuclear spins of the magnetic nuclei.

As a result, the chemical reactions, which involve free radicals or ion-radical pairs, may exhibit different reaction rates and different yields of products according to whether the reagents contain magnetic or nonmagnetic isotopes. This is known as magnetic-isotope effect. MIE is a purely kinetic phenomenon. While the classical mass-isotope effect selects isotopic nuclei in accordance with their masses, the magnetic-isotope effect selects isotopic nuclei in accordance with their nuclear spins and magnetic moments. In molecular liquids, within recent years, the magnetic isotope effects in chemical reactions have been discovered for a number of magnetic isotopes, among them H–D, ^{13}C, ^{17}O, ^{29}Si, ^{33}S, ^{73}Ge, $^{117,119}Sn$, $^{199,201}Hg$, and ^{235}U (see [5] and references therein).

On its own, MIE unambiguously evidences that the chemical reaction involves a spin-dependent rate-determining step, namely, the singlet-triplet conversion or vice versa, that is accelerated by the magnetic field of the isotope's nuclear spin. Similar spin bans arise at transitions between singlet and triplet states of molecules and macromolecules, in semiconductors, and so on. Likewise, the magnetic fields, be it exposure to the external field or the field of the magnetic isotope, are the means to change the spin state and, thus, to lift the spin ban [2–10].

14.3 Nuclear Spin Catalysis in Biomolecular Nanoreactors

All biomolecular nanoreactors, like other cell structures in living Nature, are composed from atoms of chemical elements, many of which have stable isotopes of both kinds, magnetic and nonmagnetic ones (see Table 14.1).

Of special interest is magnesium as one of the most abundant cell elements. Magnesium has three stable isotopes, ^{24}Mg, ^{25}Mg and ^{26}Mg, with natural abundance of approximately 79, 10 and 11 %. Only ^{25}Mg is magnetic (nuclear spin $I = 5/2$) whereas ^{24}Mg and ^{26}Mg are nonmagnetic (spin-less, nuclear spin $I = 0$) [11]. It is generally known that cations of Mg^{2+} serve obligate cofactor functions for the enzymes of synthesis and hydrolysis of ATP and performs many other regulatory functions in important cell processes [12, 13]. In order to search for magnetic isotope effects, we developed procedures for growing living cells in a medium containing only one magnesium isotope: either the magnetic ^{25}Mg or the nonmagnetic ^{24}Mg or ^{26}Mg [14–17].

Figure 14.2 represents experimental data of our group on influence of the different isotopes of magnesium on the kinetics of post-radiation recovery of yeast cells, *Saccharomyces cerevisiae*. The data have been obtained in collaboration with our colleagues from Department of Radiation Biology, Institute of Cell Biology and Genetic Engineering, National Academy of Sciences of Ukraine, Kyiv, and Department of Molecular and Radiation Biology, Petersburg Institute of Nuclear Physics, National Research Center "Kurchatov Institute", Russia [14, 15].

Survival of the cells transferred to nutrition agar immediately after irradiation was no more than a few percent. Upon getting so high radiation dose, most of the cells do not succeed to repair the injured genetic structures before mitosis and, as a result, nonviable daughter cells are produced. Incubation in the nutrient-free media, in which cells do not divide, provides them with more sufficient time for the repair processes and leads to the corresponding increase in survival. From the kinetics curves represented on Fig. 14.2, one can see that the cells enriched with the magnetic isotope, ^{25}Mg, are recovered essentially more effectively than the cells enriched with the nonmagnetic ^{24}Mg. It was found that the recovery rate constant was $0.058 \pm 0.004\ h^{-1}$ for the cells enriched with ^{25}Mg and almost twice less, $0.032 \pm 0.003\ h^{-1}$, for the cells enriched with ^{24}Mg (the difference between the means is statistically significant at $P = 0.02$). Thus, the enrichment of cells with the magnetic isotope gives the two-fold increase in the rate constant of post-radiation recovery [14, 15].

The magnetic isotope effects of magnesium-25 were revealed in experiments with another commonly accepted cell model, bacteria *Escherichia coli*. The length

Table 14.1 Stable Isotopes in Biological Nanoreactors

Nucleus	Nuclear spin (I), in units of $h/2\pi$	Natural abundance, atom (%)	Nucleus	Nuclear spin (I), in units of $h/2\pi$	Natural abundance, atom (%)
^{1}H	½	99.984	^{39}K	3/2	93.08
^{2}H	1	0.016	^{41}K	3/2	6.91
^{12}C	0	98.89	^{40}Ca	0	96.97
^{13}C	½	1.11	^{42}Ca	0	0.64
			^{43}Ca	7/2	0.13
			^{44}Ca	0	2.06
			^{48}Ca	0	0.18
^{14}N	1	99.635	^{50}Cr	0	4.31
^{15}N	1/2	0.365	^{52}Cr	0	83.76
			^{53}Cr	3/2	9.54
^{16}O	0	99.759	^{55}Mn	5/2	100
^{17}O	5/2	0.037			
^{18}O	0	0.204			
^{19}F	1/2	100	^{59}Co	7/2	100
^{23}Na	3/2	100	^{63}Cu	3/2	69.09
			^{65}Cu	3/2	30.91
^{24}Mg	0	78.7	^{64}Zn	0	48.6
^{25}Mg	5/2	10.13	^{66}Zn	0	27.9
^{26}Mg	0	11.17	^{67}Zn	5/2	4.12
			^{68}Zn	0	18.8
^{28}Si	0	92.21	^{75}As	3/2	100
^{29}Si	1/2	4.7			
^{30}Si	0	3.09			
^{31}P	1/2	100	^{74}Se	0	0.87
			^{76}Se	0	9.02
			^{77}Se	1/2	7.58
			^{78}Se	0	23.52
			^{80}Se	0	49.82
			^{82}Se	0	9.19
^{32}S	0	95.02	^{92}Mo	0	14.84
^{33}S	3/2	0.74	^{94}Mo	0	9.25
^{34}S	0	4.22	^{95}Mo	5/2	15.92
			^{96}Mo	0	16.68
			^{97}Mo	5/2	9.55
			^{98}Mo	0	24.13
			^{100}Mo	0	9.63
^{35}Cl	3/2	75.4	^{127}I	5/2	100
^{37}Cl	3/2	24.6			

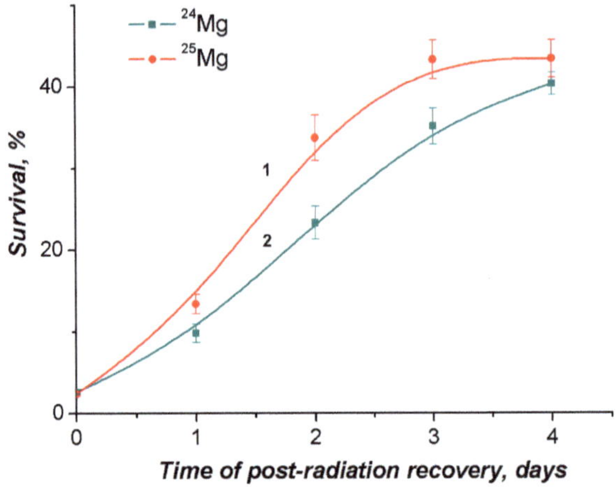

Fig. 14.2 The difference in effects of the magnetic and nonmagnetic isotopes of magnesium on post-radiation recovery of yeast *S. cerevisiae*. The cells enriched with the magnetic ^{25}Mg or the nonmagnetic ^{24}Mg were irradiated by the short-wave UV light (λ = 240–260). Survival of the cells was estimated as their ability to form colonies on nutrition agar: *1*—recovery of the cells enriched with ^{25}Mg; *2*—recovery of the cells enriched with ^{24}Mg (Compiled from [14, 15])

of adaptation (lag-phase) of the bacteria to the liquid media supplied with the magnetic isotope, ^{25}Mg, was found to be essentially shorter than the adaptation to the media supplied with the nonmagnetic isotopes, ^{24}Mg or ^{26}Mg. Hence, the bacterial cells essentially faster adapt to the growth media enriched with ^{25}Mg by comparison to the media enriched with ^{24}Mg or ^{26}Mg [16]. Furthermore, the striking effect of the magnetic isotope has been revealed after counting the colonies formed by the cells on the solid nutrient surfaces. The standard nutrient agar contains all components necessary for normal growth of cells, including magnesium. Nevertheless, the colony-forming ability of the cells, which were previously grown on ^{25}Mg, has turned out to be essentially higher in comparison with the cells which were previously grown on the nonmagnetic isotopes of magnesium. For the nonmagnetic isotopes ^{24}Mg and ^{26}Mg no difference was detected [16].

One might suggest that the differences observed in the above mentioned works were caused by different levels of impurities in the media complemented with the different isotopes of magnesium. However, according to the data of mass-spectrometry and atomic emission spectrometry, the element compositions of the media were similar for all samples with amounts of contaminants of the order of one micromole per liter or less, no matter which kind of the magnesium isotopes was used. Besides, it should be taken into consideration that the amounts of the contaminants, that were administered in the media from other basic components, have significantly exceeded the amounts of the same contaminants administered with much less additions from the magnesium isotope stock solutions.

The magnesium ions Mg^{2+} serve the obligate cofactor functions for about three hundred enzyme reactions. Among these reactions, synthesis and hydrolysis of ATP seem to be the most important [12, 13]. It is also known that adaptation of cells to novel growth conditions, especially reparation from radiation damages, requires a large variety of stress proteins to be synthesized and, correspondingly, it requires energy. Reasoning from this knowledge and the common knowledge that ATP is the main source of energy in cells, it was reasonable to suggest that the magnetic isotope effects, that have been revealed in living cells, are stemming from the higher efficiency of the "fuel-energy nanoreactors" in the cells enriched with the magnetic isotope of magnesium. In the work of A.L. Buchachenko and his co-workers with mitochondria isolated from rat hearts it was found that the ATP synthesis proceeds two-three times more effectively with ^{25}Mg than with ^{24}Mg or ^{26}Mg. The similar magnetic isotope effects were observed by the same group in their studies of creatine kinase and phosphoglycerate kinase, see the reviews [5, 18] and references therein. However, the attempts of reexamination of these pioneer works have failed [19].

Meanwhile, the ions of magnesium, Mg^{2+}, serve cofactor functions for many other enzymes including transport ATPases, DNA and RNA polymerases and so on, which catalyze hydrolysis of ATP employing the energy of this exothermic reaction. Among them, the myosin-type enzymes are responsible for myriad cellular processes such as muscle contraction, embryogenesis, intracellular cargo transport, cytokinesis, cell migration, etc. [12, 13]. Muscle myosin has received the most study. This "molecular motor" catalyzes the reaction of hydrolysis of the end phosphate bond in ATP molecule, $ATP + H_2O \rightarrow ADP + P_i$. At this, the released energy, about 0.54 eV at the physiological conditions, is used to execute muscle contraction. In essence, not a pure ATP molecule but the complex $[ATP^{4-}\ Mg^{2+}]$ is hydrolyzed into $[ADP^{3-}\ Mg^{2+}]$ and P_i in the active center of any ATP-hydrolase [12, 13, 20]. Thus, it is instructive to consider if the nuclear spin of ^{25}Mg can have an impact on operation of this "molecular motor".

We studied effects of different magnesium isotopes, the magnetic ^{25}Mg and the nonmagnetic ^{24}Mg and ^{26}Mg, on Mg^{2+}-dependent ATP hydrolase activity of the catalytic fragment (subfragment-1) of myosin isolated from myometrium muscle. The work has been performed in cooperation with Department of Muscle Biochemistry, Palladin Institute of Biochemistry of the National Academy of Sciences of Ukraine, Kyiv [21]. Three independent experiments have been done with three enzyme preparations isolated from three different animals (pigs) at different times. Figure 14.3 demonstrates the summary of the measurements.

One can see that ^{25}Mg essentially accelerates the reaction of ATP hydrolysis catalyzed by the enzyme, 2–2.5 times by comparison to the spin-less ^{24}Mg or ^{26}Mg. At that, no essential difference in the ATPase activity for the nonmagnetic isotopes ^{24}Mg and ^{26}Mg has been detected. Another important point is that the samples which contained all components of the reaction solution, except the enzyme, have manifested no magnetic isotope effect in the non-enzymatic ATP hydrolysis. Thus, we have documented the magnetic isotope effect in the enzymatic hydrolysis of ATP, the acceleration of chemo-mechanical cycle of myosin by the nuclear spin of ^{25}Mg [21].

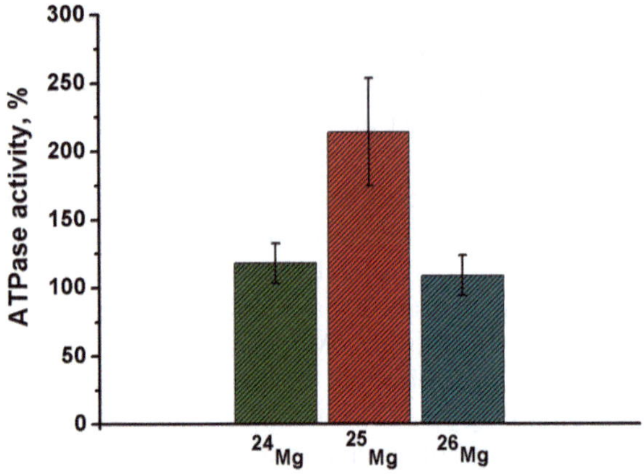

Fig. 14.3 ATPase activity of the myosin subfragment-1 in the reaction solutions supplemented with different isotopes of magnesium, 5 mM of $^{24}MgCl_2$, $^{25}MgCl_2$ or $^{26}MgCl_2$, in percentage to the enzyme activity in the reaction solution supplemented with 5 mM of "natural" $MgCl_2$ with natural isotope abundance. (Compiled from [21])

MIE means that there is a spin-selective "bottle-neck", a rate-limiting step, in the ATP hydrolysis driven by myosin. The catalytic effect of ^{25}Mg may be explained in the following way. Starting from the seventies of last century (see, for example, [22]), it was experimentally proved that ATP hydrolysis triggers electron-conformational interactions in the ATP-hydrolase's active center, thereby producing conformational changes in the enzyme macromolecule. In essence, there is the deformational excitation of the macromolecule conformation owing to the energy released from ATP hydrolysis, about 0.54 eV (see, e.g., [23] and references therein). This energy is not large enough to trigger the electron-conformational excitation of myosin into the singlet state. It is sufficient to obtain a low-level triplet state but the transition from the ground singlet state ($S = 0$) into the triplet state ($S = 1$) is forbidden by the law of spin conservation. A different situation occurs with ^{25}Mg. The ^{25}Mg isotope's nuclear spin eliminates the spin ban problem providing the necessary spin conversion from the singlet state into the triplet state. Figure 14.4 illustrates this idea. A similar mechanism has been suggested to explain the effects of magnetic fields on mobility of dislocations in solid state physics [10, 24].

Furthermore, the theoretical investigations holds that an elementary coherent deformation excitation, the so-called soliton, may arise in the quasi-one-dimensional protein molecules. The low probability of loss of energy of the ATP hydrolysis into chaotic motion (into heat phonons) is explained on the basis of high stability of solitons. Therefore, the conformational energy in the form of the soliton can be transported along the protein molecule without heat losses [25]. Besides, the conformational transitions are accompanied by the movements of the electrically charged groups. Such movements create the magnetic fields and it seems likely that interactions of the nuclear spins with local magnetic fields should be also taken into

adenosine 5'-triphosphate (ATP)

↓↑ H_2O, Mg^{2+}

adenosine 5'-diphosphate + P_i

initial conformation,
singlet state (S = 0)
↓
energy released from ATP (≈0.54 eV)
↓ ^{25}Mg
exited ("energized") conformation,
triplet state (S = 1)
↓
acceleration of ATP hydrolysis

Fig. 14.4 Nuclear spin catalysis in ATP hydrolysis driven by myosin

account. Besides, apart from the large nuclear spin moment, $I = 5/2$, the nucleus of ^{25}Mg has the quadrupole moment the value of which is rather large in comparison with other magnetic nuclei. Correspondingly, interactions of the nuclear quadrupole moment with the electrically charged groups of the protein macromolecule might be large enough to affect kinetics of the conformational transition as well as the conformational relaxation of the macromolecule.

In addition, one should view an ion-radical pair as the spin-selective "bottle-neck" in the reaction. The exothermic reaction of ATP hydrolysis with formation of ADP and inorganic phosphate (P_i) follows the basic-acid mechanism [12, 13]. Hence, in spontaneous non-enzymatic ATP hydrolysis, appearance of an ion-radical pair as the intermediate of the reaction seems to be unlikely. Indeed, no magnetic-isotope effect has been detected in our experiments on the non-enzymatic hydrolysis of ATP [21]. A different situation, however, can arise in the case of the enzymatic ATP hydrolysis. Accordingly to the quantum-chemistry and molecular mechanics calculations [20], first catalytic step in the ATPase activation of myosin during the force generating cycle is stabilization of the γ-phosphate of ATP in a dissociated metaphosphate state. At this, the myosin-bound ADP and P_i products of hydrolysis remain in close contact and only release later by myosin, upon rebinding to the actin filament. It is consistent with the well-known reversibility of ATP hydrolysis in myosin. The ATP hydrolysis reaction is reversible as long as the protein remains in the postrecovery–prepower stroke conformation [20]. Meanwhile, in the enzyme's active center under the conditions of electron-conformational excitation of the macromolecule there may be a transfer of the electron spin density onto Mg^{2+} from ADP^{3-}, or NH_2-group of Glu459, for example, or OH^- of the water molecule with formation of the relevant ion-radical pair. Due to the hyperfine coupling of the ^{25}Mg's nuclear spin with the ion-radical pair's unpaired electron, this myosin-bound intermediate ion-radical pair can be converted into the triplet state (S = 1). However, the stable ATP-Mg product is to be in the singlet state (S = 0). Thus, because of the spin ban, the nuclear spin of ^{25}Mg hampers the reverse reaction of ATP synthesis, thereby promoting the direct reaction of ATP hydrolysis.

Alternatively, one further explanation of the catalytic effect of the nuclear spin of ^{25}Mg may be considered. At conformational transitions, atomic groups of the macromolecule change their positions. At this, dehydration and rehydration of the electrically charged groups take place. Meanwhile, it is known that there are two isomers of water which differ in orientation of the hydrogen nuclear spins, namely, *ortho*-H_2O with the parallel orientation of proton spins and *para*-H_2O with anti-parallel proton spins [26]. There are the reasons to believe that *ortho*-H_2O has a preferential affinity to L-amino acids by comparison with *para*-H_2O [27]. If that is the case, when tightly bound in macromolecules, the molecules of *ortho*-H_2O are hard delivered to proper sites for proper processes. Of course, nuclear spin-rotation interaction occurs but they are weak to be effective in driving the *ortho* to *para* transitions. Again, the magnetic isotope of ^{25}Mg can make a difference eliminating the spin-ban problem and, thereby, providing the necessary conversion of the water isomers.

One way or the other, the nuclear spin of ^{25}Mg, via the acceleration of the chemo-mechanical cycle of the enzyme, helps in setting the myosin macromolecule for acceptance and hydrolysis of next ATP molecule. The detailed mechanisms, including quantum mechanics, of the nuclear spin catalysis in biomolecular nanoreactors require further investigations.

14.4 Conclusions and Outlook

In "molecular motors" which run on nonmagnetic isotopes of magnesium, the spin-catalysis functions can be served by the nuclear spins of phosphorus and protons. However, ^{25}Mg has the nuclear spin 5/2, that is five times greater than the nuclear spins of ^{31}P or ^{1}H. Furthermore, the comparatively high catalytic activity of the ^{25}Mg's nuclear spin stems from the specific localization of the Mg^{2+} ion in the enzyme active center due to which the magnesium's nuclear spin creates the comparatively high values of the magnetic field and hyperfine coupling. That is why ^{25}Mg proved to be "in the right place at the right time" to light the fine machinery of the fuel-energy cell nanoreactors.

In living Nature, apart from magnesium, there are other elements which have both kinds of stable isotopes, nonmagnetic and magnetic ones, including carbon, oxygen, calcium, zinc, etc. (see Table 14.1). In the magnetic field of Earth, the strength of which is about 0.05 mT, NMR frequencies of the nuclei fall within the range between approximately 50 and 2000 Hz. Biological effects of weak low-frequency magnetic fields are well known for a long time but poorly understood [28, 29]. Inasmuch as the nuclear spin moments of the magnetic isotopes are prone to external magnetic fields, it may bear a direct relationship to the problem of the biological effects of weak low-frequency magnetic fields. What's more, the geocosmic oscillations in processes of different nature, caused by movement of the Earth in heterogeneous and anisotropic space-time, have long been known. For example, the so-called "macroscopic fluctuations" as the anomalous scattering of the results of measuring the actomyosin enzyme activity were discovered about 60 years ago [30]. Based on the nuclear spin-catalysis background, one can suggest that the "macroscopic fluctuations" in the actomyosin and some other objects of living Nature stem from the interactions of the nuclear spins of the magnetic isotopes with the oscillating geocosmic electromagnetic fields. One can speculate that the stable magnetic isotopes may open the novel ways of control over efficiency and reliability of biosystems via the interactions of their nuclear spins with the external electromagnetic fields.

Based on the nuclear spin-catalysis background, one can further speculate that stable magnetic isotopes hold considerable promise for control over efficiency and reliability of molecular and biomolecular devices in optical communications, quantum information processing, computational schemes and the like.

Acknowledgments Magnesium oxides, ^{24}MgO, ^{25}MgO and ^{26}MgO with isotope enrichment no less than 99.8, 98.2, and 81.0 atomic percent, were purchased from Russian atomic industry. From these oxides, the relevant magnesium chloride or sulphate solutions, correspondingly, were prepared accordingly to the standard procedures. I am thankful to Dr. Vasilii K. Karandashev, head of Analytical Center, Institute of Microelectronics Technology and High Purity Materials, Russian Academy of Sciences, Chernogolovka, for measurements of the isotope contents and element analysis of solutions using ICP-MS (mass spectrometry with inductively coupled plasma) and ICP-AES (atomic emission spectrometry with inductively coupled plasma). This work was presented on 6th International Conference on Physics of Liquid Matter: Modern Problems (PLMMP-2014), Kyiv, Ukraine, May 23-27, 2014. I am deeply grateful to academician Leonid Bulavin and Professor Nikolai Lebovka who kindly invited me to give my presentation on this conference, within the precincts of Physics Department of Kyiv National University, my Alma Mater. Additionally, this manuscript benefited from the remarks of an anonymous reviewer. The Funding was provided by Russian Foundation for Basic Research (RFBR), project no. 14-04-00593.

References

1. L.D. Landau, E.M. Lifshitz, *Theoretical Physics. vol. 3. Quantum Mechanics*, vol. 3 (Nauka, Moscow, Russia, 1989)
2. A.L. Buchachenko, R.Z. Sagdeev, K.M. Salikhov, *Magnetic and Spin Effects in Chemical Reactions* (Nauka, Novosibirsk, Russia, 1978)
3. Y.B. Zeldovich, A.L. Buchachenko, E.L. Frankevich, Sov. Phys. Usp. **155**, 3 (1988)
4. B. Brocklenhurst, Chem. Soc. Rev. **31**, 301 (2002)
5. A.L. Buchachenko, *Magnetic Isotope Effect in Chemistry and Biochemistry* (Nova Science Publishing, New York, 2009)
6. V.K. Koltover, in *Nanotechnology 2010*, (Nano Science and Technology Inst, Anaheim, 2010), vol. 3, pp. 475–477
7. V.K. Koltover, in *Biomedicine*, ed. by C. Lin (InTech-Europe, Rijeka, 2012), pp. 105–122
8. P.J. Hore, Proc. Nat. Acad. Sci. USA **109**, 1357 (2012)
9. A.L. Buchachenko, J. Phys. Chem. B. **117**, 2231 (2013)
10. D. Kaplan, I. Solomon, N.F. Mott, J. de Physique Lett. **39**, 5 (1978)
11. D.M. Grant, R.K. Harris (eds.), *Encyclopedia of Nuclear Magnetic Resonance* (Wiley, Chichester, 1996)
12. L. Stryer, *Biochemistry* (Freeman, New York, 2002)
13. D.L. Nelson, M.M. Cox, *Lehninger Principles of Biochemistry* (Freeman, New York, 2008)
14. D.M. Grodzinsky, T.A. Evstyukhina, V.K. Koltover, V.G. Korolev, Y.A. Kutlakhmedov, Reports of National Academy of Sciences of Ukraine. **No. 12**, 153 (2011)

15. V.K. Koltover, V.G. Korolev, Y.A. Kutlakhmedov, *Ionizing Radiation: Applications, Sources and Biological Effects* (Nova Science Publishing, New York, 2012), pp. 117–128
16. V.K. Koltover, U.G. Shevchenko, L.V. Avdeeva, E.A. Royba, V.L. Berdinsky, E.A. Kudryashova, Dokl. Biochem. Biophys. **442**, 12 (2012)
17. V.K. Koltover, Biophysics **58**, 187 (2013)
18. A.L. Buchachenko, D.A. Kouznetsov, N.N. Breslavskaya, Chem. Rev. **112**, 2042 (2012)
19. D. Crotty, G. Silkstone, S. Poddar, R. Ranson, A. Prina-Mello, M.T. Wilson, J.M.D. Coey, Proc. Nat. Acad. Sci. USA **109**, 1437 (2012)
20. F.A. Kiani, S. Fischer, Proc. Natl. Acad. Sci. U.S.A. **111**, 2947 (2014)
21. V.K. Koltover, R.D. Labyntseva, A.A. Lul'ko, V.K. Karandashev, S.A. Kosterin, Reports of the National Academy of Sciences of Ukraine. **No. 1**, 163 (2014)
22. V.K. Koltover, L.M. Reichman, A.A. Yasajtis, L.A. Blumenfeld, Biochim. et Biophys. Acta. **234**, 296 (1971)
23. M.V. Volkenstein, *General Biophysics* (Acad. Press, New York, 1983)
24. M.V. Badylevich, V.V. Kveder, V.I. Orlov, Y.A. Osipyan, Phys. Stat. Sol. **2**(6), 1869 (2005)
25. A.S. Davydov, Sov. Phys. Usp. **25**, 898 (1982)
26. V.I. Tikhonov, A.A. Volkov, Science **296**, 2363 (2002)
27. Y. Scolnik, I. Portnaya, U. Cogan, S. Tal, R. Haimovitz, M. Fridkin, A.C. Elitzur, D.W. Deamer, M. Shinitzky, Phys. Chem.—Chem. Phys. **8**(3), 333 (2006)
28. V.V. Novikov, I.M. Sheiman, E.E. Fesenko, Biophysics **47**, 125 (2002)
29. V.N. Binhi, V.A. Milayev, D.S. Chernavskii, A.B. Rubin, Biophysics **51**, 553 (2006)
30. S.E. Snoll, Biophysics **58**, 265 (2013)

Index

L. Bulavin and N. Lebovka (eds.), *Physics of Liquid Matter: Modern Problems*,
Springer Proceedings in Physics 171, DOI 10.1007/978-3-319-20875-6

MIX
Papier aus verantwortungsvollen Quellen
Paper from responsible sources
FSC® C105338

If you have any concerns about our products,
you can contact us on
ProductSafety@springernature.com

In case Publisher is established outside the EU,
the EU authorized representative is:
Springer Nature Customer Service Center GmbH
Europaplatz 3, 69115 Heidelberg, Germany

Printed by Libri Plureos GmbH
in Hamburg, Germany